EPIDEMIOLOGY OF INFECTIOUS DISEASES

EPIDEMIOLOGY OF INFECTIOUS DISEASES

A HUMAN VIEW

Jos Frantzen

BRILL | WAGENINGEN ACADEMIC

EAN: 9789004689626
e-EAN: 9789004689657
ISBN: 978-90-04-68962-6
e-ISBN: 978-90-04-68965-7
DOI: 10.1163/9789004689657

First published, 2023

Wageningen Academic Publishers,
P.O. Box 220,
NL-6700 AE Wageningen,
The Netherlands.
www.WageningenAcademic.com
copyright@WageningenAcademic.com

Table of contents

Preface

The study of epidemics among human beings belongs traditionally to the life sciences, and more specifically the bio-medical sciences. Current textbooks in epidemiology, therefore, provide a rather strong bio-medical view on epidemics. In this textbook, the bio-medical view will be extended to a human view including insights from humanities, social sciences. This extension challenges us all the more to combine the requirement of scientific objectivity with the subjectivity inherent to human life. In addition, the bio-medical view will be deepened using knowledge of botanical epidemiology to dive into the topics of 'evolutionary dynamics of pathogens' and 'epidemic spread of pathogens'. This book is especially an invitation to bio-medically oriented students and senior scientists to reflect on the multi-dimensional, subjective, character of epidemics. Reflections that may enable appropriate, human, management of epidemics.

Reflections. We need these to deal with the intriguing interaction of, body, mind, and environment, determining human life. We distinguish these three interacting components, whereas these cannot be seen as independent entities. We are all object and subject in one, and both are determined by our social and non-social environment. And in turn, we affect the environment. We present this 'trinity of life', and we reflect on it, in Chapter 2, highlighting our immune system. A system that seems the place-to-be with respect to the triadic interaction between, body, mind, and environment. We carry the insights into, the reflections on, the trinity of life over to, especially, Chapter 7 and 9. In Chapter 7 we distinguish four domains of impact of epidemics, although these cannot be regarded as independent. A novel way of determining impact that surpasses the traditional data of mortality and morbidity. In Chapter 9, we will see the trinity of human life back as a cornerstone of appropriate, human, management of epidemics.

Reflections. We need to reflect on the wealth of pathogens as well. We present pathogens in three major categories in Chapter 3. One, the category of non-organisms that encompasses prions and viruses. Two, the category of prokaryotic organisms that includes archaea and bacteria and, three, eukaryotic organisms that encompass, protists/chromists, helminths, and fungi. Some major characteristics of each will be described, but we need to be aware of the limitations of the descriptions, as the variety of pathogens is huge in each of the (sub-) categories. The non-organisms pose specific challenges for us, because the underlying theory and supporting evidence of evolutionary dynamics are directed to organisms rather than non-organisms.

Reflections. These are inevitably to go beyond the one-dimensional view on evolutionary dynamics of pathogens, which hinges on maximisation of the basic reproductive number R_0. A view that is valid under very specific conditions only, as we will see in Chapter 5. We will extend evolutionary thinking to address multi-dimensionality. It results in models that provide profound insights into evolutionary dynamics of pathogens, but not reliable predictions. Assessments of epidemic risk are also limited in such a lack of predictability, as we will see in

Chapter 8. Insights into evolutionary dynamics also indicate the value of the mechanism of tolerance to inhibit natural selection of aggressivity, virulence, of pathogens. A mechanism that has largely been ignored in human epidemiology, so far, although employment of it contributes considerably to basic health, and, therefore, proper management of epidemics, as we will outline in Chapter 9.

Reflections. We miss these largely in the traditional human epidemiology with respect to the spatial dynamics of pathogens. It looks as though abundance of pathogens increases and decreases in time only. We also see this shortcoming in the common modelling of human epidemics using, for example, SIR-models. Inclusion of a spatial component in the dynamics of a pathogen may change drastically our view on disease epidemics, as we will recognise in Chapter 6. The consequences of a profound insight into the spatial dynamics of pathogens will be worked out further in Chapter 10. A chapter that deals with the common societal responses to epidemics.

The multi-faceted knowledge gained in the Chapters 2-7 culminates in an outline of managing epidemics, which is presented in Chapters 8-10. An outline, a direction, not more than that. Each epidemic requires tailor-made management, which we will call here appropriate, human, management. Appropriate, as it is protective with respect to people at relatively high risk while re-assuring those at relatively low risk that no specific interventions are necessary for them. In terms of diagnostics, it is management that pairs relatively high sensitivity with relatively high specificity. And human, as it needs both, to comply with The Universal Declaration of Human Rights and to address the feelings, concerns, of a whole community, as much as possible.

Reflections. We will find these throughout this textbook. It is an academic one that stimulates readers to think critically about all the facts, and fictions, provided in publications of scientific journals and books. It may enable epidemiologists to interact critically with various types of scientists in, life sciences, humanities, and natural sciences. And, of course, epidemiologists need to have a basic understanding of the various disciplines to communicate properly with them. In addition, scientists use a different scientific language across all the disciplines relevant for epidemiology. This textbook may provide some basic understanding of, and 'languages' used in, several relevant disciplines. This may encourage readers to think beyond the glimpses of epidemiology provided in this textbook. An open, critical, trans-disciplinary attitude should be the hallmark of a good epidemiologist of infectious diseases.

A book written by a single author, who will use the plural form nevertheless. It reflects the fact that a single person, or a few persons, cannot write this textbook. I was just an editor, who compiled the multi- and interdisciplinary knowledge required to arrive at a human view on the epidemiology of infectious diseases. We will explain that in Chapter 1. That chapter will also describe the need to zoom in and out in order of presenting the necessary details, while keeping the overview. This is, of course, a rather subjective, but inevitable, approach to writing a textbook on epidemiology of infectious diseases that passes the borders of bio-medical sciences. In addition, we included as much as possible references to, recent, reviews of topics to minimise the number of references. This implies that we refer to references in such publications without mentioning these explicitly. We also rely nearly exclusively on Open Access publications to guarantee, as much as possible, accessibility to the refereed publications.

Author

Jos Frantzen studied biology at the Catholic University Nijmegen, the Netherlands, with majors in geobotany, soil science and law. He did an additional study in plant pathology and he received his PhD at the Agricultural University Wageningen, the Netherlands. Spatial dynamics of (plant) pathogens became a major subject of interest. He also acquired an additional Masters in human epidemiology at the Free University Medical Centre, Amsterdam, the Netherlands. He went, subsequently, to the University of Fribourg, Switzerland, becoming a lecturer for plant ecology and epidemiology. He did his 'Habilitation' at the same university. The German word 'habilitation' is a qualification for a professorship in various European countries, like Germany and Switzerland. Back in the Netherlands, Jos was employed by the Radboud University Medical Centre in Nijmegen and, subsequently, the Free University Medical Centre in Amsterdam. He made the transition from scientist to research manager in that period. Jos, subsequently, founded his own company, Driehoek Research Support, in research management. The company is based in Leiden, the Netherlands. The interest in epidemiology remained as expressed in, for example, his textbook 'Epidemiology and Plant Ecology' (2007).

Readers

I greatly acknowledge the input of various experts in reading and commenting on earlier versions of (parts of) the book. They are in alphabetical sequence by surname,

Prof Dr Lex Bouter, Biologist-Epidemiologist, Professor Emeritus of Methodology and Integrity, Amsterdam University Medical Centers.

Bernadette Conrads, MSc Biomedical Sciences, Programme Officer ZonMw and co-founder of Bitez - Food Innovations.

Dr Els Geertman, Microbiologist, Lecturer of Natural Sciences / Applied Sciences, Fontys University of Applied Sciences.

Prof Dr Marc de Kesel, Theologist-Philosopher, Professor of Theology, Mysticism and Contemporaneousness, Radboud University.

Prof Dr Hans Metz, Mathematician, Professor Emeritus of Mathematical Biology, Leiden University.

Prof Dr Christopher Mundt, Plant Pathologist, Professor in Botany and Plant Pathology, Oregon State University.

Dr Iris Otto, Life Style Physician and Lecturer, founder of the online sustainable health platform 'Gezond Kompas'.

Dr Esther Van der Werf-Kok, Epidemiologist, Lecturer of Primary Care Infectious Diseases, University of Bristol.

Prof Dr Theo Wobbes, Physician-Philosopher, Professor Emeritus of Surgical Oncology, Radboud University Medical Centre.

EPIDEMIOLOGY IN SCIENCES

HISTORY AND CHALLENGES OF EPIDEMIOLOGY

This chapter provides a short insight into the history of epidemiology and it raises the issue of good science. We see historically a division of epidemiology into a branch directed to plant diseases and another to human diseases. Infectious diseases constitute a specific category within human epidemiology, as pathogens may disperse from one human to another. We focus on the specific role of epidemiologists to integrate knowledge of a variety of scientific disciplines into a thorough understanding of disease epidemics. An understanding that can be achieved by good science only. We will come up with an indication as to good science, and then, we will be ready to dive into the fascinating world of disease epidemics in the subsequent chapters.

1.1 Epidemiology as a scientific approach

The Greek physician Hippocrates (460-380 BC) observed disease among people (Frantzen, 2007, and references therein). He called it 'επιδεμιος', which literally means 'what is among people'. We call it 'epidemic' now. Hippocrates wondered that some young males in a gymnasium got mumps, and others did not. He also observed that disease was abundant in one year, and the less so in another year. The Greek philosopher Theophrastus (372-287 BC) extended the observations of Hippocrates to plants identifying epidemics as well. He went a step further than Hippocrates and he could attribute variation in disease abundance to environmental variation.

The Italian physician, poet, and all-round scientist Girolamo Fracastoro (1483-1553) introduced the notion of contagion observing epidemics. He introduced the term 'seed' to explain the transmission of disease from one body to another. Fracastoro indicated that these 'seeds' had the faculty of multiplying and propagating rapidly. Today, we would call it the dispersal units of an infectious agent. We may notice that Fracastoro was dealing with infectious diseases only. He described as well that some infectious diseases did occur on plants only, others on animals only, and some specifically on humans. In addition, he noticed that diseases may be organ-specific.

Epidemics have been observed and described for thousands of years. The term 'epidemiology', which is the study of epidemics, is used for the first time, as far as we know, in 1598 (Buck *et al.*, 1988). It was the Spanish physician Angelerio publishing his book 'Epidemiología' on plague. The term 'epidemiology' appeared again in Spain in 1802, when the physician Villalba published his book 'Epidemiología Española'. It describes epidemics occurring in Spain between the fifth century before Christ until the eighteenth after.

The study of epidemics got a boost in the nineteenth century, both for plant and human diseases (Frantzen, 2007, and references therein). The plant disease potato late blight, which is caused by *Phytophthora infestans*, destroyed potatoes across large areas of Europe in the 1840s. The resulting famine did not result in many deaths only, but it also led to emigration of poor people to the United States. The famous Kennedy family is an example. Anyway, the epidemic provided a strong stimulus to the science of plant pathology. Similarly, a local epidemic of cholera in London in 1848 stimulated the British physician John Snow to map cases of disease precisely in both, time and space. The epidemic could, subsequently, be inhibited rather simply by closure of a contaminated water pump. It turned out as an efficacious intervention, although the causal agent, the gram-negative bacterium *Vibrio cholerae*, was unknown at that time.

John Snow is sometimes called the father of epidemiology, because he adopted a method of quantification. If so, we should call him the father of human epidemiology, as we see a division between the study of human and plant diseases in the 19th century. This division may be explained by the fact that plant pathogens cannot infect humans, and human pathogens cannot infect plants. In addition, science, in general, got divided into various branches and disciplines keeping pace with the accelerating advance of scientific knowledge. All-round scientists, like Fracastoro, became extinct.

Epidemiology as a discipline of medicine had its own journal in 1874, the German 'Allgemeine Zeitschrift für Epidemiologie' (Zadoks, 2017). We may call it 'General Journal of Epidemiology' in English. Interestingly, the title suggests the coverage of diseases of, humans, plants and animals, whereas it was directed to human infectious diseases only. The implicit suggestion that 'human epidemiology' equals 'epidemiology' turned out to be persistent throughout the following decades. We, however, have to state precisely that human epidemiology deals with both, infectious and non-infectious human diseases. We also call the latter non-communicable diseases. In contrast, botanical epidemiology deals with infectious diseases only. It does not include non-infectious diseases. In addition, we have veterinary epidemiology, which deals with infectious diseases of animals. We may notice that a human also is an animal enabling several pathogens to infect both, man and other animals. The border between human epidemiology and veterinary epidemiology is in that sense a bit blurred.

Botanical epidemiology may be seen as arising as a discipline of plant pathology by way of the seminal works of the Swiss plant pathologist Ernst Gäumann and the South-African potato breeder James Edward Vanderplank in the 20th century. Vanderplank's statistical approach to modelling epidemics, using non-linear regression, was picked-up by the Dutch biologist Jan Carel Zadoks. He advanced it successively into, numerical, analytical, and physical-analytical modelling, respectively. His work suggests modelling is a key feature of epidemiology. Interestingly, not one of these three eminent scientists in botanical epidemiology called himself an epidemiologist, nor did they have an education in epidemiology. If we look at the history of human epidemiology, we may observe a similar phenomenon. Whereas we have a common sense on the term 'epidemic', we did not have it with respect to epidemiology. We may even wonder whether it is a scientific discipline.

The aim of sciences, in general, is to understand patterns we observe among people, animals, plants, materials, climate, or whatever (*cf.* Bod, 2020). Scientists also try to get some understanding at a specific level of organisation. It may be at, the very basic level of (sub-) atoms, molecules, cells, organisms, populations, (eco-)systems, earth, or cosmos. Epidemiology focuses on the population level. The focus is on, incidence, distribution and determinants of diseases at that level. We need, however, to pay attention to processes at the individual level, from molecule up to the whole body, and those at the ecosystem level, to really understand diseases at the population level. In addition, we need to understand not only the occurrence of illness in a community, but we also need to explain that people do not get ill. We encounter here a specific problem of human epidemiology. Ill people look for a physician, whereas healthy people, in general, do not. We may also notice that human epidemiology is limited in experimental tools to investigate infection processes in humans. It is, of course, quite different for the study of plant diseases. These, therefore, offer proper experimental models of various aspects of human infectious diseases, and especially, those of the evolutionary and spatial dynamics of the pathogens. So, we may advocate, on the one hand, the integration of medical, botanical, and veterinary epidemiology in a single discipline of epidemiology of infectious diseases to understand epidemics. On the other hand, the anatomy and physiological processes differ completely between plants and animals. These also differ among animals, although less than between plants and animals.

We may conclude from the preceding that epidemiology is hardly to see as an autonomous scientific discipline, nor to attribute to the bio-medical sciences, or any other branch of science. We see it rather as an approach to studying (infectious) diseases at the population level. We, therefore, define epidemiology here, as follows:

> *Epidemiology is a trans-disciplinary, scientific, approach to the understanding of diseases at the population level, also taking into account processes at both, the individual and ecosystem level.*

Such a definition of epidemiology enables the use of knowledge of a broad range of scientific disciplines. In addition, we may use scientific knowledge of plant diseases while studying (infectious) human diseases, as we will do in this text book. Whereas a good scientist knows a lot about a little, *i.e.* she, or he, is an expert on a very specific subject, a good epidemiologist knows a little about a broad range of subjects integrating the knowledge of various scientific disciplines to a real understanding of an epidemic. She, or he, therefore, depends completely on good science underlying the input of all those disciplines. We arrive at the topic of defining good science across a broad spectrum of disciplines. We like to stipulate that we adopt here the German concept 'Wissenschaft' that goes beyond the Anglo-Saxon notion of science. 'Wissenschaft' encompasses all disciplines of science, irrespectively these belong to the humanities or natural sciences (Brier, 2015).

REFLECTIONS

Epidemiology is regarded quite often as a bio-medical science. If so, such a view on epidemiology ignores both the multi-disciplinary character of epidemiology and the scientific knowledge we have gathered in botanical epidemiology. Knowledge that may be used in the understanding and management of human, infectious, diseases. You may reflect about a proper definition of epidemiology that fits your own experience and interests.

1.2 Good science

Scientists live in a rather confusing era. On the one hand they are embraced by society, and on the other hand they are distrusted and criticised profoundly. It is just whether findings fit specific opinions, or not. Poor quality is another reason to question science. John Ioannidis of Stanford University was the one to bell the cat with respect to the poor quality. He did it by way of a seminal publication (Ioannidis, 2005). He was able to demonstrate that most published findings in bio-medical journals are false just because of, relatively few comparable studies in a specific discipline, small-sized studies, inappropriate designs, and weaknesses in statistical analyses. We may state it in another way, most published studies are not reproducible (Begley and Ioannidis, 2015). We have a real reproducibility crisis in science. Wake-up calls for reproducibility popped up in all branches of science the last decades (*e.g.*, Munafo *et al.*, 2017; Peels, 2019). We may interpret it as one shout-out, calling for good science. We, therefore, have meanwhile a myriad of charters, protocols, guidelines, and so on, to promote good science. These are all of value, but would it be possible to provide scientists with an overall, rather elementary, understanding of good science? Say, an indication to it? We may notice that such an indication is of particular importance to epidemiologists integrating knowledge of a variety of scientific disciplines.

Let us first take a step back in time to the Carmelite Titus Brandsma (1881-1942). He was Professor of Spirituality and the History of Piety at the Catholic University in Nijmegen, the Netherlands. Titus Brandsma started to warn against the upcoming Nazism in the thirties (Hemels, 2008). He focused on warning journalists, as he felt they underestimated the danger. In addition, he detected the poor quality of journalists of the Roman-Catholic newspapers. Instead of criticizing them, Father Titus tried to guide the journalists towards good journalism. In the phrasing of Kees Waayman, who provided the introduction to Hemels (2008): "freedom of speech, orientation in such a plethora of information, and love of truth defines the triangle, in which good journalism operates". We see that Brandsma did not come up with stringent instructions to journalists, but he provided a space. A space that justifies the various characters and circumstances of individual journalists. We feel such a space is also needed for scientists. They cannot operate according to stringent directives and rules. Academic freedom is needed to explore challenging scientific questions. It should, however, not end up in sloppy and non-reproducible science. So, we need to define an operational space for scientists. A space that we may define by a triangle, as well. We will work out such a triangle further below, side by side.

We may agree on a love of truth as the driver of all good science, the basics (Fig. 1.1). But, how can we determine the truth? Does it exist? Let us take a closer look at an epidemic of influenza in the Netherlands in the flu season of 2017/2018. We use this example as the Netherlands is a relatively small, well-organised, country enabling a fairly well-monitoring of diseases. The national report on this epidemic, however, starts with a disclaimer (Reukers *et al.*, 2018). The term 'influenza-like illness' is used because the presence of an influenza virus was laboratory-confirmed in a minority of cases only. So, other pathogens may have caused the symptoms. In addition, people with 'influenza-like illness' may report it, or not. In general, we may expect underreporting, especially for those cases experiencing relatively mild symptoms. The probability of confirmation of an influenza virus increases along the way to hospitalisation. Knowing this, we may look at the actual data. About five percent of the Dutch population was classified as suffering from 'influenza-like illness' in the flu season starting in week 40 of 2017 and ending in week 20 of 2018. Two peaks were identified. One of 166 cases per 100,000 inhabitants in week 4 and one of 170 cases per 100,000 inhabitants in week 10. The number of hospitalisations due to complications was estimated at over 16,000 during the whole flu season. The all-cause mortality in the Netherlands was 9,500 higher than the average mortality in the annual flu seasons. We see in the example some points of general concern arising. One, a precise determination of the abundance of a disease is troublesome. Symptoms of various diseases overlap, like fever, and the identification of a specific cause of disease is a challenge, which comes along with costs. In addition, people have to report their illness to a care provider. Reporting may be limited in fees of consulting a care provider. Two, we may get an overall picture of a disease, but it may be more abundant in specific groups of people, or in specific areas of a country. Three, a proper determination of the mortality rate of a specific disease, a pathogen, is rather impossible. In the example above, we have about 900,000 cases of which 9,500 might have died due to an influenza virus (1%). Likely, the number of influenza-cases is largely underestimated missing the people experiencing relatively mild symptoms. In contrast, the number of deaths might be over-estimated. If so, the mortality rate will be far below 1% of the cases, which is a fraction of the whole Dutch population only. We also face here the specific problem of identifying the primary cause of mortality. If, for example, a patient, who is suffering from a severe cardiovascular disease, passes away and she, or he, also showed symptoms of flu, the cardiovascular disease will be recorded as primary cause of mortality. In contrast, a person without any known co-morbidity, but showing symptoms of the flu, influenza will be recorded as the prime cause of mortality.

Let us suppose now that the data in the example above is correct. We then have scientific findings in the sense of 5% of the population has influenza, of which 1% dies due to the infection by an influenza-virus. A scientific truth, but what does it mean? The meaning depends completely on the scientific view on it. From a medical point of view, we may consider such an influenza epidemic as a relatively mild one compared with, for example, the mortality caused by malaria in the tropics, or non-communicable cardiovascular diseases in high-income countries. We may also consider it from a point of view of population and evolutionary dynamics. Again, a mortality rate of about 0.05% of the population due to the flu may be regarded as rather negligible in the overall mortality rate of a host population. We may even go a step further questioning the role of pathogens as selection forces, in general, because of the relatively low mortality rates. The impact of pathogens has been questioned indeed with respect to shaping plant communities (Harper, 1990; Frantzen, 2007). We may also question prudently the impact of pathogens with respect to human communities from a point of view of population dynamics and natural selection. In contrast, SARS-CoV-2 clearly demonstrated the huge impact a pathogen may have on human communities. We have to go beyond the bio-medical 'facts' to really understand the impact of a pathogen.

We may conclude that 'truth' is a somewhat problematic concept. It has a context and a meaning, which are provided by one, or more, scientific disciplines. Disciplines that each inevitably encompass discussions about scientific findings. So, the love of truth is not a fact-finding exercise but, it is rather an attitude towards exploring research topics in depth. We should not stick to the simple bio-medical 'fact' of a mortality rate of 0.05% in the example above. We need to rate it adopting an inter-disciplinary scientific view. An inter-disciplinary scientific view that is the basis of the epidemiological approach as defined above.

Love of truth is a time-consuming attitude. It starts by defining the right scientific question, the hypothesis. It requires a thorough knowledge of the state-of-the-art of a specific topic. Appropriate methodology has, subsequently, to be developed, or adopted from other studies. We then need to select carefully the materials and sources to be used. The execution of the study is often a matter of failure and restart. Finally, the data has to be analysed and positioned in ongoing scientific debates. It is a tough job requiring, in general, a lot of time, besides the basic scientific skills of a scientist.

Do we have sufficient time to investigate scientific subjects thoroughly? We may doubt about it given an earlier statement in The Lancet (Chalmers and Glasziou, 2009). They estimated that 80-90% of the investment in bio-medical research is a waste of money due to all the issues of, posing the wrong, or even futile, questions and hypotheses, bad methodology, wrong analyses, and so on. We miss here grossly the love of truth. We know that love of truth cannot be put in protocols and directives. It is an attitude of each scientist. Investigating scientific topics in depth is completely different from adopting the common opinion in a discipline, or

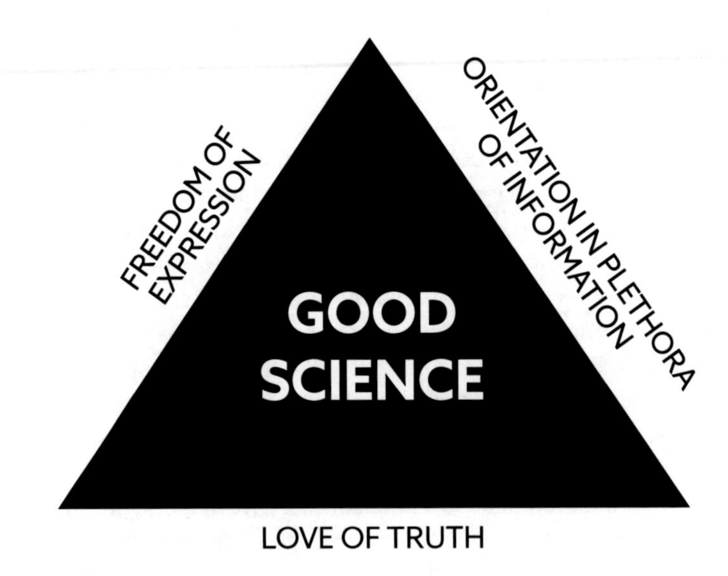

Figure 1.1. Good science seen within a triangle defined by, love of truth, orientation in a plethora of information, and freedom of expression.

the society. It also requires an open mind to other scientific disciplines, especially investigating multidisciplinary phenomena, like epidemics. Each of us faces the challenge of having sufficient time to investigate a topic in depth.

We turn now to the side of orientation in a plethora of information (Figure 1.1). The example of influenza presented in the preceding section was based on a single publication. It describes an epidemic in a single season on a relatively small spot, the Netherlands. We have seen that a scientist cannot use the results of this study straightforward. She, or he, needs to read and interpret carefully the study report. In addition, we may notice that it is just one study out of a tsunami of publications addressing influenza. It, therefore, is quite impossible for an individual scientist to keep track of all relevant literature nowadays. We faced the same problem writing this textbook. Zooming out, selectivity, and collaboration, are the keywords in addressing this problem.

Zooming out means that we step back from the flood of publications trying to arrive at an overview of a specific topic. It is modelling by mind, or computer. Such a modelling is omnipresent in science, from physics to philosophy, and all in between. The drawback of each model is that the assumptions/parameters, which underlie it, determine the outcome of the model. A selection bias is inevitable in modelling. We may notice the more complex the phenomena that are modelled, like epidemics and evolution, the larger the bias. So, we see on the one hand the need to zoom out, and on the other hand the inevitable bias that results. A good scientist is aware of the bias and she, or he, communicates it to peers and other users of scientific findings.

Selectivity is actually zooming in on a topic at, for example, the level of a single gene of a virus trying to explain, for example, an epidemic. The selection bias, which is inherent in zooming out, decreases while zooming in. In contrast, the relevance of the specific knowledge diminishes with regard to the understanding of phenomena at a higher level of organisation, as the epidemic in the example here. The specific knowledge needs to be shared with other scientists in the same discipline, or other disciplines, to increase relevance. It requires good collaboration. Irrespectively collaboration, or not, a good scientist declares the selectivity in her, or his, research preventing un-justified extrapolations of findings. For example, it is impossible to declare that a gene mutation detected in a virus will result in an epidemic without having thorough, additional, scientific knowledge at each of the levels of the individual, population, and ecosystem. We will elaborate on this, especially, in Chapter 5, which deals with evolutionary dynamics of pathogens.

Collaboration is the way ahead to orientation in a wealth of scientific information. It enables set-up of large studies, minimising the statistical bias resulting from too small studies (*cf.* Ioannidis, 2005). Execution of large studies also increases the external validity of the findings by, for example, including more populations, or using telescopes at various sites to explore the cosmos. In addition, large collaborative studies trigger open discussions about the set-up of these, methods to be used, analysis of data/sources, and interpretation of results. Openness that is pivotal in maintaining research integrity (*e.g.*, Forsberg *et al.*, 2018).

Modern technology aids the execution of relatively large and international studies. Electronic data repositories are upcoming and these become accessible by applying the FAIR guiding principles (Wilkinson *et al.*, 2016). The F stands for Findable. A scientist looking for genetic data, or historical text, should be able to discover it wherever it is. Clearly, such a search should be done online. Data, subsequently, needs to be accessible to explore the appropriateness for a specific goal. The A. We are dealing then with the so-called meta-data. It is the description of when and how data has been generated. We also need interoperability of the software, the I, to find and access data digitally. It means finding and accessing data is independent of the software used. It precludes a monopoly of a party like, for example, Microsoft. Finally, data become available for re-use. The R. Re-usable data may not enable large, international, studies only, it may also contribute significantly to reduction of the waste of research investment mentioned above (*cf.* Chalmers and Glasziou, 2009). We like to add here an additional R to FAIR. The R of reproducible. We have outlined above the ongoing reproducibility crisis in sciences. It implies that we should re-use data only that is collected in a reproducible way, as indicated by the meta-data.

We feel institutions, like universities, are responsible for the good quality, reproducibility, of data stored in repositories. The responsibility cannot be delegated to an individual scientist. Such an institutional responsibility would tie in with the Bonn PRINTEGER statement, which is directed to the role of research institutions in maintaining research integrity (Forsberg *et al.*, 2018), and a broader call for action by institutions (Mejlgaard *et al.*, 2020). In addition, institutions may excel by generating high-ranking FAIR-R data instead of producing a massive flow of publications. We may turn from publication-driven science to a FAIR-R data-driven one.

The availability of FAIR-R data does not mean that we have been arrived at Open Science. An owner of data decides to open it for use, or not. A large variety of parties may be owner of data. Universities, governmental organisations, companies like publishers and pharmaceutical ones, are free to open up their data, or not. In addition, these may set conditions for the re-use, like fees, or the need for public reporting of results. The high number, and variety, of data owners has triggered the European Commission to work on a European Open Science Cloud enabling a swift exchange of data between parties willing to make data available, at least to some extent. It is work in progress, but no scientist, no research institution, can refuse to become part of Open Science in the long term *(cf.* UNESCO, 2021). In addition, we feel commercial parties will be forced to participate in Open Science as well.

Collaboration ending up in Open Science has inevitably consequences for the staffing policy in science. We foresee the common pyramidal set-up of research groups of one high-ranking scientist on top, nowadays, will transform into a flat structure, in which people skilled in various aspects of science collaborate. In addition, such a flat research unit operates openly within an international network. Automatic systems will be used to, generate, steward, and re-use data. It implies a major role of, technicians, engineers, and informaticians, irrespectively the scientific discipline we are dealing with. It might be a scary outlook for scientists. We, however, feel it

as an opportunity for scientists who may then focus on the intellectual input, *i.e.*, designing and interpreting studies. It will also facilitate significantly trans-disciplinary approaches, like the epidemiological one.

Science funding is currently focused on the excellence of individual scientists. It does not fit our outlook for Open Science based on flatly-structured units of science in international networks. We, therefore, foresee funding will be re-directed to the higher level of a research group, department, or institute. This, in turn, may herald the transition from excellent individual scientists to excellent science institutes. If so, science institutes will provide the orientation in a plethora of information rather than individual scientists. Good scientists will then be those who are not skilled in a specific aspect of science only, but they will also be very good collaborators within, a team, an institution, and an international network.

Freedom of expression is another side of the triangle of good science (Fig. 1.1). We may wonder about its inclusion, since freedom of expression is declared as a universal human right. Its expression is, however, limited in the freedom provided by national governments and employers. National security and protection of intellectual property are, for example, common reasons to restrict freedom of expression. The freedom of scientific expression may, therefore, also be affected, although we should keep in mind that science is research, but not all research is science. So, freedom of scientific expression should be distinguished from freedom of research expression. We will explain that in the following.

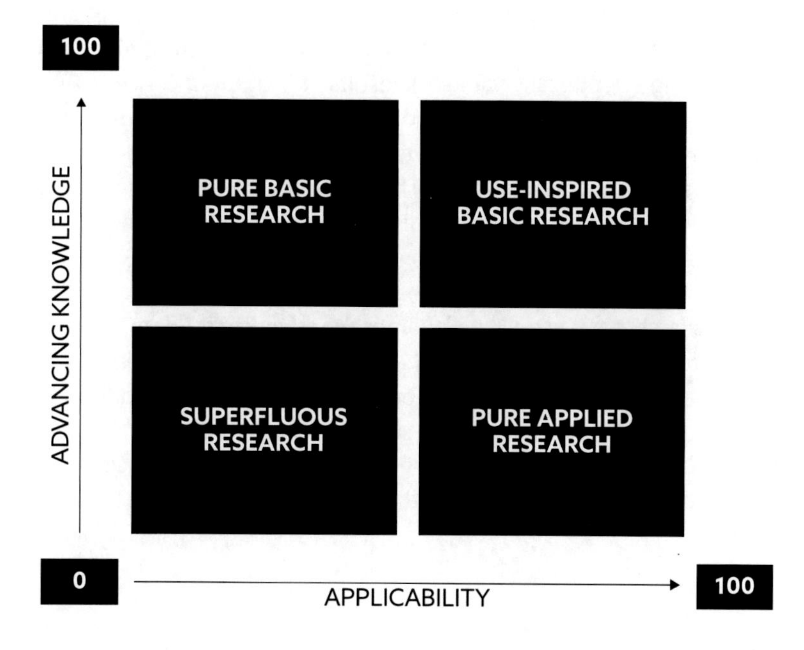

Figure 1.2. Distinguishing four types of research. The advancing of scientific knowledge is indicated on the ordinate and its immediate applicability is indicated on the abscissa. The scaling is not continuous.

We may, in general, distinguish four types of research (Chalmers *et al.*, 2014). The distinction is based on both, the advancement of scientific knowledge and its applicability to a concrete societal issue in the short term (Fig. 1.2). The advancement of knowledge and applicability both approach zero dealing with superfluous research. It is the waste of money, of which we referred to above (*cf.* Chalmers and Glasziou, 2009). Pure applied research results in relatively high applicability of the knowledge generated, whereas the contribution to the advancement of science is relatively low. In contrast, the contribution of the use-inspired basic research contributes relatively much to the advancement of science while the applicability also is relatively high. Finally, we may distinguish pure basic research, which is directed to science rather than application. Classification of research into one of the categories is not straightforward, as we can neither determine precisely the advancement of knowledge, nor the applicability. We, therefore, prefer here a division into science and applied research that is based on the starting point of the research. Is it primarily driven by scientific curiosity, or is it addressing primarily a societal request?

Freedom of expression is, in general, at quite a high level in science. National governments may provide the funding without setting stringent conditions. It is the typically primary funding of universities and (inter)national research institutes. Scientists are relatively free to investigate scientific topics of interest and they are allowed to publish the resulting data. In contrast, applied research is funded to achieve a specific goal, which is set by the funder. The funder may obstruct publication of results that do not fit the goal. The obstruction may be informal, or formally settled in an agreement. Publication of data may also be obstructed because of, national security issues, protection of Intellectual Property, political reasons, and so on. The funder, public or private, decides on the freedom of expression. Such an obstruction is less of importance concerning scientifically valueless results. It is of great concern with respect to scientific knowledge, as it may introduce a serious bias in science. A pharmaceutical company may, for example, prohibit publication of data that indicate a safety risk, or a low efficacy, of a novel type of vaccines. If so, the novel type may be erroneously seen by researchers worldwide as the way forward in treatment of infectious diseases, wasting a lot of funding in developing more vaccines of that type.

We feel applied research has a strong dominance worldwide. The dominance results from large investments in applied research rather than in science. It forces scientists the more and more into applied research, also those at universities. This trend may significantly impair the advancement of scientific knowledge. A good scientist may, however, be able to comply with conditions of a funder of applied research while maintaining freedom of scientific expression. It all starts with an attitude of being aware of the limitations of applied research from a point of view of science.

REFLECTIONS

Titus Brandsma provided a triangle, in which journalists could operate to execute good journalism. We adopted, and adapted, the triangle here with respect to science. The sides of the triangle are defined by, love of truth, orientation in a plethora of information, and freedom of expression. You may reflect about your position inside such a triangle. Alternatively, you may reflect about another concept of good science that may fit better from your point of view.

1.3 Outlook

Epidemiology is a trans-disciplinary, scientific, approach to the understanding of diseases at the population level, taking into account processes at both, the individual and ecosystem level. Good epidemiology, therefore, requires inherently a high input of scientific knowledge from a large variety of disciplines of science. Epidemiologists, subsequently, synthesise all the scientific input to an understanding of epidemics. They may also outline strategies to manage epidemics, requiring again a multi-disciplinary input of scientific knowledge to estimate the pros and cons of interventions. We like to stipulate here that an epidemiologist can never provide the truth about an epidemic and its management. He, or she, can provide a multi-faceted view on epidemics only. A human view.

We feel epidemiologists are currently limited in intellectual input and data to perform their tasks very well. We do, however, expect the upward trend in collaboration between scientists will result in an increased intellectual input. In addition, the availability of FAIR-R data will increase exponentially due to establishing Open Science worldwide. Epidemiologists will also be aided by a high degree of automatization. We may, therefore, conclude that, on the one hand, epidemiologists are facing an enormous challenge, but on the other hand, the developments in sciences will provide them with good tools to execute the job at optimum. We, therefore, expect that epidemiology will rely the more and more on good science in future.

The reader may notice that the writing of a textbook of infectious diseases epidemiology is a rather desperate task, assuming the condition of good science, as outlined above. We, therefore, intend to provide some glimpses of the various facets of infectious disease epidemiology only.

References

Begley, C.G. and Ioannidis, J.P.A., 2015. Reproducibility in science. Improving the standard for basic and preclinical research. Circulation Research 116: 116-126. DOI: 10.1161/CIRCRESAHA.114.303819.

Bod, R., 2020. Een wereld vol patronen. De geschiedenis van kennis. Prometheus, Amsterdam, The Netherlands, 474 pp. [Translation: A world full of patterns. The history of knowledge.]

Brier, S., 2015. Can biosemiotics be a "science" if its purpose is to be a bridge between the natural, social and human sciences? Progress in Biophysics and Molecular Biology 119: 576-587. DOI: 10.1016/j.pbiomolbio.2015.08.001.

Buck, C., Llopis, A., Nájera, E. and Terris, M., 1988. The Challenge of Epidemiology, Issues and Selected Readings. Pan American Health Organization, Scientific Publication 505.

Chalmers, I. and Glasziou P., 2009. Avoidable waste in the production and reporting of research evidence. The Lancet 374: 86-89. DOI: 10.1016/S0140-6736(09)60329-9.

Chalmers, I., Bracken, M. B., Djulbegovic, B., Garattini, S., Grant, J., Metin Gülmezoglu. A., Howells. D. W., Ioannidis, J.P.A. and Oliver, S., 2014. How to increase value and reduce waste when research priorities are set. Lancet 383: 156-165. DOI: 10.1016/S0140-6736(13)62229-1.

Forsberg, E-M., Anthun, F.O., Bailey, S., Birchley, G., Bout, H., Casonato, C., González Fuster, G., Heinrichs, B., Horbach, S., Skjæggestad Jacobsen, I., Jansssen, J., Kaiser, M., Lerouge, I., Van Der Meulen, B., De Rijcke, S., Saretzki, T., Tzewell, M., Varantola, K., Jørgen Vie, K., Zwart, H and Zöller, M., 2018. Working with research integrity-guidance for research performing organisations: the Bonn PRINTEGER statement. Science and Engineering Ethics 24: 1023-1034. DOI: 10.1007/s11948-018-0034-4.

Frantzen, J., 2007. Epidemiology and Plant Ecology, Principles and Applications. World Scientific Publishing, Singapore, 172 pp. DOI: 10.1142/6396.

Harper, J.L., 1990. Pests, pathogens and plant communities: an introduction. In: Burdon, J.J. and Leather, S.R. (eds.) Pests, Pathogens, and Plant Communities. Blackwell Scientific Publications, Oxford, United Kingdom, p. 3-14.

Hemels, J., 2008. Als het goede maar gebeurt. Titus Brandsma adviseur in vrijheid en verzet. Kok, Kampen, The Netherlands, 210 pp. [Translation: The Good should happen. Titus Brandsma as advisor in freedom and resistance]

Ioannidis, J.P.A., 2005. Why most published research findings are false. PLoS Medicine 2: e124. DOI: 10.1371/journal.pmed.0020124.

Mejlgaard, N., Bouter L. M., Gaskell, G., Kavouras, P., Bendtsen, A., Charitidis, C.A., Claesen, N., Dierickx, K., Domaradzka, A., Elizondo, A.R., Foeger, N., Jiney., Kaltenbrunner, W., Labib, K., Marušić, A., Sørensen, M.P., Ŝĉepanović, R., Tijdink, J.K. and Veltri, G.A., 2020. Research integrity: nine ways to move from talk to walk. Nature 586: 358-360. DOI: 10.1038/d41586-020-02847-8.

Munafo, M.R., Nosek, B.A., Bishop, D.V.M., Button, K.S., Chambers, C.D., Percie du Sert, N., Simonsohn, U., Wagenmakers, E-J., Ware, J.J. and Ioannidis P.A., 2017. A manifesto for reproducible science. Nature Human Behaviour 1: 0021. DOI: 10.1038/s41562-016-0021.

Peels, R., 2019. Replicability and replication in the humanities. Research Integrity and Peer Review 4: 2. DOI: 10.1186/s41073-018-0060-4.

Reukers, D. F. M., van Asten, L., Brandsema, P. S., Dijkstra F., Donker, G. A., van Gageldonk-Lafeber, A. B., Hooiveld, M., de Lange, M. M. A., Marbus, S., Teirlinck, A. C., Meijer, A., and van der Hoek, W., 2018. Annual report Surveillance of influenza and other respiratory infections in the Netherlands: winter 2017/2018. National Institute for Public Health and the Environment (RIVM), 140 pp. DOI: 10.21945/RIVM-2018-0049.

UNESCO. 2021. UNESCO Recommendation on Open Science. UNESCO, Paris, France, 36 pp.

Wilkinson, M. D., Dumontier, M., Aalbersberg, I. J., Appleton, G., Axton, M., Baak, A., Blomberg, N., Boiten, J., Bonino da Silva Santos, L., Bourne, P.E., Bouwman, J., Brookes, A.J., Clark, T., Grosas, M., Dillo, I., Dumon, O., Edmunds, S., Evelo, C.T., Finkers, R., Gonzalez-Beltran, A., Gray, A.J.G., Groth, P., Goble, C., Grethe, J.S., Heringa, J. 'T Hoen, P.A.C., Hooft, R., Kuhn, T., Kok, R., Kok, J., Lusher, S.J., Martone, M.E., Mons, A., Packer, A.L., Persson, B., Rocca-Serra, P., Roos, M., Van Schaik, R., Sansone, S., Schultes, E., Sengstag, T., Slater, T., Strawn, G., Schwertz, M.A., Thompson, M., Van Der Lei, J., Van Mulligen, E., Velterop, J., Waagmeester, A., Wittenburg, P., Wolstencroft, K., Zhoa, J. and Mons, B., 2016. The FAIR Guiding Principles for scientific data management and stewardship. Scientific Data 3: 160018. DOI: 10.1038/sdata.2016.18.

Zadoks, J.C., 2017. On social and political effects of plant pest and disease epidemics. Phytopathology 107: 1144-1148. DOI: 10.1094/PHYTO-10-16-0369-Fi.

INFECTION AT THE INDIVIDUAL LEVEL

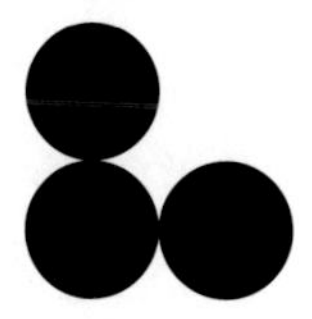

THE TRINITY OF HUMAN LIFE

We deal with the intriguing interaction of, body, mind, and environment, which determines human life, in this chapter. We distinguish these three interacting components, whereas these cannot be seen as independent entities. We are all object and subject in one, and both are determined by our social and non-social environment. And in turn, we affect the environment. We present this 'trinity of human life', and we reflect on it, highlighting our immune system. A system that seems the 'place-to-be' with respect to the triadic interaction between, body, mind, and environment. A system that also has a major role in defence against pathogens. In addition, we will explore the consequences of the trinity of life for the setting of daily health care. We touch on the population level by including the social environment, but the focus of this chapter will be at the individual level.

2.1 The human body

A body is the physical carrier of the phenomenon that we call a human. It is separated from the environment by a barrier that we may call a structural closure (*cf.* Jagers op Akkerhuis, 2010). The structural closure enables an autonomous metabolism within it. We, therefore, do not have a structural closure only, but we do have a functional one, as well. We have a so-called dual closure. It is the hallmark of organisms separating these from non-organisms. The structural closure, however, is not complete, enabling up-take of resources and excretion of waste, as needed for an autonomous metabolism. In contrast, it is closed almost completely with respect to adverse substances, like toxic ones and pathogens. The dual closure is thus characterised by selectivity. We will elaborate on this below. In addition, the human body has intrinsic reflexes, mechanisms, to survive and reproduce, like other organisms do have as well. Reflexes that also serve as responses to pathogens.

Epithelial barrier

Epithelium constitutes the structural barrier between the human body and the environment. It extends from the mouth into the stomach and gut. Epithelium also lines the respiratory tract from the mouth and nose into the alveoli in the lungs. It also covers organs with an opening to the environment to excrete waste, like the bladder. Epithelium may combine the barrier function with one of absorbing, or just excreting, compounds. The structure of epithelium depends on the specific function it exerts. If we look at the skin, it has predominantly a function of barrier. The epithelium of the skin encloses some layers of keratinised cells to make the skin quite impermeable. It inhibits both, uncontrolled loss of water and entry of pathogens, or other detrimental agents. The skin is a major constituent of the human defence against pathogens. The human body, therefore, responds immediately to a breach of the skin, a wound, to prevent entry of pathogens. The response may be supported by a medical intervention, like the use of a disinfectant, or a surgical intervention.

Breaching of the skin is, amongst others, prevented by the secretion of oil to keep it flexible and smooth. The oil is secreted along the hairs. The sebaceous glands secrete the oil in the hair follicles. Glands and follicles are lined with a specific type of epithelium enabling the secretion. Similarly, the sweat glands in the skin are coated with specific epithelium enabling the evaporation of water to cool the body. The loss of defence against pathogens due to the semi-permeability of the epithelium is counter-balanced by the movement of oil and water, respectively, from the inside to the outside of the body. These openings in the skin, however, provide some risk of pathogen entry still.

The risk of pathogen entry increases the further down the respiratory tract we go. The epithelium is relatively thin and permeable in the alveoli, compared with the upper respiratory tract, to enable the exchange of oxygen and carbon dioxide. It, therefore, is quite vulnerable to penetration by pathogens, at least with respect to those being small enough to pass down the respiratory tract into the alveoli. This passage is, however, opposed by the mechanism of muco-ciliary clearance (Bustamante-Marin and Ostrowski, 2017). The mechanism is executed by ciliated epithelial cells lining a large part of the respiratory tract. The cells are covered with a low viscosity periciliary layer of a polyanionic gel. It serves both, beating of the ciliates and restricting access of foreign particles to the epithelial cells, so far these are not trapped by the layer of mucus yet, which is located on top of the periciliary one. The mucus is secreted by specific epithelial cells, the goblet cells.

The muco-ciliary clearance mechanism is quite sensitive to the state of hydration, as this may affect the height of both, the periciliary and mucus layer, respectively. A healthy person will have a periciliary layer with a height of about 7 micrometres, which corresponds with the length of the cilia. The height of the mucus layer is in the range of 2 to 5 micrometres. All cilia beat at the same frequency, in a range of 10-20 Hertz, showing a phase-shift generating a travelling wave. The coordinated movement of the cilia results in transport of the mucus, which traps foreign particles, up through the respiratory tract leaving the body by way of the nose, or by way of the digestive tract. The velocity of muco-ciliary clearance differs between upper and lower respiratory tract (Rogers *et al.*, 2022). It may approach 15 millimetres per minute in the posterior nasopharynx of mice, whereas it may be less than 5 millimetres per minute in the trachea. This capacity of the nose to clear air of, dust, pathogens, and toxic substances, supports a view on it as the primary inlet of air. We may notice that this initial mechanism of clearance, and thus defence against pathogens, is not present in the mouth. Oral breathing, therefore, results, inevitably in a greater contamination of the air passed to the lungs compared with the one resulting from nasal respiration.

The mucus cleared from the lower respiratory tract arrives inevitably in the digestive tract. It will pass downwards even as food and drinks do. Mucus, food, and drinks, are contaminated with pathogens, which also arrive in the stomach and, subsequently, the intestine. The digestion of food is prepared along the digestive tract by the excretion of enzymes, which starts in the mouth already. We, however, need as well the help of microorganisms to digest food, like the tough fibres in the food. So, the lumen of the intestine is filled with, pathogens, commensal micro-organisms, and mutualistic ones. All these need to be inhibited to pass into the blood vessels. A mucus layer, which covers the epithelial cells, serves as a major barrier to prevent such passage (Okumura and Takeda, 2017). The micro-organisms, or the metabolites of these, in the lumen trigger the secretion of mucus by the goblet cells in the epithelium.

The mucus layer is much thicker in the large intestine than the small one as a result of a higher density of goblet cells. This fact corresponds with the higher number of micro-organisms present in the large intestine compared with the small one. We distinguish two layers of the mucus of the large intestine, a firm inner one and a loose outer one. Whereas micro-organisms may enter the loose layer, and may feed even on the mucus there, these cannot invade the inner one. The defence exerted by the mucus barrier in the small intestine, which is thus relatively thin, is complemented by chemical substances, like the anti-microbials secreted by the epithelial cells. So, we see overall in the, small and large, intestine an interesting interplay between protection and opening up by the epithelium. The latter is, of course, needed to enable absorption of nutrients and water.

The urinary tract serves the excretion of liquid waste from the human body, which we call urine. It starts with filtration of the blood in the kidney and the subsequent passage of urine to the bladder by way of the ureter. The urine is collected in the bladder and it is frequently voided by way of the urethra. The urinary tract is actually a closed system, completely lined with epithelial cells, except at the times of voiding. Bacteria, fungi, and viruses have, nevertheless, been detected in urine, especially that of women (Abelson *et al.*, 2018). Urine is not sterile. The non-sterility of urine is not problematic as long as the voiding is complete. If not, titres of pathogens may increase to a level that may enable infection of the bladder epithelium, the urothelium. The voiding may, for example, be incomplete due to obstructions in the urethra, especially among ageing men, or a loss of contraction power of the detrusor muscle, especially among women. We may notice here the relatively large differences between women and men regarding the anatomy and physiology of the urinary tract, whereas the respiratory and digestive tracts are fairly similar among the sexes, so far known.

Sexual intercourse carries a risk of breaching the epithelial integrity. The risk depends on the type and severity of the sexual activity. Intercourse, therefore, is an interesting trade-off between opening-up, as needed for reproduction, and self-protection. In contrast, the lowering of mucus production in the respiratory tract under cold and dry conditions (*cf.* Bustamante-Marin and Ostrowski, 2017) is an example of opening-up the body that is outside our control. It is a sub-optimal functioning of the epithelium due to abiotic environmental conditions.

We may conclude that epithelium is all over the human body to protect the internal, autonomous, metabolism. The type of epithelium varies according to the specific location and function. The epithelium may produce additional protectants, like the keratinised layer of the skin and the mucus within the respiratory tract. Pathogens may, however, cross the additional barriers and enter, or even pass, the epithelium, due to, wounding, dis-functioning/sub-optimal functioning of it, or using the openings needed for execution of the metabolism. In addition, pathogens may employ mechanisms to penetrate and pass the epithelium. Whatever the entrance of a pathogen may have caused, the immune system gets operationalised (Fig. 2.1).

The immune system is a complex of various mechanisms that interact. These are located in various tissues. The system operates somewhere between the epithelial barrier, and its added barriers, and the Central Nervous System (CNS), if we take the triangle-look provided in figure 2.1. It is a very schematic presentation of the human defence against pathogens. It may, however, indicate the relevance of each of the components in the defence against pathogens. The physical barriers, to which the mechanism of constitutive phagocytosis is closely linked, constitute the primary, and major, component of human defence. These protect us against the majority of pathogens that attack us daily. This fact is often overlooked in managing epidemics. We have the tendency to focus on the generation of anti-bodies by way of vaccination. Yet, the efficacy of anti-bodies hinges on the functioning of the mechanisms of, enhanced phagocytosis, inflammation and cytotoxicity, and the complement. So, we feel the role of anti-bodies is over-estimated and we, therefore, reduced it to real properties by positioning it higher up in the triangle. We put the CNS on top, although we do not know much about its' role in the defence against pathogens. It is, however, the link to both, our mind and environment, to which we return in section 2.2. and 2.3, respectively. Here, we will go firstly to the mechanism of phagocytosis.

Phagocytosis

The cellular process of phagocytosis is inextricably linked to Elie Metchnikoff (1845-1916). He was born in a village near Kharkov, or Kharkiv, in Little Russia, which we call nowadays Ukraine (Gordon, 2008). He was a zoologist and embryologist by training. He was appointed as a lecturer at the University of Odessa in 1867. Metchnikoff was just 22 years old at the time of his appointment, like most of his students. In 1882, he described the process of phagocytosis, which is actually the enclosure of extracellular particles by a cell membrane and internalising it into a vesicle, or vacuole, which we call phagosome. An Austrian professor

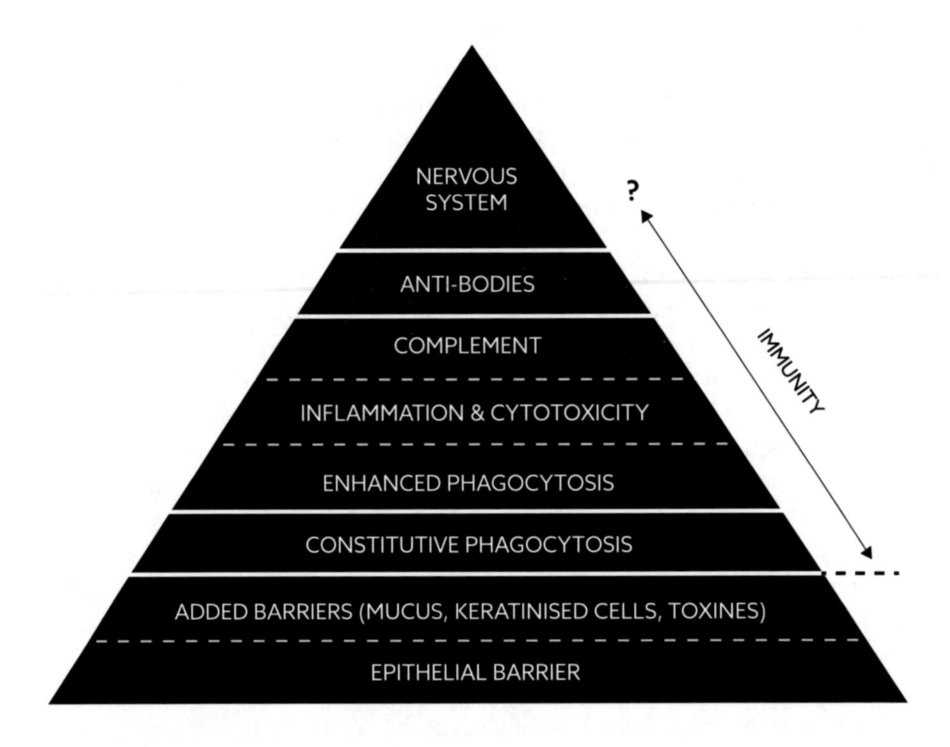

Figure 2.1. Schematic drawing of human defence against pathogens, which goes from constitutive barriers to innate and adaptive immunity, as modulated by the Central Nervous System (CNS). Lines indicate relatively easily identifiable components of defence, whereas it is less for those separated by a dashed line. The positioning of the nervous system is open for discussion.

in Zoology, Carl Friedrich Wilhelm Claus (1835-1899), suggested the general term 'phagocyte' for a cell employing phagocytosis, when Metchnikoff reported the findings to him. In 1888, Metchnikoff moved to the Institute of Louis Pasteur (1822-1895), the famous French chemist and microbiologist. He investigated there, amongst others, cellular immunity. He received the Nobel Prize in Physiology or Medicine in 1908, which he shared with the German physician and scientist Paul Ehrlich (1854-1915). Ehrlich's name is linked to the role of anti-bodies, which govern adaptive immunity, as we will outline below in the specific sub-section. In contrast, Metchnikoff focused strongly on the role of phagocytosis and innate immunity. It resulted in heated, scientific, debates between Metchnikoff and Ehrlich about the exclusivity of innate immunity, or not. Nowadays, we know that adaptive immunity, or acquired immunity, is very specific for vertebrates, and complementary to innate immunity in these, whereas innate immunity can be detected in invertebrates as well. Metchnikoff investigated both, vertebrates and invertebrates, and we may understand that he stressed the importance of innate immunity, although phagocytosis is involved in adaptive immunity as well.

Phagocytosis is a universal biological process, as determined by Metchnikoff. He had to rely on the light microscope in combination with staining of specific cell parts to investigate and describe the process. Our understanding has been deepened by the availability of molecular techniques clarifying the various roles of phagocytosis in organisms and the involvement of specific compounds in it. It serves nutrition in unicellular, eukaryotic, organisms, like amoebas, in so far as it concerns the phagocytosis of particles larger than 0.5 micrometre. This threshold is relevant in order to distinguish phagocytosis from the transmembrane translocation of other particles, substances, by way of other mechanisms. This threshold also explains that phagocytosis does not serve feeding in multicellular organisms. In these, it serves the elimination of, own, decaying cells and relatively large foreign substances, like pathogens. We will focus here on phagocytosis in man and more specifically on the defence against pathogens.

The process of phagocytosis is initiated by the detection of a target substance (Pauwels *et al.*, 2017; Uribe-Querol and Rosales, 2020). A phagocytotic cell is, therefore, equipped with receptors. The type of receptor varies according to the phagocytotic function(s) of a cell. We may distinguish non-opsonic and opsonic receptors. The non-opsonic ones detect directly molecular patterns on the target substance, whereas the opsonic ones do so indirectly by sensing proteins of the body bound to the target substance. Opsonisation means that host-derived compounds are attached to the target substance/cell enabling recognition by the host. We have two categories of non-opsonic receptors. The one encompasses those detecting apoptotic cells. Cell death is pivotal in maintaining homeostasis of the human body. Most of the body cells are, therefore, equipped with receptors to sense dead cells. It also requires that apoptotic cells are labelled as 'dead' providing the signal for phagocytosis. The other category of non-opsonic receptors includes those detecting so-called Pathogen-Associated Molecular Patterns (PAMPs). We may notice here that the non-opsonic receptors may collaborate with opsonic ones to initiate the process of phagocytosis by a cell. The opsonic receptors, which fit body-derived proteins bound to the target substance, may be classified into two categories as well. One category encompasses those initiating phagocytosis of pathogens, as governed by the complement of the human defence (Fig. 2.1). The other category includes receptors triggering phagocytosis of pathogens, as mediated by anti-bodies. We will elaborate on the complement and anti-bodies in sections below.

Recognition of a substance, and more specifically a pathogen, by receptors results in remodelling of the actin cytoskeleton and extension of membrane protrusions around the pathogen. A phagocytic cup is formed. The protrusions seal finally the phagocytic cup resulting in a protruding phagosome. The phagosome gets internalised and it starts to mature by way of fusion and fission with vesicles of the endocytic compartment. It looks like 'kiss and run'. The phagosome fuses finally with lysosomes to become a phagolysosome equipped for degradation of a pathogen, *i.e.*, a low pH, abundance of Reactive Oxygen Species (ROS), and abundance of hydrolytic enzymes. The overall maturation of phagosomes is similar amongst all cells employing phagocytosis, thus all types of phagocytes. It is, however, adapted to the specific function of phagocytes in the immune system. The pathogens are, for example, relatively rapidly and completely destroyed in phagosomes of neutrophils, the white blood cells that operate especially in the acute phase of inflammation, as we will see below. In contrast, pathogens in the phagosomes of dendritic cells are degraded partially only preserving peptides of the pathogens, which serve as antigens for T-lymphocytes operating in adaptive immunity. We pick up this topic in the part directed to the anti-bodies. We may notice here that pathogens may interfere with the maturation of phagosomes. If so, pathogens may survive in the phagocyte and these may even use it to multiply.

Phagocytosis, and especially the enhanced one, cannot be seen independently from other mechanisms of the immune system. We turn to those of inflammation and cytotoxicity (Fig. 2.1).

Inflammation and cytotoxicity

Inflammation is a very general mechanism of the body to cope with tissue injury. The injury may result from, mechanical stress, radiation, toxic compounds, pathogens, and so on. Leucocytes, white blood cells, are pivotal in the process of inflammation. Leucocytes are derived from hematopoietic stem cells in the bone marrow and may develop further, or not, in specific organs of the lymphatic system, like the spleen and nodes. Leucocytes circulate permanently in the blood system. We distinguish five types of leucocytes. The major category, in terms of quantity, consists of neutrophils. The second category regarding quantity, is the one of the lymphocytes. These are sub-divided into, B-cells, T-cells, and natural killer cells. The B- and T-cells execute the adaptive, or acquired, immune response, as we will see below in the part directed to anti-bodies. The three other categories of leucocytes are present in relatively small quantities, *i.e.*, eosinophils, basophils, and monocytes.

Injured cells express Damage-Associated Molecular Patterns (DAMPs) and pathogens express Pathogen-Associated Molecular Patterns (PAMPs). DAMPs and PAMPs are recognised by Pattern Recognition Receptors (PRRs), which are present intra- and extracellularly throughout the body (Amarante-Mendes *et al.*, 2018). Recognition results in death of cells. Three types of,

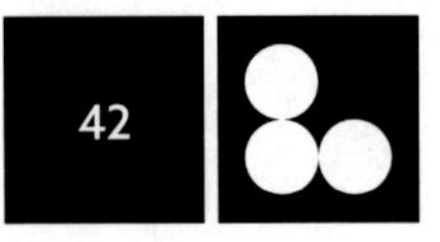

programmed, cell death may be distinguished, (1) apoptosis, (2) necroptosis, and (3) pyroptosis. The type of cell death is determined by the type of PRR recognising the DAMP, or PAMP. The programmed cell death should be distinguished from necrosis, which is an uncontrolled, accidental form of cell death caused by physicochemical insults. Pyroptosis is typically related to inflammation. The PRRs involved are able to assemble intracellularly protein complexes that we call inflammasomes. These inflammasomes initiate the process of pyroptosis, which results in both, the release of intracellular content and the pro-inflammatory cytokines IL-1β and IL-18. An inflammatory environment is created by pyroptosis. Macrophages start to phagocytose the pyroptotic cells releasing additional pro-inflammatory cytokines. Macrophages are constitutively present all over the body. These derive from monocytes in a tissue-specific process, as reflected in the names of the macrophages. We call these, for example, Kupffer cells in the liver and microglia in the central nervous system.

Blood vessels pervade tissues and leucocytes circulate in these, as mentioned before. The secretion of pro-inflammatory mediators by macrophages and mast cells initiates a process of leucocyte migration to a site of injury (Nourshargh and Alon, 2014). Mast cells are omnipresent in the body, like the macrophages. Mast cells function similar to basophils secreting, amongst others, histamine. Histamine has a vasodilatory effect increasing the blood flow, including the leucocytes, to the site of injury. The increased blood flow explains one of the typical, clinical, signs of inflammation, the redness. In addition to the increased blood flow, leucocytes start to adhere to the venular wall at a site of injury (Fig. 2.2). A process of movement of leucocytes across the endothelium and pericyte, subsequently, starts. It is called the Trans Endothelium Movement (TEM). Leucocytes fill up the intercellular space of the injured tissue. We see the second symptom of inflammation, swelling of the tissue.

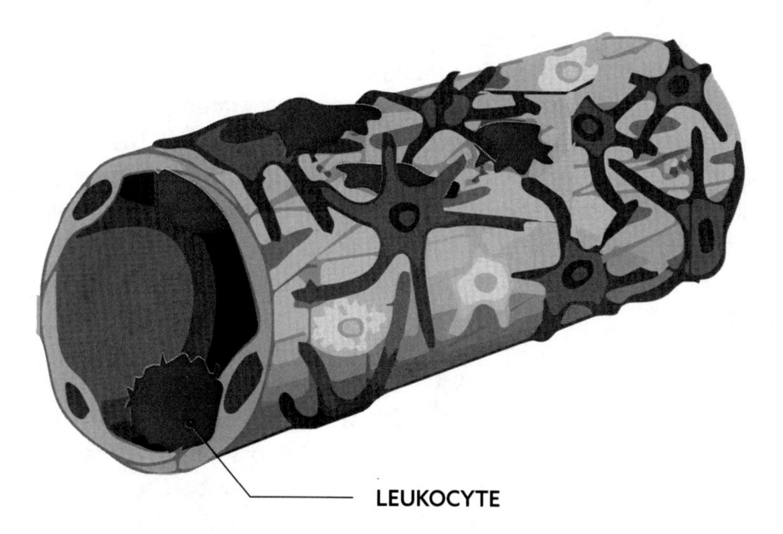

Figure 2.2. Schematic drawing of the Trans Endothelium Movement (TEM) of leucocytes (coloured) from the lumen of a blood vessel into the intercellular space of the extra-vascular tissue. A leucocyte lays above a pericyte at the end of TEM (leucocyte at the upper side of the blood vessel). See text for further explanation.

Each type of leucocyte arriving at a site of injury will execute its specific task. The abundant neutrophils act as additional phagocytes to the resident macrophages. The resident macrophages also phagocytose the neutrophils, once these have completed the phagocytosis of the target substance. Neutrophils are typically quick responders to injury and these show a short life time. Eosinophils target especially multicellular pathogens producing, amongst others, cytotoxic substances. The natural killer cells, a category of lymphocytes, do this as well.

The interstitial tissue fluid, which contains leucocytes, cell debris, blood plasma and pathogens, drains into the lymphatic system. The lymph is filtered in the lymph nodes and cleaned. Swelling of the lymph nodes, therefore, is a characteristic of inflammation. The cleaned lymph, free of pathogens, flows back into the vascular system ready to be re-used.

Inflammation may be very local, like the one following a small wound in a finger. It may become systemic in case of a severe injury caused by, for example, an abundant pathogen. If so, the abundance of pro-inflammatory mediators may, amongst others, trigger the hypothalamus to elevate the body temperature, *i.e.*, a fever. Such a fever is not a side-effect only. It may facilitate actively the control of pathogens, as many of these are rather sensitive to a relatively high temperature. Heat is the third clinical characteristic of inflammation. Inflammation may be temporary, or it may become chronic. An acute inflammation is, in general, followed by an anti-inflammatory response of the body to keep collateral side-damage as small as possible. The onset of this anti-inflammatory response may, however, be inhibited by persistence of an injurious agent, or disfunction of the inflammatory mechanisms involved. We may notice that various cells and compounds are involved in both, the pro- and anti-inflammatory response. Inflammation is a rather complicated process. And so, we arrive at a fourth characteristic of inflammation, pain. We may understand the accompanying pain of inflammation, given the abundance of sensors of the nervous system everywhere in the body. Understanding the cross-talk between central nervous and immune system, and more specifically inflammation, is a real challenge, as we will see in the part 'neuro-immune communication' below. We will turn now to another mechanism of the immune system, the complement. The name indicates the function already. It complements the mechanisms of phagocytosis, inflammation, cytotoxicity and that of the generation of anti-bodies (Fig. 2.1). We will pick up the role of anti-bodies after delineating the mechanism of complement.

Complement

We encountered the name of Paul Ehrlich (1854-1915) in the section 'phagocytosis' above already. He provided the theoretical basis of adaptive immunity, in which anti-bodies are pivotal, as we will outline below later on. He also detected 'something' that complemented the anti-bodies. The term 'complement' was born. We know now that both, the complement is part of the innate and adaptive immune system, and adaptive immunity is based on responses of the innate system.

Complement is basically a surveillance system based on a set of specific proteins (Merle *et al.* 2015). Each of the proteins is indicated by the C of complement, a subsequent number, and eventually a letter to indicate derivatives, *e. g.*, C3a. The complement involves soluble proteins, membrane expressed receptors, and regulators. The activation follows one of three pathways, in which opsonisation is pivotal. One, the 'alternative-pathway' that is based on the circulation of biologically inactive complement proteins, except one. C3 is present in an active, derived, state in a small quantity, as C3a. It is deposited on host cells and foreign ones. Host cells inactivate the protein by way of complement regulators, *i.e.*, the opsonisation 'fails' and the host cells are not targeted by the complement. In contrast, pathogens, in general, lack such regulators and the complement is triggered by way of the opsonisation. Two, the 'lectin-pathway' that also includes opsonisation, but it uses the mannose-terminating glycans of pathogens. So, this pathway is directed to pathogens only. Three, the 'classical-pathway' is based on opsonisation of pathogens using C1-protein-based immune-complexes. Interestingly, opsonisation of apoptotic host cells using the C1-protein 'fails'. Apoptotic cells are subject of 'silent', non-immunogenic, clearance.

Onset of the complement results in an enhanced phagocytosis, stimulation of inflammation and cytotoxicity, and facilitation of the adaptive immune response. The complement has a major function in enhancing the responses of the innate and adaptive immune system. In contrast, it may also alleviate the immune responses enabling the host of both, to tolerate and swich off the immune reaction, where needed. If not, auto-immune diseases may result. The complement may also have a direct, lethal, effect on specific bacteria. It, however, is a minor function of the complement only.

The complement is the final part of the so-called innate immune system. We turn now to the so-called adaptive immune system, which is specific for vertebrate animals including man. The generation of anti-bodies is key in this adaptive immunity. The execution of adaptive immunity, however, hinges on the innate system to truly eliminate pathogens. So, anti-bodies have on the one hand a major function in the adaptive system and, on the other hand, no function at all in the innate one. We, therefore, indicated the overall contribution of anti-bodies to the human defence against pathogens as relatively small (Fig. 2.1). We may, however, notice that these become pivotal in the management of epidemics by way of vaccination. Anyway, we turn now to the generation of anti-bodies and the function of these in the immune system.

Anti-bodies

Anti-bodies are produced by the B-cells of the lymphocytes, which are also called B-lymphocytes. The term 'anti-body' refers to the functional role of binding to antigens. An antigen is a molecular structure alien to the body for whatever reason, *e. g.*, part of a tumour cell, or a pathogen. We refer to the term 'immunoglobulins' from a point of view of the basic, chemical, structure of anti-bodies. The various isotopes of the basic immunoglobulin execute different functions within the immune system. The isotopes are indicated by combining the abbreviation of I(mmuno)g(lobulin) with a capital letter, *e.g.*, IgM. Immunoglobulins consist of a light and heavy chain, which are arranged in parallel like an upsilon (Fernández-Quintero *et al.*, 2021). Each of the chains consists of a stable domain and a variable one. The latter is called the V-domain and it is located at both of the upper ends of the Y-structure. The diversity of immunoglobulins regarding sequence and structure is concentrated in six hypervariable loops in the variable domain, the so-called complementarity-determining regions. These constitute the antigen-binding sites of the immunoglobulins. The complementarity-determining regions are characterised by dynamics, in which the conformation of an immunoglobulin settles finally at the time of binding with a specific antigen. We may notice that such an end-conformation of an immunoglobulin may match more than one specific antigen.

B-lymphocyte receptor genes encode the immunoglobulins (Semmes *et al.*, 2021). Recombination of the genes coding for the V-domain is a first process of the host to generate variety in the immunoglobulins. A second process is the random addition of N-nucleotides at the joints between gene segments coding for the V-domain during recombination. These two processes may together generate 10^{11} different immunoglobulin conformations already. In addition, individual B-lymphocytes may be subject of so-called somatic hypermutation upon contact with antigens, increasing the variety of immunoglobulins further. This hypermutation runs in germinal centres in secondary lymphoid organs, like the lymph nodes. In these centres, the naïve B-lymphocytes also differentiate into memory B-cells and plasma B-cells. The plasma B-cells are also called effector B-cells, as these actually interact with a pathogen, or other target object. The plasma B-cells migrate to the bone marrow to replicate, enabling a massive immune response.

Immunoglobulins may interact directly, or indirectly, with pathogens. The direct interaction is one of blocking physically a pathogen to dock onto a host cell, or an organelle, inhibiting its establishment and multiplication. The indirect interaction is by way of presenting the antigens to T-lymphocytes and, more specifically, the sub-types of CD8+ killer and CD4+ helper T-cells. The antigen-presenting function of the B-lymphocytes is similar to the one of dendritic cells. Dendritic cells have a major signalling function in the immune system, especially in tissues at the periphery of the body.

The maturation of B-lymphocytes is completed upon contact with a specific antigen. The 'recognition' is stewarded in the memory B-cells enabling a swift response upon a new encounter with an antigen, say, a specific pathogen. The maturation of B-lymphocytes in infants, *i.e.*, children up to two years, deviates from the one in older children, especially older than six years, and adults. The abundance of maternal antibodies may explain the differences. These are needed to protect the new-borns, who lack a full-grown immune system still. The maternal antibodies do, however, interfere with the maturation of the B-lymphocytes, as well. The deviant maturation may interfere with the development of long-lasting immunity in infants by way of vaccination. A vaccine is basically an antigen of a specific pathogen, in whatever form, triggering the development of specific memory B-cells preventing a full-blown immune response, and its detrimental side effects, once a pathogen passes the physical barriers of the body. Vaccination of infants is balancing between the waning maternal protection and the timely development of long-lasting immunity by the infant itself.

We have completed the description of main features of the immune system, except the role of the Central Nervous System (CNS). Should we consider it as part of the immune system? Or, does the CNS govern the system, as it does with respect to other bodily systems? We opt here for the governance and we put the CNS on top of the immune system (Fig. 2.1). The part of the triangle occupied by the CNS should, therefore, not be seen as an estimate of the magnitude of its contribution to human defence, as we did for other components of the triangle.

Nervous system

The CNS, which consists of the brain and spinal cord, is the centre of coordination of the body. The CNS communicates with all parts of the body by way of afferent and efferent nerves, the peripheral nervous system. The afferent nerves provide the sensory stimuli, to which the CNS may react by way of stimuli towards specific tissues using the efferent nerves. This is the overall picture of the functioning of the nervous system. Here, we take a closer look at the functioning of the nervous system regarding pathogens. We focus especially on the function of nociceptors.

Nociceptors, also called 'pain receptors', are the efferent nerves that sense stimuli related to potential body damage and the nociceptors transmit these to the CNS. The term is derived from the Latin verb 'nocere', of which the translation is 'to harm'. Activation of a nociceptor is associated with pain. This feeling of pain has a two-fold function. The one is that of a warning, which is needed to operationalise a protecting response of the body. The other is that of prevention. A child that has burned her hand on boiling water once, will, in general, avoid it in future. So, we learn to avoid actions that hurt by way of feeling the associated pain.

Nociceptors may sense pathogens directly, like the lipopolysaccharide of Gram-negative bacteria released during infection, or indirectly by way of sensing pro-inflammatory compounds, like IL-1β, generated by the immune system upon entry of a pathogen (Godinho-Silva *et al*, 2019; Fattori *et al.*, 2021). Induction of nociceptors, in either a way, does not result in a painful warning to the CNS only. It also liberates neuropeptides serving the modulation of inflammation. Modulation of inflammation is needed to minimise secondary damage to the body, while controlling a pathogen. The Calcitonin Gene-Related Peptide (CGRP), for example, may inhibit the activity of neutrophils, monocytes, and macrophages, while it may stimulate those of mast cells and dendritic cells. So, CGRP may both, accelerate and inhibit the control of a pathogen, and the inflammation response related to it. In contrast, another neuropeptide, substance P, has a stimulating effect only. It increases the activities of neutrophils, macrophages and mast cells, and it, therefore, enhances the control of a pathogen. All in all, the actual effect of a neuropeptide on the immune system depends on, the simultaneous release of other neuropeptides, the site of action, and the target pathogen.

Inflammatory diseases, like rheumatoid arthritis, may result from an insufficient control of inflammation. So, we see, a deliberate balancing of the body between pathogen control and avoiding secondary damage to the body due to inflammatory immune responses. We also see that balancing in the daily practice of physicians. Control of inflammatory diseases by way of, for example, administration of analgesics may result in compromised responses to pathogens, as nociceptors are inhibited.

Neuro-immune communication does not hinge on neuropeptides only. Neurotransmitters and hormones also are players in this communication. Noradrenaline, which is also called norepinephrine, is a special one. It is a compound that functions as both, a (adrenergic) neurotransmitter and a hormone. Noradrenaline is involved in various bodily processes including those touching on mental ones, like the one of generating anxiety. Here, we focus on the bodily processes and more specifically the one of haematopoiesis, the differentiation of hematopoietic stem cells in the bone marrow in various types of cells including ones of the immune system, like the lymphocytes. An impairment of the adrenergic innervation of the bone marrow may, therefore, result in a compromised immune system, as we may, for example, observe frequently in ageing people.

We outlined human defence here as a relatively simple modular and bottom-up system (see fig. 2.1.). It is certainly not. It is neither a system governed hierarchically by the central nervous system. It is a multi-dimensional system, in which we see all kinds of interactions, vertically and horizontally, if we take the triangular view presented in the figure. We, however, adopted this particular triangular view to stress the primary role of physical barriers in human defence against pathogens and the secondary one of innate immunity, whereas adaptive immunity has, in general, a tertiary role only. The term 'immunisation' is closely-related with adaptive immunity. An initial exposure to a pathogen, or a vaccine, prepares the immune system for a subsequent exposure and it may react quicker. So, strictly speaking, people are not immune, but their response time to a subsequent exposure to the pathogen is shortened, minimising the collateral damage of an immune response. The term 'pre-disposition' would actually be a more accurate description of this phenomenon than 'immunisation'. We will, however, continue to use the term 'immunisation' as it is very commonly used.

We presented an overall picture of human defence to pathogens here. We see, however, a huge variation in immunity among people and within an individual's life (Liston *et al.*, 2021). Genetic variation, ageing, life style and environmental effects do inevitably result in differences in innate and adaptive immunity among people and among various stages in one's life. We may safely assume that such a variation also exists with respect to the physical barriers.

We positioned the nervous system in a fourth position, as we are just at the start of exploring its role in human defence. The neuro-immune communication, and more specifically the sensation of pain, does, however, indicate that body and mind interact with each other. We may also refer here to the example of patient Anna O. of Sigmund Freud (1856-1939). This Austrian physician, and neurologist, was the founding father of the psychotherapy, which was based on his extensive experience with patients consulting him about their mental problems. Anna O. consulted Freud about her mental problems while she was disabled physically. She couldn't walk. Surprisingly, she was able to walk after a prolonged period of psychotherapy, when her

mental problems were alleviated. And, the case of Anna O. was, and is, not exceptional in the practice of psychiatrists. So, time to turn to the mind, as a second side of human life, in the next section.

REFLECTIONS

The human body faces the trade-off between protecting its autonomous functioning as much as possible and opening itself sufficiently to take up resources and to excrete waste. Defence mechanisms, like the muco-ciliary clearance and immunisation, serve the opening of the body while keeping out detrimental effects of pathogens. We may reflect about the strengths and weaknesses of the structure and functioning of the body. Could we envisage significant improvements of it?

2.2 The human mind

Pain. Pathogens may cause pain as we have seen in the previous section. At least, we know that nociceptors are triggered directly, or indirectly, by pathogens invading the body. These, subsequently, send a signal to the central nervous system. But then? How do we notice pain, become conscious of it? It remains, so far, a mystery, of which we do neither know the mechanism nor where it is located. We may remind that the human body is delineated clearly by the epithelium enabling an autonomous functioning of our metabolism, provided a sufficient exchange of resources and waste with the environment. The body has an identifiable dual closure. We cannot define such a dual closure for the mind, or in a broader sense, the psyche. We do, nevertheless, attribute specific mental characteristics to a specific body completing it to a person, *e. g.*, she is a pretty person, or he is an angry person. This attribution is generated by a so-called third-person perspective. Interestingly, we also look at ourselves from a certain, mental, distance. A position that is, at least virtually, external to the body. It is the second-person perspective. And then, we have the first-person position. The position at which we experience body and mind at the same position.

Pain. A young girl is falling and she certainly feels pain. The first-person perspective. She does, however, not start to cry immediately. First, she looks whether her mother, or another beloved person, is nearby. If so, she starts crying. If not, she continues playing. The second-person perspective. If a parent is present, the reaction of her, or him, determines whether the girl stops crying quickly, or not. A parent getting upset may prolong crying, whereas one staying calm and appeasing the girl may stop the crying relatively quickly. The parent, however, is influenced by the common, societal, opinion regarding the severity of a child falling. The third-person perspective. Now, let us imagine that a subject is requested to indicate pain on a scale of 1 to 10, due to, for example, the bladder pain syndrome. Which perspective, or perspectives, does the resulting score reflect? Is it the one of feeling pain as a first-person perspective subject? Or, at some virtual distance in the second-person perspective? Or, does it include the envisaged opinion and expectations of the physician/researcher, the second-persons' perspective? Or, the common opinion in the social environment? The third-person perspective. In addition, the ratio between the effects of the three perspectives on the pain score varies inevitably among subjects. So, we may doubt, in general, the objectivity of such a score, although we assume it in medicine.

Psychology is the science of the psyche. Psyche, the goddess of soul in the Greek mythology. The soul as an expression of the non-tangible part of a person. The non-tangibility that we see in the mind and behaviour of people still today. Psychology is directed to these phenomena, which emerge consciously and unconsciously. The non-tangibility of the psyche results from its dual character, being attributed to a physical body, while it is simultaneously going around in the air like a 'ghost'. We may call the psyche a 'multi-dimensional phenomenon', which means that we cannot measure it simply using a single, one-dimensional, metric (Desmet, 2018). We have seen pain already as an example of such a multi-dimensional phenomenon having various perspectives, dimensions. Psychology, therefore, faces the challenge of expressing the multi-dimensional, subjective, psyche in objective, one-dimensional, parameters. It, in general, fails adopting a reductionistic approach, in which subjects are reduced to objects. We should not wonder about that. Actually, we are all subjects having, our own life history, our own environment, our own body. That all generates a rather unique, multi-dimensional, person of each of us. How to deal then with objectivity, which may be seen generally as the paradigm of science, in psychology?

Psyche is expressed in a 'stream of consciousness', the flow of psychical experiences that wells up continuously in our mind. It goes from verbal thoughts to sensory impressions, emotions, affects, and bodily sensations (Desmet, 2019). It arises from associative structures, memory networks, crystallising in lived experience. We may consider the lived experience as a narrative, a life history. A narrative that encloses clues to explain mental disorders, and eventually accompanying bodily expressions, like neuroses and psychoses, but also clues to our daily behaviour, like a dominant one, or just a humble one. A psychologist, or a psychiatrist, has the challenge to analyse a narrative and to support a patient in perceiving the narrative in a different way. It sounds a bit like freestyle talking. It is not, adopting, for example, the model of the French psychiatrist and psycho-analyst Jacques Lacan (1901-1981). Lacan provided a purely graphical model that is called the 'graph of desire'. The model completes, and extends, the pioneering work of Sigmund Freud (1856-1939), to whom we paid some attention already at the end of section 2.1 above.

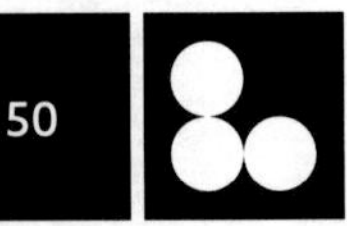

The basic idea of the graph of desire is that the life history, narrative, of a subject is determined by others, or better, the way of perceiving others by the subject (Desmet, 2019). The mother is, in general, the first other that a newborn encounters. She cares for the baby and the baby adapts to her. We may say that all desires of the child are fulfilled by the mother, up to about six months after birth. The baby is, therefore, copying her, or somebody else who is the first other, one to one. It starts to change as a child discovers itself at about six months after birth. It is the start of the so-called mirror stage lasting until the age of about three years. The child senses that it is not the same as the first other. This also opens up space for another other, which is, in general, the father. The others are seen as perfect at this stage still. That changes seriously at the oedipal stage, which is the one between three and six years old. A child discovers the more and more the imperfections of the others. Imperfection should be interpreted here as unfulfilled desires. So, a child starts to discover that others also have desires. Desires that may be hidden from the consciousness of the others even. The child finds itself in a situation of having its own desires, recognising the desires of others, and wishing to fulfil the desires of others. Fulfilling the desires of ourselves, and others, may be seen as a basic mental attitude of humans.

The build-up of a narrative of a subject continues, of course, after six years. The experiences of young childhood will, however, be carried on, especially unconsciously. These, therefore, determine largely the narrative structure, in which new experiences need to be fitted in. The graph of desire expresses the unconscious, former, experiences of the subject with others in the so-called 'lower part'. Signals of the 'lower part' may pop up in the so-called 'upper part' of the graph, which expresses the consciousness of experiences with others. Three types of experiences may be distinguished, (1) the symbolic (2) the imaginary, and (3) the real. Symbolic are the experiences that we express in language. We should notice that the meaning of each word depends on the context and the style of speech. In addition, we are in short of words always to express what we really desire. Imaginary refers to the experiences expressed in images and the real ones are those expressed by the body.

A subject is a multi-dimensional phenomenon whose actual status is determined by past experiences with others, if we adopt the model of Lacan. Most people are able to insert properly new experiences in their narrative of life. Some do not and they may suffer from mental and behavioural disorders. They need support to rewrite their narratives. We stress the term 'rewrite'. The past experiences, the words, remain the same, but the meaning may change just by putting these in another perspective, another sequence. Psychotherapy may provide such a support. It is based on the rewriting of a narrative by the subject itself. The main function of the psychotherapist is to stimulate a patient to speak freely about experiences, especially those hidden in the unconsciousness. In addition, the therapist provides the patient with subtle clues to elaborate on essential parts of the speech. Provision of these subtle clues is pivotal in psychotherapy. The graph of desire guides the therapist in recognising the right clues. We may abstract the aim of psychotherapy as supporting a patient to move detrimental experiences that are stored in the unconsciousness, the lower part of the graph of desire, into the consciousness, the upper part of the graph, and to take these back properly to the unconsciousness. A process of detoxification of past experiences.

The beauty of Lacan's model is providing some logic to something that seems complete irrational as, for example, a neurosis. It provides an objective space, in which subjectivity in all its dimensions can be analysed. An analysis that is per definition retrospective. The graph of desire does not indicate, deterministic, predictions. It explains the chaos of subjectivity retrospectively using just a single graph. A graph that has not been presented here, as it would require too much explanation to understand it fully. We may refer to Desmet (2019) for a full explanation of the graph.

The validity of Lacan's model has been demonstrated in the daily practice of psychotherapy. The focus there is on alleviating existing mental and behavioural disorders. The common focus on therapy may mask the fact that two major aspects of Lacan's model may be relevant for people in general, including those who do not suffer from disorders. Each of us encounters others throughout life determining our personality, our self. That is one. We also have the intrinsic desire to fulfil the, unspoken, desires of those others we encounter. That is two. The fulfilment of a desire may become even that strong that we sacrifice our body. We see it among, terrorists, suicides, soldiers, anxious people, missionaries, staff of emergency services, people donating a vital organ, and so on.

Our ratio is based in the psyche. We use it to bring some order to the chaos of life. The Greek called it 'logos'. The irrational also has a basis in our psyche, 'mythos' as the Greek called it (Armstrong, 2009). The use of myths is another way of coping with the chaos of life. It is actually transcending the chaos. We put events, which we cannot understand, into a higher order, accepting the non-understanding. People are generally in need of myths, whether we call these, religion, ideology, humanity, or just simply love. The desire for transcendency seems as old as *Homo sapiens*. In addition, myths set us into action creating, parts of, these in reality. Mysticism has received specific attention throughout ages (Underhill, 1911). People get a notion of the, transcendental, 'true reality' and some of them, the pure mystics, get merged with the Absolute by an extreme extension of the consciousness surpassing the sensory system of the body. The mystics are, in general, not able to express their experience in human language. They use symbols, which are often related to a religion.

The British novelist Evelyn Underhill (1875-1941) became an expert in mysticism without having any higher education. She distinguished, (i) the body generating sensory experiences, (ii) the mind generating psychological experiences, and (iii) the soul generating spiritual experiences and it is the soul that merges finally with the Absolute. The merge with the Absolute is the final stage of the spiritual path that has to be followed. It starts with the phase of awakening, becoming aware of the transcendental. This phase may be initiated by a disruptive event, like a burnout, or a beloved-one passing away. People may then enter a phase of purification, in which they try to get rid of the illusions of our sensory world. It is a heavy, painful, phase followed by a phase of enlightenment. We get a view on the Absolute. Glimpses of the Absolute are quite common among artists, scientists, and just ordinary people. These are the moments that you transcend the 'selfish', the borders set by human time and space. Few people will enter, and pass through, the next phase of darkness. The phase that the final parts of the 'selfish' need to be removed in a process of real distress. You need to be fond of the Absolute to pass this phase, which may take years and years. If so, you will arrive in the last phase of Unity. You are a real mystic.

A mystic is not necessarily a person being in permanent contemplation. It may be the contrary even. They are very active people whether we are dealing with Hildegard of Bingen (c. 1098-1179), Teresa of Ávila (1515-1582), Mahatma Gandhi (1869-1948), Albert Schweitzer (1875-1965), or Dag Hammarskjöld (1905-1961). We may also notice some interesting medical phenomena among mystics. They do not respond to any stimulus of the environment during periods of trance. They don't hear, or see, anybody, they do not even feel pain of, for example, needles pushed into their skin. The body cools down and the heartbeat goes down. In contrast, they are completely awake during the trance, according to their own opinion. In addition, such episodes of trance, which may last for hours and days, may be followed by symptoms of a burn-out underpinning the intensity of the experience.

The experiences of mystics show some similarities with near-death experiences. These are well-known in medicine. The near-death experiences received relatively little scientific interest until the Dutch cardiologist Pim van Lommel initiated a relatively large, prospective, epidemiological study in the eighties. Patients who resuscitated after cardiac arrest were eligible for the study (van Lommel *et al.*, 2001). The patients were included consecutively to avoid selection bias. Three-hundred-and-forty-four patients were included during about four years. These were admitted to 10 Dutch hospitals in total. The majority of patients who resuscitated within a hospital had a circulatory arrest of less than two minutes and the unconsciousness lasted less than five minutes. In contrast, the majority of patients resuscitated in an ambulance experienced a circulatory arrest longer than two minutes and unconsciousness longer than 10 minutes. All 344 patients were interviewed shortly after resuscitation. The majority of these, 82%, had no memory at all about the period of near-death. Others had memory that ranged from some recollection to a very deep near-death experience. Awareness of being dead and positive emotions were reported most frequently as near-death experiences. A completely dysfunction of the body seemed to be associated with an extension of the consciousness.

Near-death experiences may have long-lasting effects on life. Matched cohorts of subjects reporting near-death experiences (35 subjects), or not (39 subjects), were interviewed again two years after resuscitation focusing on changes in their life. Those with a near-death experience reported positive changes in their life significantly more often than those without. The changes encompassed the attitude to, society, religion, death, meaning of life, and just ordinary daily matters. Changes seemed to persist, as indicated by subsequent interviews eight years after resuscitation. The numbers of subjects in both cohorts, however, were quite low by 23 subjects and 15 for near-death experience, or not, respectively.

Van Lommel used the clinical finding to explore the phenomenon of 'consciousness' further in a more philosophical sense. He postulated that consciousness may remain after the death of the body. A kind of separation of body and mind *in extremis*. Van Lommel admits that his axiom is rather impossible to demonstrate. It is like the zero hypothesis in statistics.

We may infer that our psyche may be the basis of two narratives, the rational one focusing on the physical world and the mythical one focusing on the transcendental world. The two narratives are actually expressed in one narrative encompassing various proportions of rationality and, as we might call it, irrationality. We, therefore, have as many narratives as we have people on earth. Each narrative has its own dynamics and each narrative determines the way of coping with, for example, infectious diseases, in both, a bodily and behavioural sense.

Each narrative of life is based on experiences. Experiences with pathogens, or other living organisms, the biotic environment. Experiences with others, the social component of the biotic environment. But, also experiences with the a-biotic environment, like, drought, heat, cold, or whatever. Experiences that are primed in the body and the mind and these, thus, predispose us to new experiences with our environment. So, it's time to turn to the third side of human life, the environment.

REFLECTIONS

Psyche is a complex, multi-dimensional, phenomenon encompassing conscious and unconscious experiences that determine our mind and behaviour, our narrative of life. The narrative is shaped by experiences with others triggering the question of what the 'self' of us is? We may reflect about the potential answers to this question. In addition, we may wonder whether the body is governing the psyche, or that the reverse is true. Let us just reflect about the consequences of the one option, or the other, regarding health and disease.

2.3 The human environment

We may define the environment at various scales, from the very micro one, in which an individual lives, up to the macro-scale of the world. The environment is also composed of a variety of abiotic and biotic factors, like temperature and pathogens, respectively. A special category of the biotic environment is the one of our social contacts. We are part of a family, have neighbours, colleagues, are citizens of a country, belong to a race, a specific socio-economic class, live on a continent, in a climate zone, in an urban environment, or just the country side. These are relatively stable factors. The environment also encompasses very dynamic factors, like the weather, traffic accidents, loss of a beloved one, birth of a child, and so on. The environment of man is, all in all, highly diverse and dynamic. The probability of encountering specific pathogens, therefore, depends on the specific environment, in which we are living at a certain moment. People living in the tropics, for example, have a high probability of being infected by one of the species of *Falciparum*, the cause of malaria. And so have deprived people a higher probability of suffering from cholera, which is caused by *Vibrio cholerae*, homosexual men are more likely to catch the HIV-virus causing AIDS, children have a higher probability of getting measles than adults, and the risk of the flue is higher in winter than in summer.

We sense our environment, like every organism, in various ways. We use our eyes, ears, nose, and skin to pick up stimuli and to react to these, or not. If so, the reaction may be immediate, especially, in case of danger to our body. Such a reaction may happen in a split second without any reflection. We may also reflect firstly on a reaction delaying it. The reaction may also require a continuous process of adaptation over generations resulting in, for example, a dark skin to protect the body against high UV-radiation. The duration of a reaction, therefore, depends as well on the presence of a stimulus for a shorter, or longer term. If longer, the adaptation gets fixed. The fixation concerns our genetic make-up in case of a bodily reaction, like the various responses of our immune system. It is in our culture, in case of behavioural responses, like washing our hands to reduce the exposure to pathogens.

We are bombed with environmental stimuli, 24 hours a day, 7 days a week. We are filtering these continuously, but many remain that trigger a conscious, or unconscious, response of us. A response, action, that results from a certain tension, stress, that is caused in our body by the stimulus. We may, therefore, regard such triggering stimuli as 'stressors'. The response triggered by a stressor is one of counter-balancing the impact of the stressor on the functioning of our body and mind. We call it resilience of body and mind. We see it, for example, in the down-regulation of our body temperature to about 37 °C after an inflammatory response of clearing a pathogen, or recovering after a burn-out.

We elaborate on resilience here adopting a psychological view. We, therefore, assume that psychological and bodily responses are intertwined. It is likely, as we highlighted it for pain above. We, thus adopt a model proposed by IJntema *et al.* (2023), although it was explicitly presented as one of psychological resilience. The model encompasses two mechanisms, tolerance and constructing narratives, and six possible outcomes of (non-) resilience (Fig. 2.3).

Tolerance is just not responding to a stressor. You sense the stressor, and its' potential impact on your life, but you do not defend yourself against it. You are able to bear it mentally. Pain caused by a stressor, for example, is tolerated by one person better than by another one, although they may experience the same injury of the body, or mind. We will elaborate on the specific tolerance of pathogens in the Chapters 5, 9 and 10. Tolerance is not an on/off-mechanism, although it is suggested in figure 2.3 for ease of presentation. The outcome of tolerance may show various levels of enduring a stressor, *e. g.*, tolerating a bit of pain, some, moderate, or heavy pain. The magnitude of tolerance may depend on so-called pre-adjustments, say previous experiences with a specific stressor. Tolerance in turn affects the outcomes of the second mechanism of resilience, the construction of narratives. A typical mental mechanism. It is the ability of a person to make sense of experiencing a stressor, or in terms of Lacan's model, to detoxify the experience consciously passing it, subsequently, to the unconsciousness.

The construction of narratives may follow the route of assimilation. Assimilation is the incorporation of the experience with a stressor in the existing narrative of one's life without re-writing it. In contrast, accommodation is re-writing the narrative using the experience with the specific stressor. A racing driver, for example, may assimilate a car accident during a race in his narrative of life, whereas a common car driver may need to accommodate it. She, or he, does, in general, not anticipate an accident, whereas the racing driver does.

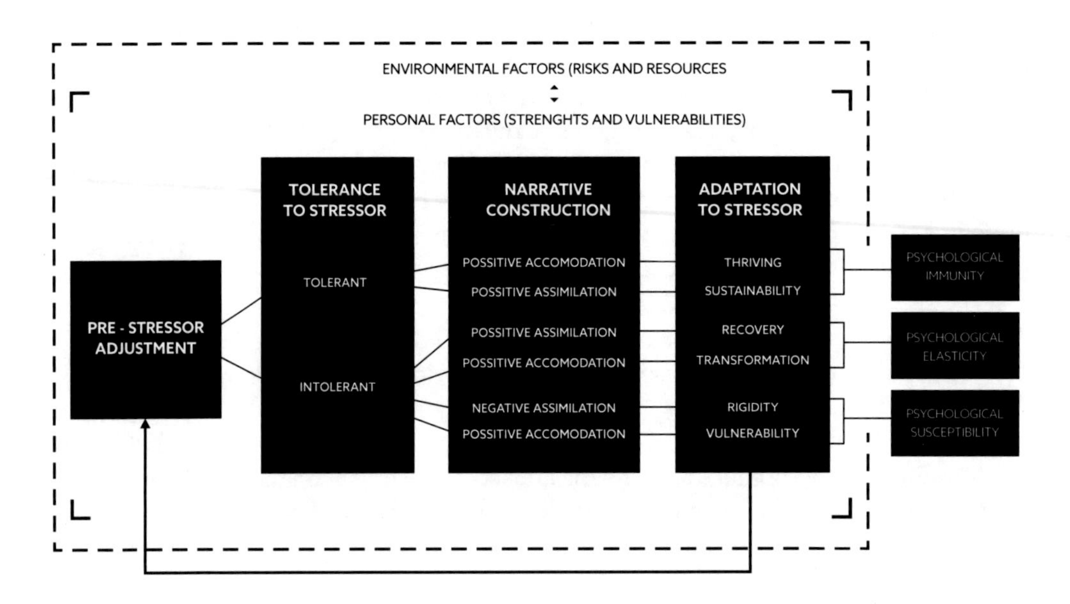

Figure 2.3. A model of psychological resilience. Tolerance is indicated as a dichotomy for ease of representation. See text for further explanation.

A (re-)construction of the narrative may result in various outcomes depending on the level of tolerance in combination with assimilation, or accommodation, of a stressor in one's narrative of life. We may assign the outcomes to three major categories, (i) immunity, (ii) elasticity, and (iii) susceptibility. Immunity means a surplus of the personal functioning after exposure to a stressor protecting a person against a subsequent exposure to a similar stressor. It fits the bodily immunity regarding pathogens, if we restrict it to adaptive immunity. Immunity may be achieved by way of thriving, a positive accommodation of a stressor in the narrative, or sustainability, a positive assimilation of the stressor. We may see immunity as a surplus of resilience. The personal functioning is better after the stressful experience than before. The category of elasticity is resilience in the strict sense. The functioning of a person bends back into the one preceding the exposure to the stressor. The overall effect of elasticity is zero. Elasticity may be achieved by way of recovery, a positive assimilation of a stressor in a narrative, or transformation, a positive accommodation of it. Susceptibility, the third category, is actual a lack of adaptation, a lack of resilience. The stressor is negatively assimilated in the narrative in the case of rigidity, or negatively accommodated in the case of vulnerability. Vulnerability means that a subsequent encounter with the stressor will result in additional damage. It is the worst-case scenario.

We adapt our body and mind continuously to the environment. It is one way of dealing with the diverse and dynamic environment, in which we live. We also did, and we do, the opposite. We adapt the environment to our desires. An adaptation exerted by consecutive generations for millennia. We domesticated plants and animals to assure our feeding, protected ourselves by shelters and houses, we developed engines to execute the heavily physical work, we developed the world-wide web to communicate with anybody worldwide, and so on. We were that successful in adapting the environment that we became the dominant animal species on earth. The exponential growth of *Homo sapiens* did, also, result in an exhaustion of natural resources, a toxification of our environment, an accelerated climate change, and a loss of resilience. Our behaviour towards the environment is meanwhile threatening our own health (Inauen *et al.*, 2021). The threat is that severe that a new (sub-)discipline of science emerged, environmental health psychology. It takes into account the reciprocal effects of man and environment investigating human behaviour that may be beneficial for both, our environment and health. It surpasses the common setting of healthcare, in which this reciprocity is hardly addressed, if at all. We will turn to that common healthcare in the next section.

REFLECTIONS

We have to cope with a diverse and dynamic environment, as did our ancestors before. We developed, and develop, bodily and mental responses to cope with all the stressors, biotic and abiotic ones, of the environment. Mankind did not adapt only to the environment. It also started to change the environment, which accelerated considerably over the last century. We may reflect about the pros and cons of both strategies of coping with the environment regarding our health at the individual level, our own life.

2.4 The setting of healthcare

We look here at the perspectives of the first person, the second, and the third one, which we mentioned earlier in section 2.2. We feel our body and sense the environment from a first persons' perspective. We, also, look virtually from a certain distance at ourselves, it is a kind of second persons' perspective, and we start to interpret our feelings, our experiences. We notice consciously, or unconsciously, the fragility of ourselves. This notice may result in some concern. We may also develop fear of, or anxiety for, threats, which may be real or imaginary, to our body and mind. We cannot mark sharply the borders between, concern, fear, anxiety, and panic as an extreme of anxiety.

Fear is a primary emotion that is common among animals (Heeren, 2020). Fear is directed to a clearly identifiable threat that is present in one's environment. The response to the threat is one of 'fight-or-flight' and it encompasses the typical signs of, elevated blood pressure, pupil dilatation, and so on. The response exhausts the energy reserves and it should, therefore, be as short as possible. Continuous fear may result in mental disorders, like phobias. In contrast, anxiety is a future-oriented emotion. It anticipates a less identifiable and uncertain threat. Anxiety generates a long-term response characterised by physical tension and avoidance of situations, in which the threat might occur. Excessive anxiety may result in mental and physical disorders, like the obsessive-compulsive and cardiovascular ones. We may, roughly, see an increase of severity of expressing the awareness of our fragility going from no concern at all to panic (Fig. 2.4). The predictability of an expression shows just an opposite trend, at least from a point of view of a third person. We may understand the concerns of somebody else. We may imagine fear of another person, although we do not experience a similar fear. We get out of understanding, and therefore predictability, in the case of extreme fear and anxiety. If we look

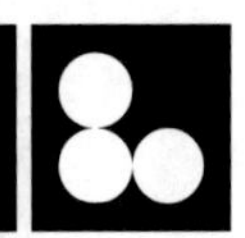

at pathogens, epidemics, we notice a hybrid character, as those are identified by experts, but for most of the people these remain invisible. So, even if we may identify a specific pathogen as a threat, it may feel for most people as a less identifiable, and uncertain, one. Pathogens may cause extreme fear and anxiety among people.

We may replace the term 'fragility' by the term 'health'. We have, concerns, fears, and anxieties regarding our health, especially adopting the definition of the World Health Organisation (WHO). It defines health as "a state of complete physical, mental and social well-being and not merely the absence of disease or infirmity". Reading this, and especially the term 'complete, it seems impossible to call anybody healthy. The weakness of the WHO-definition has been recognised and alternative definitions, or we may better state 'concepts' here, have been proposed (van Druten *et al.*, 2022). A proper one-size-fits all concept of health seems rather impossible. Health is a, subjective, multi-dimensional phenomenon. We may deal with the lack of a common concept of health as long as we follow the common trajectory in healthcare starting at the individual who feels ill. It changes as soon as we arrive at the population level. We are then talking about public health and policies to improve it. Or, do we mean simply control of a specific disease? If so, we may take measures to take control of a specific disease, like a lockdown, that are harmful with respect to the prevention and treatment of other diseases, and thus public health. We, therefore, do not reduce the role of epidemiology to provenance of knowledge for barely control of a specific disease here. We will take a broader view. Epidemiology should facilitate appropriate, human, management of epidemics serving public health at maximum. We will develop some feeling for 'public health' as we go from chapter to chapter in this book rather than coming up with a new concept of health.

Let us assume that a subject is healthy as long as she, or he, feels so. It changes as soon as any physical, or mental, complaint comes to the mind. The subject feels ill, as we call it. The perspective of the first person in collaboration with the virtual second-person perspective. She, or he, consults an expert, if not able to deal with the complaint in the daily setting of life.

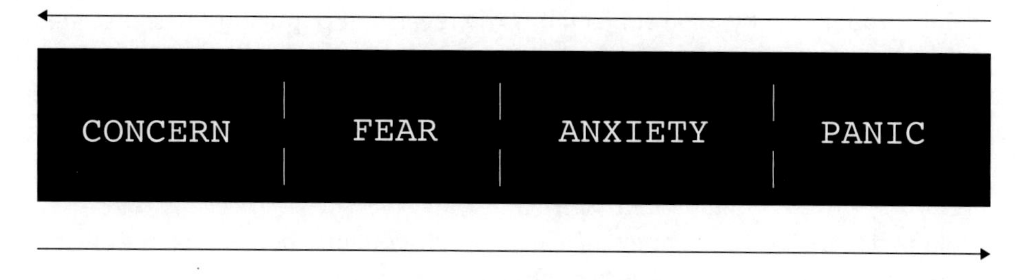

Figure 2.4. Simplified scheme of expressing, and observing, feelings about our fragility from a first- and second-persons' perspective, respectively. See text for further explanation.

The real second-person perspective comes into play. Such an expert may be a physician, a medicine man, or whoever, depending on the society, the culture, in which a subject is living, and the nature of the complaints. We will refer further to an expert as 'physician', thus in a very broad sense. The consultation of a specific 'physician' depends not on the type of complaints only. An ill subject also needs to trust a 'physician'. A trust between physician and subject that is needed to arrive at a proper diagnosis of the complaints. It is such an essential condition that the freedom of consulting 'physicians' is commonly part of national legislation. This freedom is extended to the acceptance of a treatment by a subject, or not. People cannot, in general, be forced to accept a treatment proposed by a 'physician'.

The diagnostics of complaints starts, in general, with a confidential conversation between subject and 'physician'. The subject presents its' narrative of the complaints. The 'physician' will trigger the narrative in such a way that it fits the own profession. A psychiatrist, for example, needs to hear a different type of narrative than an internist. This guidance is, in general, the Achilles heel of health care. The choice of the 'physician' determines largely the outcome of the diagnostics. It is relatively straightforward in cases of obvious disorders, like a broken leg. It is less so for general complaints, like fatigue.

A 'physician' analyses the complaints, the narrative, of a subject and the 'physician' arrives at a diagnosis using, eventually, additional tests, like a physical examination. The 'physician' gets part of the narrative of a subject and the subject turns into a patient with a specific disease. So, we see that the original illness, in the first- and virtual second-person perspective, turns into a disease diagnosed by a 'physician', a real second person, who uses the knowledge of its' peers, who provide the third-person perspective in the process of diagnostics. The diagnosis of a disease is an abstract of a narrative that fits a category agreed among 'physicians' dealing with the specific complaints. The 'physicians' themselves are, however, affected by the society, in which they are living. The society may declare subjects as 'sick' regarding their body and mind. 'Physicians' may, subsequently, pick up such societal signals creating a 'disease', or accommodating an existing disease typology. The Attention Deficit Hyperactivity Disorder (ADHD) and Erectile Dysfunction (ED) are some examples of a 'sickness' turning into a 'disease' of specific 'ill' people. The process of diagnosing, however, remains the same, starting with complaints brought up by an ill subject. It is the reverse in population-based screenings and vaccination campaigns. The aim of these is diagnosing and preventing hidden illness, respectively, or in other words, declaring people sick. The freedom of participation is, in general, guaranteed legally. Society may, however, exert a powerful, mental, force on an individual to participate in a screening.

Diagnosing a disease is one. Treatment of a disease, subsequently, depends on the perception of the treatment by the patient. The perception is affected first of all by the relationship between patient and 'physician'. Is it trustful, or not? The perception also depends on a weighing of the concerns/fear regarding the disease and the side effects of the treatment. Do, for example, the concerns/fear of an overactive bladder (OAB) outweigh the concerns/fear of an injection of botulinum toxin in the bladder muscle? The narrative of life of the patient, including eventual

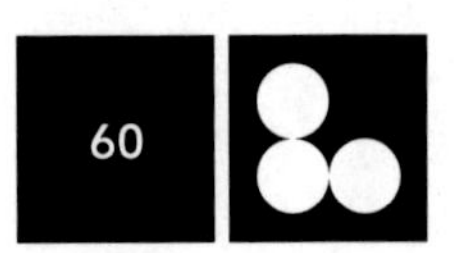

experiences with OAB and injections, affects the answer to this question. In addition, the social environment influences the view of the patient on the proposed treatment. Does it, for example, favours medical interventions, or just not? Finally, the severity of the disease symptoms, the prognosis of the treatment, and the availability of an alternative to deal with the disease, determine whether the patient accepts the treatment, or not. Accepting a treatment becomes more complicated even in disease prevention, like the vaccination against SARS-CoV-2, as we do not have the disease then, or not yet. We need to estimate the risk of getting ill. We feel, that is not easy for a subject in a world full of framing and opinions.

Addressing properly the questions around acceptance of a treatment, or not, is first of all a matter of the fundamental right to freedom of life, as laid down in the Universal Declaration of Human Rights in 1948. A treatment needs to fit the life, as valued by a patient. An elderly person may, for example, experience a demanding surgery quite differently from a young adult. The freedom of an individual is, however, not unlimited. The Declaration also stresses the social context of enabling the life of an individual. An individual, therefore, has duties regarding the social environment, as expressed in a partner, children, friends, neighbours, and so on. A patient with a poor prognosis may, for example, be requested to abandon a graft of a scarce organ in order to aid a patient with a better prognosis.

A deliberated decision regarding a treatment is also needed from a point of view of efficacy. We enter the topic of the effects of placebo and nocebo on health (Pardo-Cabello *et al.*, 2022). A placebo is, in general, included in Randomised Clinical Trials (RCT), in which new medical treatments are tested. A new treatment is applied to one group of trial subjects, whereas a new treatment without the presumed active component, the placebo, is applied to another group. Subjects are randomly assigned to one group, or the other, and the subjects are, preferably, blinded regarding the treatment. Blinding is relatively easy with respect to testing drugs. It is less easy regarding, for example, surgical interventions and psychotherapy. We may notice that the term 'sham treatment' is also used to indicate a placebo in studies not directed to drugs. We keep the term 'placebo', in a broad sense, here. Analyses of data of RCTs and other studies including a placebo indicate that the health status of subjects in a placebo arm also improves. This effect is referred to as a 'placebo' one, in reference to the verb 'placere' in Latin, which means to please. So, the term 'placebo' refers to both, the pseudo-treatment and the type of effect resulting from the pseudo-treatment. In addition, a placebo may result in a nocebo effect. It refers to the verb 'nocere' in Latin, which means to harm. Subjects, who receive a pseudo-treatment in a study, may show side effects related to the treatment tested in the study. Such side effects are reported, but without the notion of being a nocebo effect often.

An unequivocal determination and quantification of placebo, and nocebo, is rather difficult. The inclusion of a 'non-treatment' arm would be required in a randomised study design. Blinding of the non-treatment to the subjects is then per definition impossible. In addition, a non-treatment may be unethical, depending on the severity of the disease and the availability of an efficacious treatment, other than the one under investigation. We may notice that the use of a placebo may be impossible as well in such a situation. Evidence of placebo and nocebo is, therefore, provided by rather descriptive data. This data, however, provides substantial evidence of abundance of placebo and nocebo in treatments of a variety of disorders, especially as we get meanwhile more grip on the underlying mechanisms. We may, therefore, conclude that, amongst others, a trustful relationship between subject and 'physician' is a basis of generating placebo effects rather than nocebo-ones in a treatment.

The setting of healthcare varies considerably around the world. We cannot identify a common setting, or focus, of healthcare. The setting may be directed especially to bodily complaints, or to mental ones, but the setting may also enable a more integrated view on health (Fig. 2.5). An integrated approach to health seems to be the characteristic of a setting of more traditional medicine. In contrast, a clear separation between mental and physical disorders is, in general, a major characteristic of healthcare in Western countries. In addition, the environment may be regarded either as a pool of health threats only, or one providing opportunities for resilience as well. An individual experiences inevitably various stressors in its professional and private daily life, as outlined above. We may reduce the severity of the stressors, the threats. It is one way to deal with stressors, but it is per definition insufficient to cope with all daily stress. We, therefore, need each day a period, or periods, with relatively little stress in order to build up some resilience. We may imagine a lot of ways of to build up resilience, like sleep without disturbing noise, fresh air, some relaxing physical exercise, pleasant social contacts, and so on. A healthy environment offers such opportunities to build up resilience.

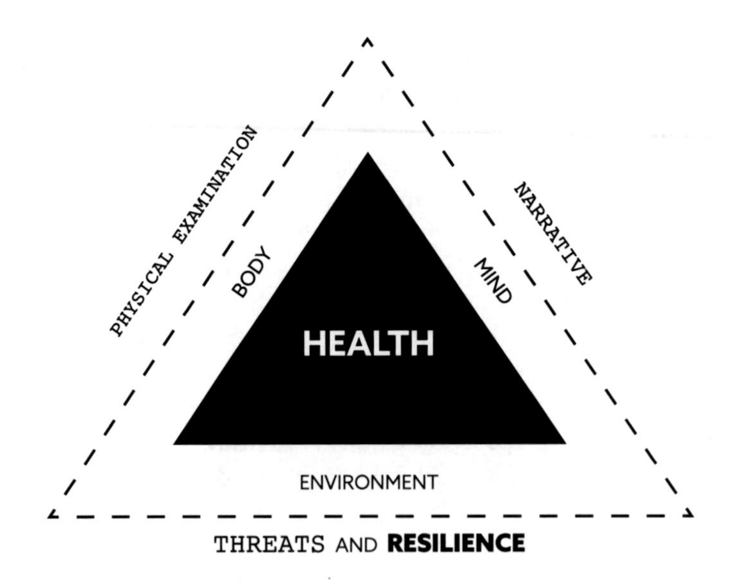

Figure 2.5. Interaction between body, mind and environment determining human life. The setting of health care is expressed in the physical examination of the body, including all kinds of tests, the narrative generated by mind, and the environment, being both, a pool of threats to health and an opportunity to strengthen resilience. See text for further explanation.

The funding of healthcare also varies considerably among, and within, countries, from completely public to completely private, and all in between. Companies have, in general, a prominent position in healthcare by way of providing, insurance, drugs, medical devices, equipment, staffing, and so on. Hospitals may also be run by private parties. Healthcare is a global market meanwhile. Clearly, the products can be bought only by people, governments, having sufficient funding. In addition, healthcare was, and is, a major issue in politics. It sounds great. It, however, is not according to the Italian philosopher Giorgio Agamben (Reader, 2022). He saw human life reduced to bare, bodily, life at the onset of the COVID-19 epidemic in Italy, due to a governmental policy of social distancing and closing down of public facilities and events. He wondered about the willingness of Italian people to minimise their narrative of life to a simple survival of the physical body. He accused the Italian government of 'bio-politics', forcing citizens to give up their mental life, their life, just for saving the physical body, while serving the benefits of industry. Agamben also denounced the obligation of wearing face masks as a killer of public life and debate. He sees participation of citizens in politics at risk. He warns against a turn from humanism to barbarism even. Agamben may overstate the case, but he highlighted some logical consequences of subordination of public health to control of a single disease. In addition, he questioned actually the dominant position of healthcare in society.

The Dutch physician and philosopher Marli Huijer also addressed the topic of bare, physical, life. She did it in a context of people approaching the end of, bodily, life (Huijer, 2022). The people encompass those who suffer from a terminal disease, or elderly people passing the age of 65-70 years. People who enter, in general, a stage of life without professional employment and a relatively high degree of suffering with respect to body and mind. She addresses the question whether these people need to start a fight to prolong their bodily life without complaints, as much as possible, while sacrificing the potential meaning of the end of life? An end that is inevitable anyway. Huijer sees the meaning of the latest stage in life in enrichment and, especially, share of the narrative of the own life with beloved ones. A meaningful end of life that may facilitate those beloved ones to cope with the final loss of the elderly person. We may notice that Huijer wrote her book during the COVID-19 epidemics in the Netherlands. She was struck by the observation that young people suffered mentally from all kinds of restrictions imposed by the Dutch government trying to save the lives of, especially, elderly people. People who were at the end of life anyway, at least most of them.

REFLECTIONS

We are all aware, consciously and unconsciously, of the fragility of our life. It triggers concerns, fears, anxiety, or even panic, regarding the environment, in which we are living. These all trigger mental and bodily responses, which may be experienced as complaints. We feel ill and we enter voluntarily a process of disease diagnosis and treatment. It is the basis of healthcare worldwide, irrespectively the setting. We may reflect about this basis, taking into account the societal pressure exerted on individuals in times of epidemics affecting their voluntary health behaviour.

2.5 Outlook

Subjectivity is inherent to human life. Each of us has a unique narrative of life, as it is determined by the dynamics of the interactions between, body, mind, and environment. The challenge of healthcare is to address properly the uniqueness, subjectivity, of an individual. If so, we arrive at an efficacious healthcare. We use the term 'efficacious' rather than the one of 'efficient' here. Efficiency is primarily directed to minimising costs. This stimulates 'one-size-fits-all' approaches to health, like blockbuster drugs and nation-wide interventions by governments. In contrast, efficacy is primarily directed to maximising the outcome, *i.e.*, the quality of healthcare, while minimising the costs related to this goal. It enables tailor-made approaches to health, as we may see it with respect to personalised treatments in, for example, oncology, psychotherapy, and lifestyle medicine. It also takes into account the triadic interaction of, body, mind, and environment.

We feel efficacious healthcare is achievable in societally well-developed societies. Societies in which people can live meaningfully by way of, social contacts, culture, religion, nature, or whatever. Meaningful life implies that we have fewer incentives to use professional healthcare, saving time and money that we may attribute to those unfortunate subjects with relatively severe, or chronic, disorders. Similarly, well-developed societies also offer ample opportunities of resilience in daily life, reducing the need for professional healthcare, as well. So, we may conclude that well-developing societies may turn their systems of healthcare from high-quantity into high-quality ones.

References

Abelson, B., Sun, D., Que, L. *et al.*, 2018. Sex differences in lower urinary tract biology and physiology. Biology of Sex Differences 9: 45. DOI: 0.1186/s13293-018-0204-8.

Amarante-Mendes, G.P., Adjemian, S., Migliari Branco, L., Zanetti, L.C., Weinlich, R. and Bortoluci, K.R., 2018. Pattern recognition receptors and the host cell death molecular machinery. Frontiers in Immunology 9: 2379. DOI: 10.3389/fimmu.2018.02379.

Armstrong, K., 2009. The Case for God. Knopf Doubleday Publishing Group, New York, United States, 432 pp.

Bustamante-Marin, X.M. and Ostrowski, L.E., 2017. Cilia and mucociliary clearance. Cold Spring Harbor Perspectives in Biology 9: a028241. DOI: 10.1101/cshperspect.a028241.

Desmet, M., 2018. The pursuit of objectivity in psychology. Borgerhoff & Lamberigts, Ghent, Belgium, 109 pp.

Desmet, M. 2019., Lacan's logic of subjectivity: a walk on the graph of desire. Owl Press, Ghent, Belgium, 484 pp.

Drunen van, V.P., Bartels, E.A., van de Mheen, D., de Vries, E., Kerckhoffs, A.P.M., and Nahar-van Venrooij, L.M.W., 2022. Concepts of health in different contexts: a scoping review. BMC Health Research Services Research 22:389. DOI: 10.1186/s12913-022-07702-2.

Fattori, V., Ferraz, C. R., Rasquel-Oliveira, F. S. and Verri Jr, W. A., 2021. Neuroimmune communication in infection and pain: friends or foes? Immunology Letters 229: 32-43. DOI: 10.1016/j.imlet.2020.11.009.

Fernández-Quintero, M.L., Georges, G., Varga, J.M. and Liedl, K.R., 2021. Ensembles in solution as a new paradigm for antibody structure prediction and design. mAbs 13: e1923122. DOI: 10.1080/19420862.2021.1923122.

Godinho-Silva, C., Cardoso, F. and Veiga-Fernandes, H., 2019. Neuro-immune cell units: a new paradigm in physiology. Annual Review of Immunology 37: 19-46. DOI: 10.1146/annurev-immunol-042718-041812.

Gordon, S., 2008. Elie Metchnikoff: father of natural immunity. European Journal of Immunology 38: 3257-3264. DOI: 10.1002/eji.200838855.

Heeren, A., 2020. On the distinction between fear and anxiety in a (post)pandemic world: a commentary on Schimmenti *et al.* (2020). Clinical Neuropsychiatry 17: 189-191. DOI: 10.36131/cnfioritieditore20200307.

Huijer, M. 2022., De toekomst van het sterven. Uitgeverij Pluim/de Groene Amsterdammer, Amsterdam/Antwerpen, The Netherlands/Belgium, 150 pp.

Inauen, J., Contzen, N., Frick, V., Kadel, P., Keller., J., Kollmann, J., Mata, J. and Valkengoed van, A. M., 2021. Environmental issues are health issues: making a case and setting an agenda for environmental health psychology. European Psychologist 26: 219-229. DOI: 10.1027/1016-9040/a000438.

Jagers op Akkerhuis, G.A.J.M. 2010., Towards a hierarchical definition of life, the organism, and death. Foundations of Science 15: 245-262. DOI: 10.1007/s10699-010-9177-8.

Liston, A., Humblet-Baron, S., Duffy, D. and Goris, A., 2021. Humane immune diversity: from evolution to modernity. Nature Immunology 22: 1479-1489. DOI: 10.1038/s41590-021-01058-1.

Lommel van, P., Wees van, R., Meyers, V. and Elfferich, I., 2001. Near-death experience in survivors of cardiac arrest: a prospective study in the Netherlands. Lancet 358: 2039-2045. DOI: 10.1016/S0140-6736(01)07100-8.

Merle, N.S., Noe, R., Halbwachs-Mecarelli, L., Fremeaux-Bacchi, V. and Roumenina, L., 2015. Complement system part II: role in immunity. Frontiers in Immunology 6: 257. DOI: 10.3389/fimmu.2015.00257.

Nourshargh, S. and Alon, R., 2014. Leukocyte migration into inflamed tissues. Immunity 41: 694-707. DOI: 10.1016/j.immuni.2014.10.008.

Okumura, R. and Takeda, K., 2017. Roles of intestinal cells in the maintenance of gut homeostasis. Experimental & Molecular Medicine 49: e338. DOI: 10.1038/emm.2017.20.

Pardo-Cabello, A.J., Manzano-Gamero, V. and Puche-Cañas, E., 2022. Placebo: a brief updated review. Naunyn Schmiedebergs Archives of Pharmacology 395: 1343-1356. DOI: 10.1007/s00210-022-02280-w.

Pauwels, A-M., Trost, M., Beyaert, R. and Hoffmann, E., 2017. Patterns, receptors, and signals: regulation of phagosome maturation. Trends in Immunology 38: 407-422. DOI: 10.1016/j.it.2017.03.006.

Reader, J., 2022. Review of Giorgio Agamben (2021). Where are we now? The epidemic as politics. Postdigital Science and Education 4: 590-594. DOI: 10.1007/s42438-021-00247-3.

Rogers, T.D., Button, B., Kelada, S.N.P., Ostrowski, L.E., Livraghi-Butrico, A., Gutay, M.I., Esther Jr, C.R. and Grubb, B.R., 2022. Regional differences in mucociliary clearance in the upper and lower airways. Frontiers in Physiology 13: 842592. DOI: 10.3389/fphys.2022.842592.

Semmes, E.C., Chen, J-L., Goswami, R., Burt, T.D., Permar, S.R. and Fouda, G.G., 2021. Understanding early-life adaptive immunity to guide interventions for pediatric health. Frontiers in Immunology 11: 595297. DOI: 10.3389/fimmu.2020.595297.

Underhill, E., 1911. Mysticism. A study in the Nature and Development of Man's Spiritual Consciousness. Translated in Dutch and annotated by J-J. Suurmond, 2022, Skandalon Press, Middelburg, The Netherlands, 508 pp.

Uribe-Querol, E. and Rosales, C., 2020. Phagocytosis: our current understanding of a universal biological process. Frontiers in Immunology 11: 1066. DOI: 10.3389/fimmu.2020.01066.

IJntema, R.C., Schafeli, W.B. and Burger, Y. D., 2023. Resilience mechanisms at work: the psychological immunity-psychological elasticity model of psychological resilience. Current Psychology 42: 4719–4731. DOI: 10.1007/s12144-021-01813-5.

A WEALTH OF PATHOGENS

This chapter provides insight into the wealth of pathogens that may cause disease among people. We will first of all reflect on the term 'pathogen' and we arrive at a definition of it. We will, subsequently, assign pathogens to some major categories. One, the category of non-organisms that encompasses prions and viruses. Two, the category of prokaryotic organisms that includes archaea and bacteria and, three, eukaryotic organisms that encompass, protists/chromists, helminths, and fungi. Some major characteristics of each will be provided, but we need to be aware of the limitations of these characteristics, as the variety of pathogens is huge in each of the (sub-)categories.

3.1 Defining pathogens

The common view on pathogens is one of microorganisms invading a body, the reproduction of these within that body, and the subsequent transmission to another body. Pathogens, therefore, are the causal agents of infectious diseases of humans, as well as other organisms. This common view may be questioned in various ways. First of all, the term 'microorganism'.

Life is attributed to organisms as the key ontological unit of biology (Jagers op Akkerhuis, 2010). Organisms are, therefore, the research subjects of biology. We do not elaborate on the definition and meaning of life here, but an organism has the ability of autonomous life by way of a dual closure in order to keep itself intact as such. An organism has a structural closure separating it from the environment by way of, for example, a membrane. We may notice that a closure should not be complete to enable uptake and secretion of substances. An organism also has a functional closure enabling an autonomous metabolism. We may infer from the dual closure concept of an organism that prions and viruses cannot be considered as organisms (see sections 3.2 and 3.3 below). In addition, we may question the term 'micro' as we look, for example, at parasitic worms that may be several metres long. So, we may prefer here a general term like 'agent' instead of 'microorganism'.

Secondly, we look at the phrase 'invading the body'. Epithelium is the a-vascular tissue providing the structural closure of the human body. It covers all bodily surface that may come into contact with foreign substances, like the food in our gut and dust in our respiratory tract. It may also protect the body from detrimental waste products, like urine in the bladder. The epithelium of the skin and mucous membranes, therefore, is a necessary barrier between inside and outside, as we described in Chapter 2 already. Actually, it also is the first line of defence of the body against all kinds of agents that may cause disease. We, however, know that agents may cause disease, although these are located outside the human body. The bacterium *Vibrio cholerae*, for example, colonises the gut, but the bacterium itself does not enter, or pass, the epithelium. It secretes a toxin that causes the severe diarrhoea and accompanying disease symptoms by way of interference with the secreting function of epithelial cells. We, therefore, like to stress that an agent does not necessarily need to invade the body in the narrow sense of the term in order to cause an infectious disease.

Thirdly, the term 'reproduction' needs a closer look. It is in itself a relatively neutral term, but we associate it commonly with the reproduction of an organism. We can, however, not see prions and viruses as organisms, as outlined above. We, therefore, prefer here the term 'multiplication' to avoid any restriction to organisms. We also like to stipulate that the multiplication of an agent may be in the human body, as well as on it. The addition of the term 'on' is relevant with respect to agents like *Vibrio cholerae*.

Fourthly, we zoom in on the crucial phrase of 'transmission to another body'. We feel this to be the major characteristic of infectious, communicable, diseases. One human may infect another human, or an animal a human in case of zoonotic pathogens, as it is phrased commonly. It is, of cause, the agent that passes from one body to another. The term 'transmission' suggests, in our opinion a rather active process of passing from one body to another. It may be correct looking at, for example, *Plasmodium falciparum* causing malaria. Mosquitos pass it from one human body to another in a rather directional process, although *Plasmodium falciparum* itself is passive inside the mosquito. In contrast, other agents pass like particles from one body to another by way of rather stochastic processes, like air currents, or water streams. The term 'dispersal', therefore, is used in botanical epidemiology to cover all kinds of ways that agents pass from one host to another (*cf.* Zadoks and Schein, 1979). It emphasises the rather stochastic character of the passing of an agent from one host to another. We will adopt the term 'dispersal' here, and so, we have finally arrived at an operational definition of the term 'pathogen' with respect to man:

> *A pathogen is an agent that is, (i) able to cause disease among man, (ii) able to multiply on, or in, the human body, and (iii) equipped for dispersal from one human body to another, either by way of another host species, or not, and being able to cause disease of the subsequent body again.*

We include the term 'able' specifically, as a pathogen does not cause necessarily disease upon contact with a host. It depends on the status of the host whether disease is the result. A body needs to be in a sensitive, predisposed, state to get diseased (Casadevall and Pirofski, 2014). We see, in general, that pathogens cause disease among a minority of a host population only. The term 'able' expresses that a pathogen and host have a basic compatibility, but the actual states of, pathogen, host, and environment, determine whether disease results indeed from a contact between a host and pathogen (Zadoks and Schein, 1979). We do not need to ditch the term 'pathogen', as proposed by Casadevall and Pirofski (2014), by inclusion of the term 'able' in our definition of a pathogen.

We turn now to the description of major categories of pathogens. We use a classification that is based primarily on the structure of the pathogens. We start with the simplest structures, those of the prions, and we go up to the complex ones of the helminths. We, subsequently, describe each (sub-)category further with respect to, (i) the mode of multiplication, (ii) the disease caused among man, and (iii) the mode of dispersal from one human body to another. We are aware that each classification is arbitrary and that we are dealing with a huge variety, even in a sub-category (*cf.* Parish and Riedel, 2020). We, however, feel a concise overview pinpointing major differences among the (sub-)categories is needed to deal appropriately with epidemiology of infectious diseases.

REFLECTIONS

We use the term 'pathogen' frankly in daily life. Everybody seems to know what is meant. In contrast, the common definition does not cover properly the wealth of pathogens. We may reflect about the validity of the definition provided here. Do we agree with it? Could it be improved?

3.2 Non-organisms

3.2.1 Prions

The American physician and chemist Stanley Prusiner won the Nobel Prize in Physiology/Medicine in 1997 for his ground-breaking science on, fatal, neurodegenerative diseases. He addressed the own research, and that of others, on the 'proteinaceous and infectious' agent, which he called in short 'prion', in his Nobel-lecture (Prusiner, 1998). Prusiner described more than 25 years of research to identify the pathogen causing fatal neurodegenerative diseases, and more specifically Kuru and Creutzfeldt-Jakob. Diseases that are characterised by a formation of plaques in the central nervous system resulting in lesions. The odyssey passed along the scrapie disease of sheep and mad cow disease. It started by considering the diseases as degenerative ones, *i.e.*, no involvement of a pathogen. The diseases would then result from a degeneration of the body by way of external, non-pathogenic, factors, or inherited ones. This hypothesis was based on the failure of identifying foreign DNA and RNA in diseased bodies. But how can you demonstrate that something is not present? It is scientifically impossible. One may reason always that the method of detection is not sensitive enough. A scientist has to demonstrate that something exists. We have to add here that the infectious character of the diseases had been demonstrated already at that time. So, Prusiner and others knew a pathogen had to be involved. The discovery of an infectious protein, however, was really ground-breaking. All the pathogens we knew until then contained RNA and/or DNA, even the simplest virus (viroid). Another 30 years later, we now have sufficient evidence that prions may be quite common (Carlson and Prusiner, 2021).

A prion consists of a host-encoded protein only. The protein has a conformation, a folding, that deviates from the 'normal' one. So, a host produces a protein using the genetic information in the DNA/RNA. That protein folds either in a normal, healthy, conformation, or not. The significance of folding with respect to healthy, or diseased, was first shown for the so-called Prion Protein (PrP). It is a common protein in mammals, which is encoded by the PRNP-gene.

Figure 3.1. Schematic, simplified, representation of the Prion Protein in the 'healthy' (left) folding and the 'diseased' one (right).

The protein in the healthy conformation is indicated as PrP^{C}. The protein with the deviant conformation is indicated as PrP^{Sc} (Fig. 3.1). It differs from PrP^{C} by its β-sheet folding. It is clear that the protein sequence, which is based on the amino acid sequence of the DNA, does not dictate the conformation. The host DNA may, however, affect the susceptibility of a host with respect to prion formation, and its disease associated, as we will elaborate below under the heading 'multiplication'.

Prion formation is not restricted to the PRNP-gene resulting in the so-called PrP prion. The Aβ prion also has a conformation deviant from that of the 'normal' Aβ peptide. This peptide results from endoproteolytic cleavage of the Amyloid Precursor Protein (APP). Interestingly, mutations in the encoding gene APP may increase the level of a Aβ prion, although the mutations are outside the coding sequence of the peptide itself. Mutations in the PSEN1 and PSEN2-gene, which encode for proteins involved in the processing of APP, may also affect the formation of Aβ prions.

Table 3.1. Four types of prions and, the encoding genes in humans, the associated diseases, and the likelihood of passing from one human body to another.

PRION	GENE	DISEASE(S)	HUMAN-TO-HUMAN
PrP	PRNP	Kuru, Creutzfeldt-Jakob	Possible
Aβ	APP	Alzheimer	Possible
Tau	MAPT	Alzheimer, chronic traumatic encephalopathy, frontotemporal dementia-tau	Unknown
α-Synuclein	SNCA	Parkinson, multiple system atrophy	Unknown

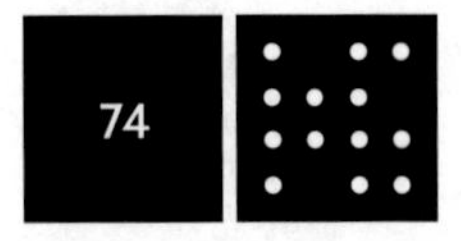

Two other categories of prions are the tau-prions and the α-synuclein-prions. Tau-prions are deviant conformations of the Microtubule Associated Protein Tau (MAPT). A protein that is encoded by the MAPT-gene. The α-Synuclein-prions result from deviant conformations of α-synuclein, which is encoded by the SNCA-gene.

Multiplication

We have convincing evidence for the existence of four types of prions (Table 3.1). Interestingly, we see variants of the conformation of a prion as well (Ghaemmaghami, 2017). So, a prion shows phenotypical variation with respect to the conformation. This variation may result from, (i) mutations in the encoding gene, (ii) entry of a foreign prion, and (iii) unknown internal factors. Once emerged, a variant of a prion multiplies without initiating transcription of the encoding gene itself. A kind of chain reaction starts, in which the prion uses the correctly folded protein as substrate. The prion serves as a template for an unmodified protein, which, subsequently, turns into the modified form. The multiplied and modified protein passes then from cell to cell along neurological pathways.

We are talking about strains of a prion taking into account both, the phenotypic variation and the ability to multiply exponentially, like we do see for bacteria and fungi. We may even extend the analogy with these microorganisms. A process of natural selection has been determined with respect to the PrP prions (Ghaemmaghami, 2017). Drugs applied to clear disease plaques in cell cultures, or mice, did result in selection of PrP^{Sc} strains insensitive to the drugs. The prion-population turned into a drug resistant one. So, the variation in prion conformation enables the prion to cope with adverse conditions, like a drug treatment. It is fascinating, and perhaps also alarming, that a non-organism may obey biological processes. Anyway, it justifies that we deal with prions as pathogens.

Diseases

All prions known are linked to neurodegenerative diseases (Table 3.1). Kuru was a common disease among the Fore-people living in a remote area of New Guinea (Nelson, 2020). The word kuru means shivering, or trembling, in their language. It indicates the loss of control of the muscles, which exaggerates and results finally in death. Death is between 3 and 24 months after onset of the symptoms. Pathology of the brain indicates the characteristic, non-inflammatory, spongiform lesions. We like stress here the term 'non-inflammatory'. The lack of inflammation triggered researchers for a long time to think about a non-pathogenic cause of the disease. Kuru resembles very much scrapie of sheep.

The early symptoms of Creutzfeldt-Jakob are rather atypical. Fatigue, insomnia and loss of muscle control are followed by a fast deterioration, resulting in dementia and other adverse neurological events. The average time to death is about six months. Brains of the patients deceased show the typical spongiform lesions, without any sign of inflammation. Evidence of an inherited autosomal disorder is detected in about 10% of the cases. A point mutation in the

PRPN-gene is most likely. It implies that Creutzfeldt-Jakob does not necessarily result from the PrP prion. In contrast, a variant of Creutzfeldt-Jakob disease has been linked to the PrP prion of Bovine Spongiform Encephalopathy (BSE), the mad cow disease.

Some other neurodegenerative diseases are associated with prions, as misfolded protein aggregates are involved. Alzheimer is one of these. It is a major cause of dementia (Long and Holtzman, 2019). It is predominantly diagnosed among people of 65 years and older. Alzheimer may, however, be present 15 to 20 years before the onset of cognitive impairment. It results from an autosomal dominant inheritance in about 1-2% of the patients. The pathogenesis proceeds most likely along an initial deposition of Aβ prion-aggregates in the brain, the amyloid fibrils. This, subsequently, leads to deposition of tau-prion aggregates, which results in loss of neurons and synapses. So, we see here an interaction of two major types of prions, and each of it having the potential of the typical prion multiplication described above.

Chronic traumatic encephalopathy resembles closely Alzheimer. Tau-prions are involved in the pathogenesis (Fesharaki-Zadeh, 2019). The pathology of brains of patients with this encephalopathy shows tau-filaments with a β-helix region containing a cavity, which is absent in Alzheimer. The cavity contains a co-factor that might be involved in the propagation of the tau-prion. Chronic traumatic encephalopathy is linked to sportsmen exposed to head blows, like boxers and American football players.

Frontotemporal dementia-tau is a group of dementia syndromes with an onset at a relatively young age (Hofmann *et al.*, 2019). The common characteristic is the frontotemporal lobar degeneration, which differs, in general, from that of Alzheimer-patients. The pathology of frontotemporal lobar degeneration may be divided into three categories reflecting the major proteins involved. One of these is defined by tau-proteins/prions.

Parkinson's disease is characterised by a degeneration of dopamine neurons in the substantia nigra in the midbrain (MacMahon Copas *et al.*, 2021). The loss of dopamine results in the primary motor symptoms, like tremor and postural instability. The α-synuclein prion may have a role in the pathogenesis, as an aggregation of it has been demonstrated in astrocytes resulting in increased neuronal death. The prion, however, seems to operate in a minority of patients with Parkinson only. Interestingly, the α-synuclein prion strain involved in Parkinson's disease differs from the one involved in multiple system atrophy (Peng *et al.*, 2018). The difference between the strains, and the related pathology, is determined by differences in the intracellular environment, in which these multiply.

Dispersal

The diseases listed in Table 3.1 are referred to as prion diseases by Carlson and Prusiner (2021), indicating these as infectious diseases. In contrast, Nelson (2020) does list Kuru and Creutzfeldt-Jakob as such only. Other authors, who are referred to above, indicate the proteins that are misfolded without using the term 'prion'. A reason for avoiding the term 'prion' may be that

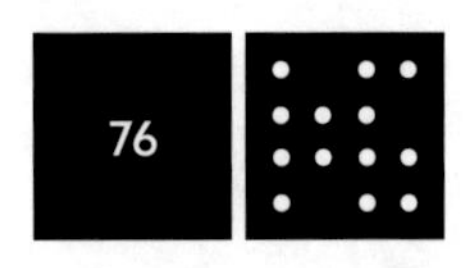

the misfolding of a protein within a body may initially result from three sources. It may result from, (i) a mutation in the encoding gene, (ii) other internal factors, and (iii) a misfolded protein entering the body. These are called according to the origin as, (i) genetic, (ii) sporadic, and (iii) infectious. We may notice that, actually, distinguishing these three types is a challenge. The use of the term 'prion' is, strictly speaking, proper with respect to the infectious origin only. So, we need to look at the dispersal of a misfolded protein to declare it as a prion, or not.

Oral digestion of prion-contaminated substances is one means of dispersal. Consumption of contaminated brains of relatives passed the PrP prion directly from one body to another among the Fore-people, which caused Kuru (Nelson, 2020). Kuru gradually declined by banning cannibalism in the 1950s. The Australian missionaries and settlers, who banned cannibalism, did unintentionally an efficacious intervention to halt Kuru-epidemics. Cannibalism, however, is not the only means of passing prions directly from body to body. Grafts of prion-contaminated dura mater and the use of human growth hormone prepared from contaminated pituitary glands have been identified as causes of Creutzfeldt-Jakob disease. In addition, digestion of prion-contaminated meat has been discovered as mode of dispersal of PrP prions between cattle, suffering from the mad cow disease, and man. This interspecies dispersal causes a specific variant of Creutzfeldt-Jakob, as mentioned above.

Dispersal has been demonstrated for PrP prions only, so far. The other types of prions, Aβ, tau and α-synuclein, do, however, have the ability of dispersal as well. One, inoculation of these types of prions in animals did result in the corresponding disease (Carlson and Prusiner, 2021). Two, medical interventions may introduce the prions, as demonstrated by the use of the human growth hormone. It triggered the fear of passing prions iatrogenically, for example, by way of blood transfusion (Nelson, 2020). Three, the mechanisms of multiplication and dispersal inside a body are quite similar among the various types of prions and homologues of each of the encoding genes are present in other animals. It all points to the ability of all major types of prions to pass among bodies, intra- or inter-specifically.

3.2.2 Viruses

The discovery of viruses started actually in 1886 (Kuhn, 2021). The German agronomist Adolf Eduard Mayer demonstrated that sap of diseased tobacco plants transferred to healthy tobacco plants caused disease, indicating an infectious agent. He could, however, not detect bacteria in the sap using a light microscope. Bacteria were the smallest pathogens known at that time. The Russian plant physiologist Dmitri Iosifovich Ivanovsky, subsequently, concluded that the infectious agent should be smaller than a bacterium. He passed the tobacco plant sap through so-called Chamberland-filters, but the sap remained 'infectious'. These filters had been designed to filter out bacteria. One may notice that we are dealing with sizes of bacteria of micro-metres, whereas those of viruses are in nano-metres. So, Ivanovsky concluded that the infectious agent of the tobacco mosaic disease, as the disease was called, was 'ultra-filtrable'. The Dutch microbiologist and botanist Martinus Beijerinck arrived at the same conclusion as Ivanovsky in 1898 using the Chamberland-filters. A nice example of reproducibility 'avant la lettre'. Beijerinck called it a contagious living fluid. We know now that the term 'living' is not appropriate, as we are not dealing with an organism. Scientists started to call the unknown, ultra-filterable, infectious agent a 'virus'. It is the Latin name for poison. It reflects the invisibility of the agent in combination with the detrimental effects on the host, like a real poison.

The American chemist Wendell Meredith Stanley continued the quest for the identification of the unknown virus. He was able to crystallise the virus causing the tobacco mosaic disease. He demonstrated that the crystals consisted largely of proteins. Stanley was awarded the Nobel Prize in Chemistry in 1946 for his research. It was another Nobel Prize winner, the German physicist Ernst Ruska, who paved the way to the final identification of a virus. He was awarded in 1986 for his research on the electron microscopy, which came available in Germany in 1931. It was his brother Helmut, who applied, together with Gustav-Adolf Kausche and Edgar Pfankuch, electron microscopy to the tobacco mosaic virus in 1939. The resulting image of the virus provided a definite demonstration of the existence of viruses. We may conclude that the worries about a disease of tobacco resulted finally in the discovery of a completely new type of pathogens. Viruses of which was already known at that time that these may also cause disease among animals including man. The electron microscope turned out as an essential tool to visualise viruses. The seminal work of Watson and Crick in the fifties, subsequently, started the era of discovering DeoxyriboNucleic Acid (DNA) and RiboNucleic Acid (RNA) in viruses. The carriers of the genetic information.

The core of a virus consists of DNA, or RNA, surrounded by a protein coat (Parish and Riedel, 2020). The core is called the 'virion' and the protein coat the 'capsid'. The DNA and RNA may be double-stranded, or single-stranded. Various shapes of capsids exist, *e. g.*, helical and filamentous ones. The virion may also be enveloped by a lipo-protein coat, the envelop, which is acquired from a host cell membrane.

Systematics of viruses is a real challenge (Kuhn, 2021). The current systematics reflects on the one hand the methodological hurdles in the past and, on the other hand, the seemingly lack of barriers separating viral species. We pointed out already that viruses cannot be considered as organisms, as these lack a dual closure. Viruses do neither have a structural closure separating these from the environment, except the ones with an envelope, nor a functional closure, which enables an own metabolism. We feel the relatively simple and open structure may explain the abundance of horizontal gene transfer among viruses. In addition, the mutation rates are relatively high. It altogether obscures phylogenetic relationships, if present at all.

The International Committee on Taxonomy of Viruses (ICTV) is leading in the systematics of viruses. It adopted a bottom-up approach rather than a hierarchical top down, as it is common in the systematics of organisms. So, everybody may come up with a systematics proposal for a newly detected virus, or a re-positioning of a known virus. The ICTV strives at adopting the binary system of organisms, *i.e.*, the first name of the genus (*e. g. Homo*) and the second

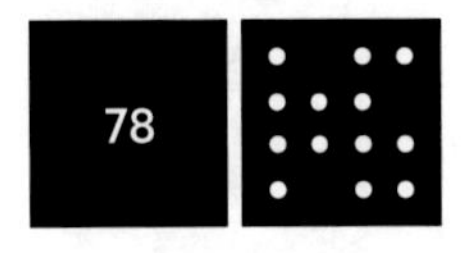

name of the species (*sapiens*). Adoption of such a system, however, implies some notion of the phylogenetical evolution of viruses. A quest that just has been started. It faces serious, methodological, challenges (*e.g.*, Malik *et al.* 2017).

The size of a virus genome varies between about 3000 and 2.8 million base pairs (Yin and Redovich, 2018). The Pandora virus has the largest genome, so far known. It is, therefore, called a giant virus. The genome is even larger than the one of some bacteria. The number of virus proteins correlates approximately linearly with the genome size (Fig. 3.2). A virus contains, on average, 1 protein per 1000 base pairs (kb).

Multiplication

Viruses need obligatorily a host cell to multiply (Yin and Redovich, 2018). We see a sequence of events of, (1) a virus particle adhering to a cell wall, (2) up-take of the virus particle, or the genome only, by the host cell, (3) replication of the virus genome and transcription/translation

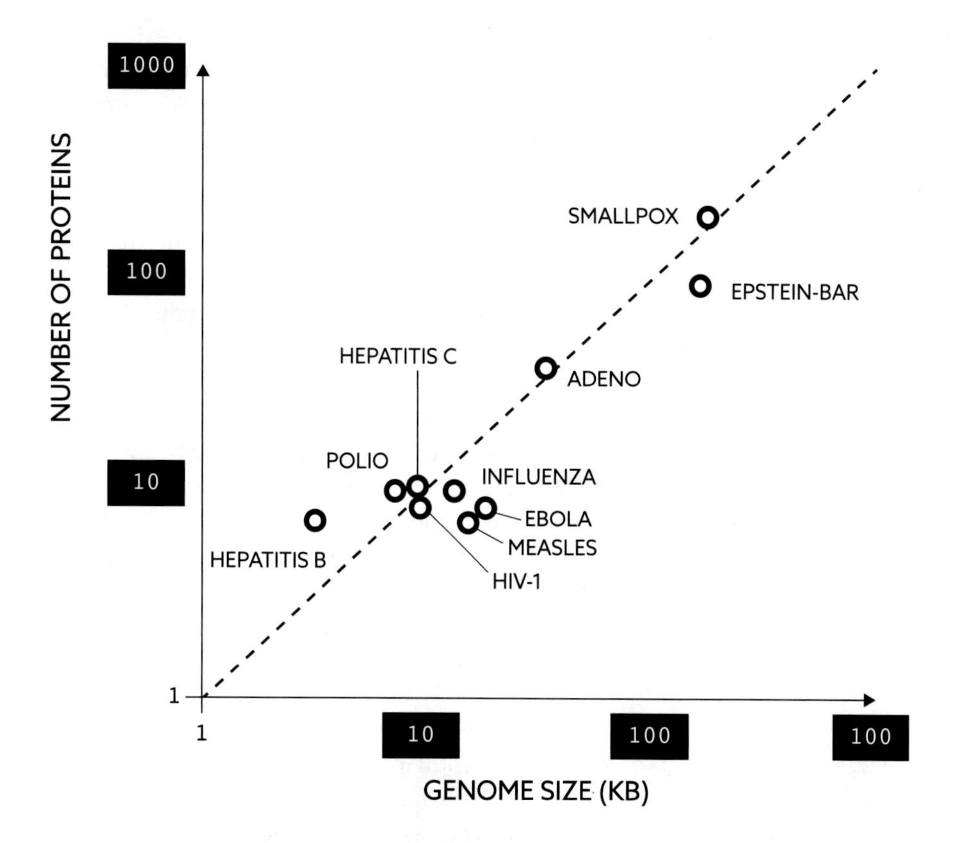

Figure 3.2. Association between size of virus genome, as expressed in 1000 base pairs (kb), and the number of virus proteins. Entries are various viruses causing diseases among animals, including man, and bacteria.

of the genome into virus-proteins, (4) assembly of the proteins in pro-capsids, (5) packaging of the replicated virus genome in capsids using the pro-capsids, and (6) release of the new virus particles from the host cell. Proteins of both, virus and host are involved in the adherence and up-take. The virus genome entered into a cell starts to use the multiplication mechanisms of its host. The multiplication is completely in the cytoplasm of prokaryotic cells, which are missing a nucleus. It may be, partially, in the nucleus for viruses entering eukaryotic cells. The release of the new virus particles is by way of exocytosis, or lysis of the complete host cell.

The use of computational models turned out to be essential in investigating the multiplication of viruses. The cellular processes at the microlevel are so complex already that the use of solely biochemical methods fails. Empirical studies had, and have, to go hand in hand with computational modelling to get grip on the intra- and intercellular processes involved in virus multiplication. We encounter the intriguing question whether we develop a computational model to fit biochemistry, or do we fit biochemistry to the model, like it happens in the so-called 'synthetic biology'. We may rephrase this question in a broader sense. Do we adapt computational modelling to reality, or do we adapt reality to modelling? A question that we will face at any scale of epidemiology, from the molecular level up to the ecosystem level. We feel adaptation of the computational modelling to reality is embraced in science trying to understand phenomena. In contrast, adaptation of the reality is common in applied research. For example, computational modelling is used in managing epidemics at the community level. Parametrisation of the models, however, determines the outcomes. It is inevitably prone to selection bias, as we have partial knowledge of epidemics only. A biased parametrisation, therefore, determines the resulting management. Management that shapes ultimately a community, in which an epidemic occurs. Reality is adapted to the computational model. So, we feel computational modelling is a valuable tool to gain scientific knowledge. The use of it as a predictive tool in policy-making seems tricky.

We return to the use of computational modelling to investigate multiplication of viruses (Yin and Redovich, 2018). It started by studying bacteriophages. These are relatively simple viruses infecting bacteria. The models have been extended to viruses infecting man. The term 'burst size' has a prominent position in the modelling. It is the number of virus particles coming out of a host cell infected by a virus. The burst size depends strongly on the growth conditions of the host. For example, a bacteriophage T4-infection of slow-growing *Escherichia coli* had a burst size of 1, whereas it was a hundred-fold in *E. coli* growing under conditions at optimum.

The discrepancy in burst size between a poor and well-nourished host triggered, amongst others, modelling of the energy costs of constructing a virus in a host cell (Mahmoudabadi *et al.*, 2017). The costs at the cellular level were expressed in the number of Adenosine Tri-Phosphate (ATP) hydrolysis events, a common measure of cellular costs. The costs were calculated for the whole process of attaching of a virus particle to a host cell until leaving it by way of the new particles. The energy costs of multiplication were modelled for both, the T4-bacteriophage

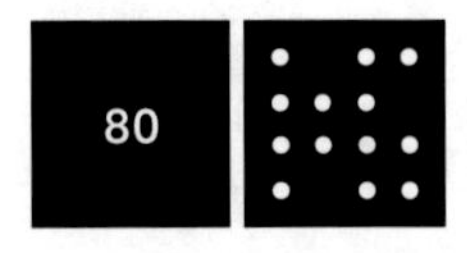

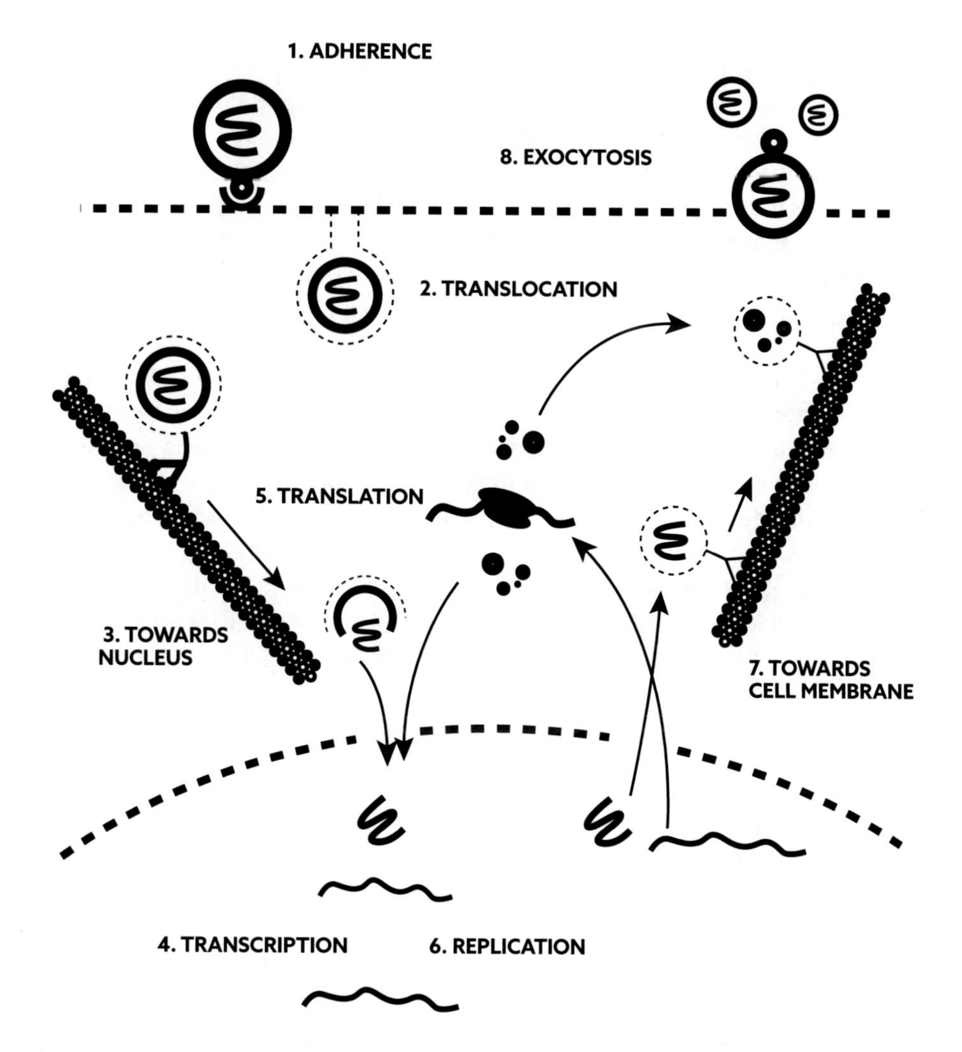

Figure 3.3. Schematic drawing of the process of uptake of a virus by a host cell, the subsequent multiplication of it by way of the molecular pathways of the host, and the release of the virus particles by way of exocytosis. Viral DNA is replicated in the nucleus, as indicated here, viral RNA is replicated in the cytosol. Virus particles may be released by way of exocytosis, as indicated here, or by lysis of the whole host cell.

infecting a prokaryotic host cell and an influenza virus infecting a eukaryotic host cell. The T4-bacteriophage has a double-stranded DNA-genome and the influenza virus a single-stranded RNA one. Figure 3.3 illustrates the multiplication process of the virus.

The energy costs of producing one new virus particle did not differ between the T4 and virus. Quite a surprising result looking at the rather different genome of DNA and RNA, respectively. It is less surprising, if we know that most of the estimated energy costs were related to the translation of genetic information into proteins. In contrast, the energy costs differed significantly taking into account a burst size of 200 and 6000 for T4 and influenza, respectively. The costs for the host with the influenza-infection exceeded those of the host with T4 by far. The relative costs, however, were low, *i.e.*, 1% of the estimated energy budget of the host cell. In contrast, T4 uses for its replication about a third of the energy of the host cell. In addition, we like to stipulate that influenza particles exit the cell by way of exocytosis, keeping the cell intact, whereas T4-particles burst of the cell, *i.e.*, the host cell is destroyed. So, the impact of the influenza virus on its host is far less than that of a T4 bacteriophage on its host.

The burst size of a virus depends on the growing conditions of the host, as indicated above. In addition, we need to take into account that multiplication may results in defective virus particles (Yin and Redovich, 2018). The defective particles have deletions in one, or more, essential genes with regard to the ability to multiply. Interestingly, these defective particles interfere with the production of complete ones, in case of co-infection of a host cell, by way of using cell resources. It is called the von Magnus phenomenon. So, we see an additive, negative, effect of defective particles on the effective burst size, *i.e.*, the number of virus particles leaving a host cell that are able to multiply in another host cell. Interventions increasing the number of defective virus particles thus offer opportunities for anti-viral therapies (*e. g.* Li *et al.*, 2021).

Diseases

The number of diseases caused by viruses is substantial among man. The diseases may be classified in various ways. We highlight here some rather than providing a systematic overview (Table 3.2). The selected samples may, however, illustrate a bit of the variety of viral diseases.

Table 3.2. Examples of viruses causing diseases among man and the organs affected. The taxonomy of viruses is continuously updated. The current taxonomy/nomenclature of the viruses can be checked on the website of the ICTV (https://ictv.global/).

AFFECTED ORGAN(S)	DISEASE	VIRUS
Central nervous system	Rabies	Rabies lyssavirus
Respiratory tract, systemic	Measles	Measles Morbillivirus
Respiratory tract	COVID-19	SARS-CoV-2
Immune system	AIDS	Human Immunodeficiency Virus
Immune system, systemic	Ebola virus disease	Zaire Ebolavirus
Skin, mouth, throat	Smallpox	Variola virus
Liver	Hepatitis B and C	Hepatitis B virus and Hepacivirus C
Gatro-intestinal tract	Gastroenteritis	Norwalk virus

The central nervous system, and especially the brain, is quite well protected against virus infections. The Rabies lyssavirus is a well-known exception. This RNA-containing rhabdovirus causes rabies among mammals. The disease presents in humans, in general, in three phases, (1) the prodromal one, in which the symptoms are non-specific, (2) a phase of acute neurologic manifestations, in which the virus enters the central nervous system, and (3) a phase of coma of the patient resulting in death (Riccardi *et al.*, 2021). Diagnosis may be blurred by the rather unspecific symptoms even in the neurological state of pathogenesis. In addition, rabies is relatively rare in Western countries, which results in un-awareness of the disease among healthcare providers. Rabies is fatal for nearly all patients, once clinical symptoms appear. An efficacious treatment is lacking still. Prevention by way of vaccination is possible, even after being wounded by an infected animal. Implementation of vaccination is, however, limited in un-awareness of rabies and costs. Vaccination of at-risk subjects takes priority currently.

The respiratory system is relatively sensitive to viral infections. The barrier between inside and outside is relatively thin in the bronchial tree and alveoli, enabling the exchange of gases, like oxygen and carbon dioxide. The Measles Morbillivirus is a well-known virus entering man by way of the respiratory tract (Rota *et al.*, 2016). It passes from epithelial cells into lymphatic tissue becoming systemic. Multiplication starts in the epithelial cells of the respiratory tract already. Systemic infection is mostly visible in the typical rash. The rash results from a virus-specific immune response. It diminishes during clearance of the virus by the immune system. Children are especially prone to measles, but older people may also become diseased. New-borns are protected by maternal antibodies until about a half a year after birth. Mortality results from complications in any organ affected by the virus, *e. g.*, pneumonia. Complications may be present among about 40% of the patients. Cases with impaired cellular immunity, like those with AIDS, are very sensitive to complications. The case-fatality rate varies between 1 and 5% depending on various factors, like vaccination, access to health care, and nutritional status. We may, therefore, state that both, the incidence and case fatality rates of measles are very low in most Western countries compared with, for example, sub-Saharan African countries. Case-fatality rates may go up to 30% under specific, adverse, conditions, like in camps of refugees. Man is the only reservoir for the Measles Morbillivirus. In addition, a rather efficacious vaccine is available. These two facts, and knowing the success of eradicating smallpox (see below), triggered the WHO to aim at eradication of measles, which has not, however, been achieved so far.

The Severe Acute Respiratory Syndrome-related CoronaVirus number two (SARS-CoV-2) also enters man by way of the respiratory tract. The single-stranded RNA-virus causes COVID-19. COVID-19 is characterised by a majority of people having relatively mild symptoms, or even no clinical symptoms at all. In contrast, a minority of cases suffers severely, resulting in hospitalisation and the need of intensive care. The majority of severe cases is characterised by fragility due to, either co-morbidity, or ageing. Estimation of a case-fatality rate is rather troublesome due to the identification of symptomless, infected, people as case, or not. We, therefore, turn to estimations of the infection-fatality rate (Ioannidis, 2021). The estimations were based on data of seroprevalence of antibodies coupled to mortality data of the catchment population. The estimations of the infection fatality rate of COVID-19 were in the range of 0 to 1.63% based on publications retrieved until September 2020. The median of the estimates was 0.23%. It was 0.05% for infected people younger than 70 years, indicating very old people are especially at risk of severe COVID-19. The actual infection-fatality rates may be even lower, as people suffering from COVID-19 are more likely to be tested on anti-bodies. The Achilles heel of estimating the infection fatality rate, however, is attributing death primarily to SARS-CoV-2-infection, or not. We mentioned the difficulties in attributing death of patients suffering from co-morbidity, or ageing, to a specific disease in Chapter 1 already.

Acquired ImmunoDeficiency Syndrome (AIDS) is caused by the Human Immunodeficiency Virus (HIV), which is a double stranded RNA-virus. It enters cells of the immune system, especially T4-lymphocytes (Nelson and Celentano, 2020). HIV was detected among humans in Congo back in 1890. It seemed to remain in Africa until the eighties. It was then identified in the USA and, subsequently, worldwide. Two variants of HIV are known, type 1 and 2. Infection with both types of this retrovirus may result in AIDS. AIDS may, subsequently, enable all kinds of opportunistic infections and tumours. Patients cannot be cured from HIV, but its multiplication can be reduced using various anti-viral drugs. The case-fatality rate is rather difficult to assess as both, the latent period may be decades and AIDS is intertwined with other diseases.

The Zaire Ebolavirus is one of the Ebola viruses causing Ebola Virus Disease. It is a single-stranded RNA-virus belonging to the filoviruses. The typical symptom is one of bleeding by patients in an advanced stage of disease (Baseler *et al.*, 2017). The name of the disease, Ebola, was adopted from the Ebola River in the Democratic Republic of Congo. The first outbreaks of Ebola were reported from the area around the river back in the seventies. The Democratic Republic of Congo was called Zaire at that time. The virus enters its host by way of epithelial cells of mucous membranes, or breaches in the skin. It, subsequently, passes into the lymph system and the blood circulation. It may enter finally all types of organs. The typical bleeding results from a disruption of the coagulation. The second week of disease is crucial for the outcome. It is the week that vital organs get infected, or not. If not, the immune system is able to cope with the viral infection and patients will recover. We may notice that the burden of disease results from both, the virus multiplication and the violence of the immune response. The case-fatality rate went up to 90% in the past, as an efficacious therapy was lacking. Case fatality rates of more recent outbreaks, however, were substantially lower, *i.e.*, about 40%. The lower rates might be attributed to both, early detection of cases and improved supportive care. Isolation of cases and their corpses, if passing away, is a common method of prevention. In addition, approved vaccines are available since 2019.

Smallpox was caused by the Variola-virus, a double-stranded DNA-virus, until it was eradicated. Smallpox was characterised especially by pustules on the skin, which became scabs that felt off (Breman and Henderson, 2002). Children and elderly people were, especially, vulnerable to the Variola virus. This poxvirus has two variants, major and minor, indicating the size of the pustules caused. The case-fatality rate was about 30% among patients infected by the major variant. Death could result from hypotension and toxaemia related to the immune response.

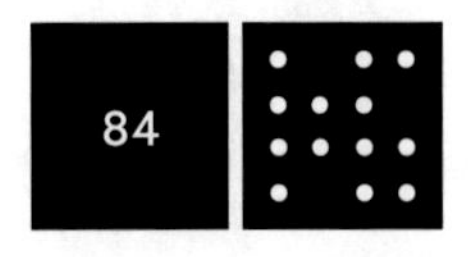

Illness caused by the minor variant was, in general, mild. The virus was effectively eradicated by a worldwide vaccination campaign in the sixties and seventies. The World Health Organisation declared the world free of smallpox in 1980. Besides a very efficacious vaccine, the success may be explained by the lack of a reservoir of the virus outside man. Nowadays, the virus is stored solely in a US- and Russian laboratory, so far known.

The eradication of smallpox was the result of 1000 years of trial and error (Breman, 2021). Traditional medical practitioners inoculated people with material from the pustules, scabs, by way of the nose, or skin, to protect them. The procedure of inoculation was also called 'variolation', which explains the name assigned to the virus later on. The practice of using infectious material to protect people became more common in the 18th century. The British country physician Edward Jenner promoted the use of infectious material of cowpox for protection against smallpox. A strategy of initiating cross-resistance. Vaccination got, subsequently, established in industrialised countries around 1900. You may notice that the term 'vaccination' was used instead of the earlier terms of 'inoculation', or 'variolation'. The term 'inoculation' is in use in science still to demonstrate the infectivity of an agent according to Koch's postulates (see section 3.2.2.). Anyway, the Soviet Union proposed a programme of global smallpox eradication to the WHO in 1959. The programme did not really take off. The call to action was repeated by the USA in 1966. The WHO, subsequently, took the lead in standardising and improving vaccines worldwide. In addition, the administration of the vaccines was improved. It resulted finally in the eradication of smallpox in a relatively short period of time. The only memory of smallpox is the 'Jennerian pustule' on the skin of those, who were vaccinated back in the seventies, or earlier.

Hepatitis is an inflammation of the liver caused by various pathogens and factors other than pathogens. Various viruses may cause hepatitis, but we focus here on the viruses causing hepatitis B and C. These are the double-stranded DNA Hepatitis B Virus and the single-stranded RNA Hepacivirus C. Infection by the Hepatitis B Virus (HBV) may be transient, causing acute symptoms, or chronic (Castaneda *et al.*, 2021). Symptoms may include, fatigue, nausea, and vomiting preceding jaundice. Responses of the immune system may clear HBV. If not, patients enter a phase of chronic infection after about six months and they may remain free of disease symptoms for years. These patients are, however, pre-disposed to liver cirrhosis due to prolonged parenchymal inflammation and fibrosis. In addition, chronic HBV is correlated to hepatocellular carcinoma. Drugs for viral suppression therapy are available. Tolerance of the drugs by patients, however, is rather poor. Cure is not possible, except by performing a liver transplant. Liver transplant is the ultimate therapy for patients with acute liver failure, decompensated liver cirrhosis, and hepatocellular carcinoma. A vaccine to prevent HBV-infection is available. A vaccine to cure HBV-infection even seems to be upcoming.

The majority of people infected by Hepacivirus C (HCV) do not show disease symptoms, or symptoms are not specific for hepatitis. Those presenting the disease have the common symptoms of hepatitis. Responses of the immune system may clear HCV. Up to 50% of infected people may clear HCV. If not, patients enter a phase of chronic infection after about three months. Chronically infected patients may develop cirrhosis up to 30 years after the initial HCV-infection. Patients developing cirrhosis do also have an elevated risk of developing hepatocellular carcinoma. Various drugs are available for anti-viral therapy and treatment of HCV-infection is expanding. A vaccine to prevent, or cure, HCV-infection is not available yet.

The Norwalk Virus causes gastroenteritis among man. It is a single-stranded RNA virus. The name of the virus resulted from an outbreak of the virus in the city of Norwalk (USA) in 1968 (Robilotti *et al.*, 2015). Illness due to the virus was described in 1929 already as 'winter vomiting disease". The attribution to the virus had to wait until 1968. Healthy adult male prisoners were inoculated with filtrate from rectal swabs of diseased individuals. Rectal swabs of the diseased volunteers were, subsequently, used to inoculate other volunteers confirming the presence of a pathogen. Subsequent investigations of clinical specimens resulted in the detection of the virus. The virus is called Norwalk according to the ICTV, whereas others call it the (human) norovirus (Robilotti *et al.*, 2015). Dealing with the systematics of viruses is a challenge. Anyway, vomiting and diarrhoea are the common disease symptoms caused by the Norwalk virus. Symptoms last, in general, one to two days. Disease may be more severe among aged people and those with co-morbidity. In addition, neonates may develop necrotising enterocolitis. Treatment of the gastroenteritis is rather supportive by reversing dehydration and loss of electrolytes.

Dispersal

Looking at the viruses described here, it indicates the large variety of modes of dispersal of viruses already. Some viruses infect humans by way of a vector, like the Rabies lyssavirus. Bats and dogs introduce the virus into humans by way of wounding. The virus cannot disperse from one human body to another infecting the latter. Viruses entering man by way of the respiratory tract disperse by way of the air contained in droplets and aerosols. Aerosols are liquid, or solid, particles suspended in a gas, like oxygen and water vapour. A particle inside an aerosol has a size of about 1 micrometre, or less. Aerosols, which include virus particles, may be generated by infected man. The aerosols may, subsequently, disperse to another person entering its respiratory tract.

Viruses may also disperse by way of sticking to all sorts of materials. We call the contaminated material a fomite. Contaminated needles are well-known fomites, transmitting viruses, like HIV and Hepatitis B. Transmission of viruses by way of fomites is a common problem in hospital settings. Food is a way of getting infected by the Norwalk virus. The virus may also disperse and infect by way of water, like it happens by drinking contaminated tap water. Dispersal by way of blood, or other bodily fluids, is another common way of passing infection directly from one person to another. The Zaire Ebola virus provides a well-known example of such a dispersal and subsequent infection. Direct contact among people by way of kissing and sexual intercourse may also pass viruses from one body to another. The HIV-virus and herpesviruses present well-known examples of such a dispersal.

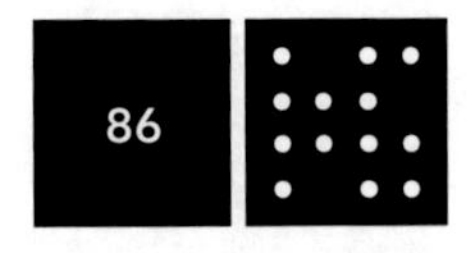

We feel determining dispersal of a virus is a challenge. We are dealing with ultra-small, non-living, particles that, therefore, disperse in a rather stochastic way, except dispersal is by way of direct contact among people. In addition, a virus may disperse by way of various modes including a non-human host as well. Dispersal may also occur at scales from micro-metres up to thousands of kilometres. Investigating virus dispersal is often like searching for a needle in a haystack.

REFLECTIONS

The prions and viruses bring us at the border of biology, as these are non-organisms. A matter that is commonly overlooked. We may reflect on the consequences with respect to detection and classification of prions and viruses, respectively. In addition, we may reflect about the consequences of treating diseases caused by these.

3.3 Prokaryotic organisms

3.3.1 Archaea

Organisms, living entities, are characterised by a dual closure, as outlined in section 3.1. An organism has the ability of autonomous life by way of this dual closure. It has a structural closure separating it from the environment by way of, for example, a membrane. An organism also has a functional closure enabling an autonomous metabolism. Organisms are divided into pro- and eukaryotes. The Greek word 'κάρυον' (karyon) means kernel. It indicates the presence of the nucleus in cells of the eukaryotes, whereas it is lacking in the prokaryotes, which are believed to have preceded the eukaryotes. The prokaryotes encompass two major domains, the Archaea and the Bacteria, whereas the eukaryotes do encompass the domain of Eukarya only.

Systematics of all organisms is based on the binomial system, *i.e.*, the first name is provided by the genus, *e.g.*, *Homo*, and the second name by the species, *e.g.*, *sapiens*. The (light) microscope was pivotal in taxonomy of the prokaryotes, until the introduction of molecular tools. The possibility of DNA-sequencing enabled identification of prokaryotes without the cultivation of these. It revealed several new lineages in the tree of life (Castelle and Banfield, 2018). The domain of the prokaryotic Archaea seems phylogenetically closer to the eukaryotic Eukarya than the prokaryotic Bacteria, at least from a molecular point of view (Figure 3.4). It may even be that the origin of the eukaryotes is within the domain of the Archaea (but see Zhu *et al.*, 2019). Interestingly, no microorganisms belonging to the domain of Archaea are known to cause diseases of man yet, although these are that closely related to eukaryotes (Lurie-Weinberger and Gophna, 2015). In contrast, pathogens of humans are abundant in the domain of the Bacteria. Did we overlook pathogenic microorganisms belonging to the Archaea so far? It might be. We, however, know that Archaean microorganisms are abundant in the microbiome of the human gut, subgingival dental plaques, and on the skin. So, we do have awareness of the Archaean organisms. We are left with unanswered questions. Why do Archaean organisms not cause disease among man? Are there biological constraints? Are we limited in clinical assessment tools regarding the domain of Archaea? Anyway, we may focus further on (pathogenic) bacteria in the following section.

3.3.2 Bacteria

We cannot deal with bacteriology, the science of bacteria, without acknowledging the German physician-scientist Robert Koch (1843-1910). He was the first one to link a specific bacterium with a specific disease (Blevins and Bronze, 2010). It concerned the attribution of anthrax to *Bacillus anthracis* by elucidating the life cycle of the bacterium. He discovered that the bacterium could survive harsh conditions by way of resting spores. He published the life cycle in 1876. Interestingly, he published it in a botanical journal, Cohns Beiträge zur Biologie der Pflanzen, although *B. anthracis* causes disease among mammals, including man, only. The journal was published by the renowned German botanist Cohn. He seemed that much impressed by the work of Koch that he liked to publish it in his own journal. The 'golden age' of bacteriology, and its impact on medicine, started thus in a botanical journal.

Microscopy was essential in the research of Koch. He improved both, the technique of (light) microscopy and the slides of the bacteria by using various dyes. He provided the first (light-microscope) photographs of bacteria. These were published again in a botanical journal, the 'Beiträge zur Biologie der Pflanzen'. Koch did not stick to microscopy only. He also developed methods to culture bacteria purely on various media. One of his post-doctoral assistants came up with the use of agar as a medium to culture bacteria. We are using the medium, and variations of it, still. Koch published his so-called plating technique in 1881. It enabled observations of bacterial growth on plates, as affected by various inhibiting, or stimulating, factors.

Robert Koch went on to identify the cause of tuberculosis. He could identify *Mycobacterium tuberculosis* by trial and error with dyes to stain it in infected tissue. His next challenge was to culture *M. tuberculosis*, which he was be able to do on coagulated blood serum. Finally, he inoculated animals with the cultured, pure, *M. tuberculosis* to demonstrate it as the pathogen of tuberculosis. He presented his finding in 1882, paving his way to international fame. He was awarded the Nobel prize in Physiology or Medicine in 1905.

The name of Koch is inevitably associated with his postulates. Interestingly, a former teacher of Koch at the University of Göttingen, Jacob Henle, provided the concept of the postulates. Robert Koch, however, was the first one to validate the postulates with respect to, anthrax,

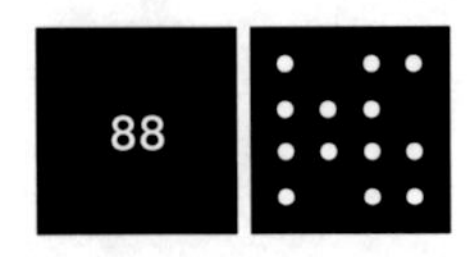

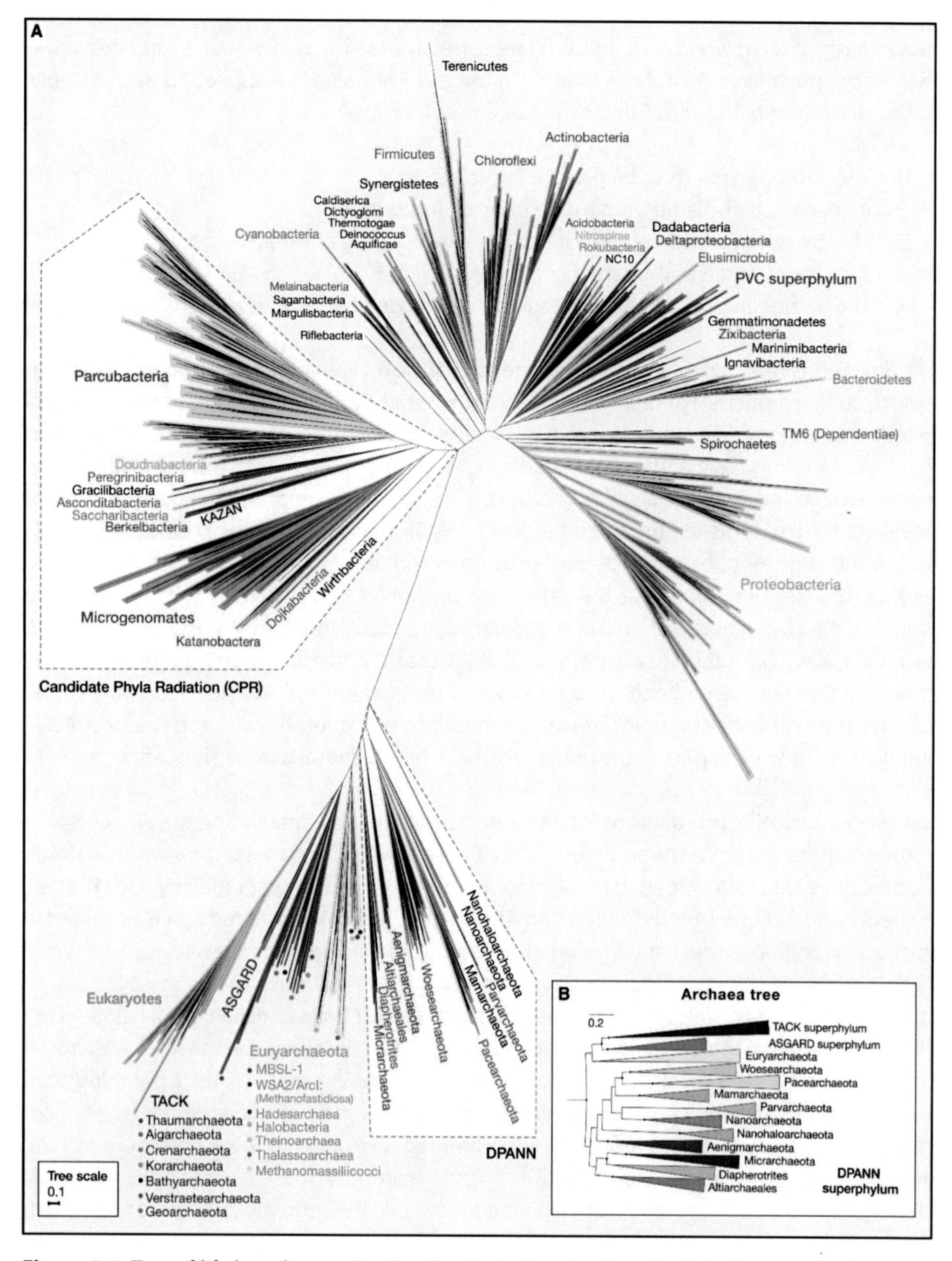

Figure 3.4. Tree of life based on molecular data including the domain of the Bacteria, Archaea and Eukarya. With permission of Castelle and Banfield (2018).

tuberculosis, and cholera. So, we should state correctly here the Henle-Koch postulates rather than Koch's postulates. Postulates that are in use still. These are not applied to bacteria only, but to all kinds of pathogens. The postulates are, as follows,

1. The pathogen needs to be present in each case;
2. It needs to be isolated from a case to be cultured purely;
3. The disease needs to be reproduced upon inoculation of a healthy body, or animal, with the cultured pathogen;
4. The pathogen needs to be isolated from the inoculated host.

The limitations of the Henle-Koch postulates are obvious. How to deal with pathogens that cannot cultured purely, like many viruses? And what about pathogens that may have very long latent periods, like HIV? How to handle with co-infections of pathogens? What to do if a pathogen does have a human host only, *i.e.*, an appropriate animal model is missing for inoculation? A reconsideration of Koch's postulates has been proposed including the aid of novel, molecular, tools (Fredricks and Relman, 1996). The disadvantage of such diagnostic tools is the huge number of microscopic organisms and non-organisms we discover, of which we have to assess whether these are pathogenic, or not (*cf.* Lipkin, 2009). We actually replace empirical data by an observational one to determine a causal relationship between pathogen and disease. We have then to deal with probabilities and potential confounding. So, we may conclude that the Henle-Koch postulates are of great value still. We need to apply these, whenever possible from a point of view of methodology and ethics. If not, the postulates should warn us with respect to the pitfalls in establishing a causal relationship in another way.

Bacteria are unicellular organisms lacking a nucleus, which is a membrane-enveloped space containing the DNA (Parish and Riedel, 2020). The DNA is, in general, present in a single chromosome that is free floating in the cytoplasm of the cell. A membrane envelops the cell, guaranteeing its autonomous functioning. The bacterial cell is surrounded by a wall, except in bacteria belonging to the genus *Mycoplasma*. The cell wall includes the specific peptidoglycan. The thickness of the peptidoglycan layer determines the response to the so-called Gram-staining, which was developed by the Danish bacteriologist Hans Christian Gram (1853-1938). The Gram-negative bacteria have a relatively thin layer, which cannot keep the dye very well. It colours red/pink. In contrast, the Gram-positive bacteria have a relatively thick peptidoglycan layer keeping the dye rather well. It appears blue/purple. So, we have now already three major categories of bacteria, (i) Gram-positive, (ii) Gram-negative, and (iii) the wall-less ones of the genus *Mycoplasma*. Mycobacteria constitute a fourth category. These bacteria have a cell wall, but it does not colour using the Gram-staining procedure. The term 'myco' refers to the growth that resembles mycelial growth of fungi. Presence, or absence, of a cell wall and response to staining are two criteria to classify bacteria. A third one is the shape of the cells. These may be, (i) round, or spherical, like those of *Staphylococcus aureus*, (ii) rod-shaped like those of *Bacillus anthracis*, or (iii) curved and spiral like those of *Borrelia burgdorferi*. A fourth criterion is the need to live obligatory, or facultatively, under aerobic, or just anaerobic conditions.

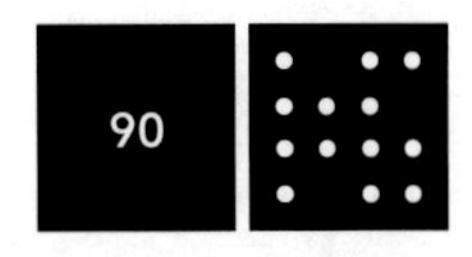

Plating and staining of bacteria in combination with the use of the light-microscope was the primary method of investigating morphology and physiology of bacteria for more than a century. Systematics was, therefore, based on this method. New tools like Scanning Electron Microscopy and DNA-sequencing enabled both, a refinement of systematics and the discovery of new species (*cf.* Fig. 3.4).

Some bacteria are able to produce spores to survive harsh conditions. A spore is produced inside the cell. It is liberated as soon as the cell decays due to, for example, a lack of nutrients. Such bacterial endospores are very resistant to, chemicals, heat, drought, frost, and so on. The endospore-producing bacteria remain dormant until the conditions change positively. If so, the spore germinates and a new bacterial cell emerges.

Multiplication

Multiplication of bacteria is asexual by way of fission. DNA, organelles and cytoplasm double while the cell membrane and the cell wall, if present, enlarge. The enlarged cell containing the doubled content is, subsequently, split into two, the more, or the less, identical cells. We call the period between start of the replication until the actual fission the lag-phase. It needs to be distinguished from the stationary phase of a bacterium, in which the whole metabolism is actually halted due to a shortage of external resources (Bertrand, 2019). In contrast, the lag-phase is initiated by sensing new resources, a new environment. In addition, bacteria show phenotypic plasticity with regard to the duration of the lag-phase. We see variation among cells of the same genotype under similar conditions, but also variation in response to varying conditions. The former enables the bet-hedging principle in an unpredictable environment, *i.e.*, some cells profit in the short term from the prevailing conditions, whereas others may profit from different conditions later on. A matter of spreading the risk. Adapting responses to changes in the environment enable bacteria to cope with adverse conditions. For example, the lag-phase may be prolonged under exposure to anti-biotics. A prolonged lag-phase means a delay in passing into the stage of cell division, in which a bacterium is rather sensitive to anti-biotics. So, the bacterium may avoid the exposure to anti-biotics in that stage by way of adapting the duration of the lag-phase.

Sensing the biotic and a-biotic environment is essential for each organism to survive, grow and multiply. Social sensing is essential for organisms living closely to other individuals of the same species, or other species. Bacteria are, in general, such social organisms living in colonies and interspecific communities. The human microbiome is a striking example of a microbial multispecies community. The social sensing of bacteria is predominantly based on quorum sensing in a broad sense (Grandclément, *et al.* 2016). Bacterial cells produce molecules that can be received, sensed, by other cells. The quantity of a molecule increases as the number of bacterial cells increases by way of multiplication. It will pass a threshold, a quorum, at a certain density of the bacterial cells initiating negative feedback to bacterial cells reducing their activity. The population, or community, density regulates the gene expression of individual cells. The molecules involved are called quorum sensing signals. A large diversity of such signals exists. Signals that do not perceive cellular density only. The signals may control bacterial genes belonging to four major functional categories, (i) cell maintenance and multiplication, (ii) external cell performance, like motility and adhesion, (iii) horizontal gene transfer, and (iv) interference with other organisms, like that with a host, or a competing bacterium. Sensing a competitor may result in the production of anti-biotics to outcompete the other. The signals may, however, be cleared and inactivated by a host, or competitor. We call such a process 'quorum quenching'. Production of the quorum sensing signals may also be inhibited by various natural and synthetic compounds. We may notice that bacteria balance between living too isolated, which exposes these to all kinds of adverse conditions, and living in an over-crowed setting of insufficient resources resulting in decay. Pathogenic bacteria, for example, living together in a well-developed bio-film, which can be produced by bacteria, are less sensitive to anti-biotics than those without such a film. Clearly, interfering with the quorum sensing of pathogenic bacteria is favourable from a point of view of humans. In contrast, it is detrimental for our health with respect to, for example, mutualistic bacteria of our intestinal microbiome.

We see a vertical gene transfer during the multiplication of bacteria, in which the original cell may be seen as the 'parent' and the resulting two cells as 'off-spring'. Phylogenetic associations are based largely on the assumption of such a vertical gene transfer indeed. Horizontal gene transfer, however, is rather common among bacteria passing species barriers (Arnold *et al.*, 2022). Three major mechanisms of horizontal gene transfer exist. One, bacteria may take up bare DNA from the environment, which is incorporated in the own genome after passing the cell membrane and eventually the cell wall. It is called transformation. Two, an infecting bacteriophage brings in the foreign DNA, which we call transduction. Three, DNA may pass directly between two bacterial cells by way of a pilus connecting both cells. It is called conjugation. We may notice here that the concept of species is based on the absence of interspecific gene transfer, which does evidently not hold for bacteria.

Diseases

Bacteria cause a variety of diseases among man. We highlight just a few here rather than providing a systematic overview (Table 3.3). We feel the selected samples may provide some insight into the variety of bacterial diseases among people.

Cholera is caused by a Gram-negative staining bacterium, *Vibrio cholerae*. We mentioned the bacterium in Chapter 1 already. An outbreak of cholera in London resulted in the seminal observations of John Snow and his subsequent intervention without knowing the cause of the epidemic. It was Robert Koch who identified *V. cholerae* unequivocally as cause of cholera later on. We also had to enlarge the definition of 'pathogens' in section 3.1 due to organisms like *V. cholerae*. These do actually not enter the host, but these exudate toxic substances damaging host cells. It happens for *V. cholerae* in the human gut, excreting the toxin that causes severe diarrhoea with accompanying symptoms. We, however, need to see it especially as an ectoparasite of various water organisms (Lutz, *et al.* 2013). We may even state that it lives predominantly in an aqueous environment. Actually, two out of the *c.* 200 strains known of *V.*

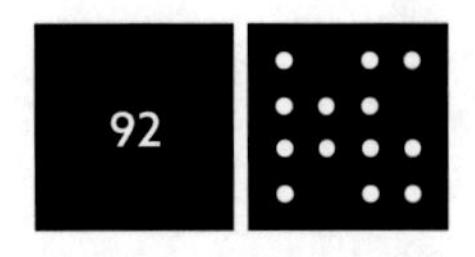

cholerae are pathogenic for humans only, *i.e.*, these produce the cholera toxin. We have seen above that such a toxin serves as a quorum sensing signal. So, the production of it seems to be related to the specific, multi-species community, in the human gut. Good sanitation is the way to prevent cholera. Epidemics are, therefore, observed only in areas missing proper water supply due to, poverty, war, or natural catastrophes, like an earthquake. Treatment of severe cholera largely relies on re-hydration of patients, orally, or intravenously. If not, death is inevitable.

The Gram-positive staining *Staphylococcus aureus* is, like many bacteria, both a commensal organism and a pathogen. The site of abundance determines its' character with regard to a host. The bacterium lives under aerobic conditions, but it may, facultatively, inhabit anaerobic sites as well. It is present among about a third of people (Tong *et al.*, 2015). A variety of diseases is related to *S. aureus* infections like, bacteraemia, endocarditis, cutaneous abscess, osteomyelitis, septic arthritis, pneumonia, meningitis and the toxic shock syndrome. Infections are often hospital acquired due to, patients being in a susceptible state, invasive treatments, and the relatively easy adherence of *S. aureus* colonies to medical devices including prosthetics. Treatment of infection is limited in an increasing abundance of anti-biotic resistant strains of *S. aureus*. Abundance of Methicillin-Resistant *S. Aureus* (MRSA) in hospitals is well-known.

Another Gram-positive staining bacterium is *Mycobacterium leprae*. It causes leprosy, which is also called Hansen's disease. The Norwegian physician Gerhard Hansen (1841-1912) linked the bacterium to the disease, although he was not able to culture it purely. *In vitro*-culture of *M. leprae* is not possible still (Lowe, 2021). Leprosy is characterised by peripheral nerve damage and skin lesions. Leprosy may result in disabilities, if not treated on time. Treatment is based on a multi-drug strategy. The Bacillus Calmette-Guérin vaccine, which prevents tuberculosis, may also provide some degree of protection against *M. leprae*. Tuberculosis is caused by *M. tuberculosis*, a bacterium closely related *M. leprae*. Leprosy, which is thus caused by *M. leprae*, is indigenous in tropical countries. The *M. lepromatosis* bacterium may cause a very similar disease, which is called diffuse lepromatous leprosy (Singh *et al.*, 2015). Both bacteria share presumably a common ancestor some 14 million years ago. Cases of diffuse lepromatous leprosy have been detected in Mexico only, so far.

The sexually transmitted disease gonorrhoea is caused by the Gram-negative staining bacterium *Neisseria gonorrhoeae*. It is an obligate human pathogen (Hill *et al.*, 2016). It is a very common disease. The bacterium adheres to epithelial cells using pili, especially in the urogenital tract. It may survive outside a cell, or invade it. Symptoms of gonorrhoea are, in general, mild. The symptoms may become more severe in a few percent of cases, especially in women. It may result in sterility, ectopic pregnancy, septic arthritis and eventually death, if not treated adequately. Treatment of gonorrhoea is limited in an increased abundance of resistance to anti-biotics among *N. gonorrhoeae*. Development of vaccines is limited in the rather weak immune response to an *N. gonorrhoeae*-infection. In addition, vaccines that are developed so far, do not impart the memory of the, adaptive, immune system.

Table 3.3. Examples of bacteria causing diseases among man and the organs affected.

AFFECTED ORGAN(S)	DISEASE	Bacterium
Intestine	Cholera	*Vibiro cholerae*
Skin, mucosal membranes	Various	*Staphylococcus aureus*
Nerves, skin	Leprosy (Hansen's)	*Mycobacterium leprae*
Urogenital tract	Gonorrhoea	*Neisseria gonorrhoeae*
Lymph nodes, lungs, vascular	Plague	*Yersinia pestis*
Skin, systemic	Lyme	*Borrelia burgdorferi*
Respiratory tract	Respiratory diseases	*Mycoplasma pneumoniae*

The Gram-negative staining bacterium *Yersinia pestis* causes the plague disease. The bacterium may enter mammals, including man, by way of a bite of a flea (Dean *et al.*, 2018). Fleas are ectoparasites sucking blood from hosts. A typical example is *Xenopsylla cheopis*, the oriental rat flea. Bacterial cells that enter the human skin pass to the regional lymph nodes, like those in the armpits, in which these multiply causing the typical, painful, swellings, the buboes. It is, therefore, called bubonic plague. The bacterium may also pass into the vascular system causing septicaemic plague. Pneumonic plague results from either direct inhalation of *Y. pestis*, or secondary by way of the lymphatic, or vascular, system. A primary pneumonic plague may also result in septicaemic plague. Tissue necrosis is a major feature of plague and it, therefore, has been known as 'Black Death' in the Middle Age. The skin and extremities are especially prone to necrosis. Treatment of plague is based on anti-biotics (Nelson *et al.*, 2020). The case-fatality rate is between 7 and 30%, even among patients treated with anti-biotics. The rate depends on, type of anti-biotic, type of plague, and age. A timely application of an anti-biotic seems pivotal in a successful treatment, especially with respect to bubonic plague. We may notice that some anti-biotics, like gentamicin, cause severe side effects prompting to a careful use. Plague is indigenous in remote areas, in which *Y. pestis* causes, relatively, small outbreaks and cases from time to time.

The spirochaetal, gram-negative staining, bacterium *Borrelia burgdorferi* causes Lyme disease, which is also called Lyme borreliosis, among man. Pathogenesis starts with a bite of a tick belonging to the genus *Ixodes* (Radolf *et al.*, 2021). Ticks are ectoparasites taking up blood of a host, like the fleas mentioned above. An infected tick passes *B. burgdorferi* to the host during uptake of the blood. It starts to multiply upon establishment in the skin. The multiplication results in a migration of *B. burgdorferi* along the plane of the skin and towards the blood vessels. We may see the typical expanding red rash at the bite spot, which is called 'erythema migrans'. If so, it may develop within a period of three to thirty days. The migration is at a relatively fast rate of about 4 micrometres per second. The bacterium becomes systemic upon entering the blood circulation. Detection of *B. burgdorferi* in an early stage of Lyme disease is a challenge due to the relatively low abundance in the blood, *i.e.*, about 1 cell per 10 ml blood. The bacterium may, subsequently, enter various organs. The heart, joints, and nervous system seem to be the organs entered most frequently. Accurate data, however, is missing due to, ignorance of the early stage of disease, an incubation time of the severe symptoms going up to years, and limitations of diagnostic tools. In addition, Lyme disease may be caused by other bacteria than *B. burgdorferi*. Anyway, Lyme disease becomes severe upon attacking vital organs. We are talking then especially about (i) Lyme neuroborreliosis upon infection of the central nervous system, (ii) Lyme carditis upon infection of the heart, and (iii) Lyme arthritis upon infection of the joints. Observation of Lyme arthritis among children back in the seventies in the USA resulted finally in the detection of *B. burgdorferi* as cause of Lyme disease in the early eighties. Death due to Lyme disease caused by *B. burgdorferi* appears to be rather exceptional. Treatment is based on various anti-biotics depending on the stage of pathogenesis. We have an ongoing debate about the duration of anti-biotic treatments, which pivots on the persistence of *B. burgdorferi* in the human body. Introduction of a vaccine was a commercial failure. The majority of people, who are infected by *B. burgdorferi*, recover, but a minority suffers from chronic symptoms that last for years.

The cell-wall lacking *Mycoplasma pneumoniae* causes various diseases of the respiratory tract. Species of the genus *Mycoplasma* are the smallest living organisms, so far known. These do pass even microbial filters that cannot be passed by other bacteria. It is, therefore, that mycobacteria, which constitute the class Mollicutes of the Bacteria, have been considered previously as viruses. The term 'myco' refers to fungi because one species, *M. mycoides*, and one only, shows a growth resembling the one of mycelial, fungal, growth (Waites and Talkington, 2004). The term 'mycoplasma', therefore, is not appropriate, but we keep on using it being that common. In contrast, the term 'Mollicutes' is appropriate, as it indicates 'soft skin', the lack of a cell wall. The lack of a cell wall, may explain the relatively easy adherence of *M. pneumoniae* to epithelial cells in the respiratory tract without being removed by the muco-ciliary clearance mechanism. The adherence is crucial for survival of *M. pneumoniae*, as it does not enter an epithelial cell to take up resources. It is, in a narrow sense, an ectoparasite, as the epithelium is the outer tissue layer. Some species of the genus *Mycoplasma*, also the ones colonising humans, may be regarded even as commensals. It all points to the relatively mild symptoms of disease caused by *M. pneumoniae* in general. The clinical symptoms are primarily caused by the adherence of *M. pneumoniae* to an epithelial cell, which results in an oxidative stress response of the host cell. It may damage the cilia of the respiratory tract. In addition, an immune response may cause illness. The stronger the response, the more severe the illness. It is a reason to consider an immune-modulatory therapy for patients rather than one based on anti-biotics. Pharyngitis, tracheobronchitis, or pneumonia, requiring treatment and, eventually, hospitalisation occur in a minority of infected people only. People may also carry the bacterium without any, significant, disease symptom. We also have some evidence that *M. pneumoniae* infection may pave the way for settlement of other pathogenic bacteria and pathogenic viruses.

The bacterium *M. pneumoniae* has been detected occasionally outside the respiratory tract. It has been associated with, for example, neurological disorders and renal failure. We do not know whether *M. pneumoniae* is the cause of these disorders indeed. We encounter here a potential weakness of a diagnostic tool with a relatively high sensitivity. It may detect such a variety of organisms, and at such low quantities, that establishment of a causal relationship becomes troublesome, if it exists at all. We see it for diagnostic tools, like the ones based on the Polymerase Chain Reaction (PCR). These are very sensitive, detecting a single cell even. We feel molecular tools provide us, in general, with a wealth of data, of which the interpretation is a real challenge.

Dispersal

The bacteria described here in detail indicate already the large variety of modes of dispersal of bacteria from one host to another. We see the so-called oral-faecal route, in which humans ingest water, or food, contaminated with bacteria and release these by way of faeces again in the environment. These may then pass to other individuals, whenever sanitary conditions are insufficient. Existence of, and persistence in, environmental reservoirs outside man may be pivotal in survival of these bacteria. A striking example is *Vibrio cholerae* (Lutz *et al.*, 2013). Its adaptability to the (water) reservoirs is expressed in the name. The term 'vibrio' is derived from the Latin word 'vibrare', which may be translated as 'moving'. The bacterium is able to move in water using its flagella. So, it is able to pass actively from one host to another in water. Man is just one potential, unnecessary, host. People may also be a dead-end host, as it is for *Borrelia burgdorferi*. A human infected by *B. burgdorferi* will not liberate the bacterium later on, as the probability that a tick may take up *B. burgdorferi* from a human is infinitesimally small. So, the bacterium passes usually between mammals in the wild, like rodents, by way of tick bites.

Dispersal of *Yersinia pestis* may involve an animal vector, or not. Fleas function as the vector in case of the bubonic plague. The bacterium is picked up from rodents by the fleas and, subsequently, transmitted to people. So, rodents like rats are the natural reservoirs of *Y. pestis*. Man is a side-host only. Interestingly, *Y. pestis* seems to be evolved from the non-pathogenic *Y. pseudotuberculosis* (Spyrou *et al.*, 2018). A major adaptation with respect to pathogenicity was the one of becoming able to colonise arthropods, and more specifically, fleas. The production of an appropriate bio-film seemed to be essential in this colonisation (Hinnebusch and Erickson, 2008). Bubonic plaque may develop into pneumonic plague, as we outlined above. If so, *Y. pestis* may disperse to the lungs of another person within aerosols. So, we see for *Y. pestis* two modes of dispersal, an animal vector-based one and an air-based one. An immediate passing from one person to another may result from sexual intercourse, as exemplified by *Neisseria gonorrhoeae*.

Bacteria may encounter adverse conditions, under which neither survival in a host nor dispersal as a functional bacterium to another host is possible. Some bacteria may, however, produce spores, as we have seen above. These spores may, rest in the soil, be ingested by animals, or be taken up by water streams. Production of spores by a bacterium, therefore, is a mode of dispersal, as well. The cause of tetanus disease, *Clostridium tetani*, is an example of a bacterium that produces spores serving both, survival and dispersal.

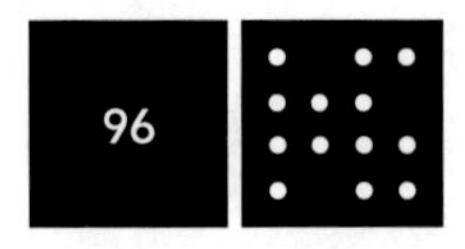

REFLECTIONS

The postulates of Henle-Koch were based on research directed to determining the pathogenicity of bacteria. The applicability, however, spans all agents that are nominated as pathogens. The execution of the postulates may face methodological constraints. We may reflect about the execution when such constraints are faced. How to apply the postulates properly in an adapted form?

3.4 Eukaryotic organisms

The domain of Eukarya encompasses a huge variety of organisms, from unicellular yeasts to whales. It includes plants, animals, and all around. The major, and common, feature of all these is a cellular nucleus containing DNA. A two-super kingdom and seven-kingdom classification, with sub-categories, was proposed by Ruggiero *et al.* (2015a, b) for living organisms supporting the Catalogue of Life (www.catalogueoflife.org). We dealt already with two of the kingdoms, Archaea and Bacteria, constituting the super kingdom Prokaryota. The other five kingdoms, Protozoa, Chromista, Fungi, Plantae, and Animalia constitute the super kingdom Eukaryota. The Catalogue of Life is intended to be reviewed and revised at five-years intervals. It is not a strict phylogenetic tree, although it is a hierarchical one. The catalogue intends to reflect common sense among scientists, at a specific moment, incorporating unsolved issues of systematics.

Eukaryotic pathogens of humans may be assigned to three categories of organisms, (1) protozoa, (2) helminths, and (3) fungi (Parish and Riedel, 2020). The helminths belong to the kingdom Animalia, which is also known as kingdom Metazoa. Research on pathogenic protozoa and helminths belongs historically to medical parasitology, whereas that of fungal pathogens to medical microbiology. We may notice that several species assigned to the Kingdom of Protozoa by Parish and Riedel (2020) belong to the Kingdom of Chromista according to Ruggiero *et al.* (2015a). The demarcation between the kingdoms of Protozoa and Chromista is troublesome. Various species have, however, been transferred from the kingdom of Protozoa to the one of Chromista, which seems to be missed by Parish and Riedel (2020). We will treat the human pathogens belonging to the Protozoa, or Chromista, in one section accounting for both, the use of the term 'protozoa' in systematics as a specific kingdom and its popular use in a broad sense, which includes protozoa, protists, and chromista, chromists.

3.4.1 Protozoa/Chromista

The kingdom of Protozoa includes unicellular organisms only. The kingdom of Chromista encompasses both, unicellular and multicellular organisms. Protozoa and Chromista differ in membrane topology and protein targeting (Cavalier-Smith, 2018). The difference may have resulted from a different evolutionary path. Protozoa are assumed to have evolved ancestrally and mono-phylogenetically from a prokaryotic organism enslaving symbiotic purple bacteria. Chromista would have been evolved from a eukaryotic-chimaeras that arose by symbiotic enslavement of a eukaryote. Each of the kingdoms encompasses a huge variety of organisms in terms of, structure, multiplication, and dispersal. In addition, species pathogenic to humans are relatively rare in these kingdoms. We, therefore, do not provide general characteristics describing Protozoa and Chromista, like we did for prions, viruses and bacteria. Some typical human pathogens among Protozoa and Chromista are listed in table 3.4, of which we will describe four in detail below.

The ectoparasite *Giardia lamblia* causes giardiasis among mammals, a diarrheal disease. It is also known as *G. intestinalis*, or *G. duodenalis*. The life cycle of *G. lamblia* encompasses the three stages of, trophozoite, cyst, and excyzoite (Bernander *et al.*, 2001). A trophozoite cell is flagellated enabling movement within the intestine. The cell is binuclear. A trophozoite cell multiplies by way of binary fission. The nuclei, therefore, cycle between a diploid and tetraploid state. It implies that a trophozoite cell cycles between a state of tetraploid and octoploid during vegetative growth of *G. lamblia* in the intestine. The nuclear state of a trophozoite cell changes from binary to tetrad upon turning into a cyst. A cyst is a cell-walled survival and dispersal unit.

Table 3.4. Examples of pathogenic species belonging to the kingdom of Protozoa and Chromista, respectively, referring to the diseases these cause and the organs affected.

AFFECTED ORGAN(S)	DISEASE	SPECIES
Small Intestine	Girdiasis	*Giardia lambia*
Blood, lymph, brain	Trypanosomiasis	*Trypanosoma brucei*
Urogenital tract	Trichomoniasis	*Trichomonas vaginalis*
Intestine, systemic	Amoebiasis	*Entamoeba histolytica*
Brain	Primary amoebic meningoencephalitis	*Nagleria fowleri* ▲
Blood, liver	Malaria	*Plasmodium spec* ▼
Intestine, systemic, brain	Toxoplasmosis	*Toxoplasma gondii*
Gastrointestinal tract	Cryptosporidiosis	*Cryptosporidium spec.*
Blood	Babesiosis	*Babesia spec.*

Assignment of species to kingdom according to Catalogue of Life (www.catalogueoflife.org) 2022; species upwards the arrow belong to the Protozoa and those downwards to the Chromista.

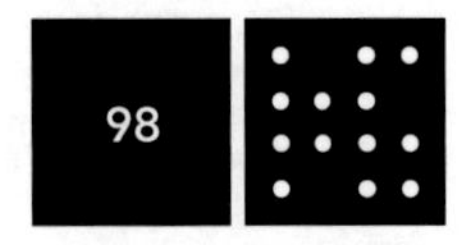

Each of the nuclei inside the cyst is tetraploid. So, we have 16 copies of the *G. lamblia* genome in a cyst. The cysts leave the intestine of a host within the faeces. A cyst may, subsequently, be ingested by a host arriving in the intestine. A process of encystation starts, in which four mature trophozoites are released. The life cycle is closed. The release of four trophozoites by one cyst may explain the relatively high infection efficiency of *G. lamblia*. We will elaborate on the term 'infection efficiency' in Chapter 4. A trophozoite cell includes multiple peripheral vacuoles (Cernikova *et al*., 2018). These are used to take up resources and exudate 'waste' of the cell. A vacuole fuses temporarily with the cell membrane, opening it up. Extracellular compounds are taken up, sorted out in the vacuole, and the beneficial compounds are passed to the connected endoplasmic reticulum. It is called the 'kiss-phase'. Useless compounds are released into the extracellular space upon the next fusion between vacuole and cell membrane, the so-called 'flush-phase'. The change from sporozoite into cyst seems to be triggered by a shift from acidity towards alkalinity, depletion of lipids, and a high density of trophozoites. The first two factors may mimic changes in the environment along the gastro-intestinal tract whereas, the third may make sense in an overcrowded intestine that restricts uptake of resources. The thick-walled cysts leaving the host may remain viable for weeks at room temperature, up to several months at relatively low temperature. A cyst does not have flagellates. It disperses passively in flows of water. The prevalence of giardiasis among man ranges between 2% in high-income countries and 30% in low-income countries. The symptoms of giardiasis are, in general, mild and patients clear the pathogen with 2-3 weeks. If not, severe disruption of the epithelial barrier function may result. Treatment with anti-biotics is based on the organic compound 5-nitro-imidazole.

The life cycle of *Trypanosoma brucei*, which causes trypanosomiasis, is completely within hosts (Schuster *et al*., 2021). A survival/dispersal unit, like the cyst of *G. lamblia*, is missing. The tsetse-fly injects *T. brucei* in metacyclic form into the blood of a mammal taking a meal (Fig. 3.5). The metacyclic *T. brucei* cells become slender upon starting to multiply. The multiplication ends and slender cells turn into stumpy ones. These may be picked up from the blood by a tsetse-fly. The stumpy cells turn into pro-cyclic cells within the midgut of the fly. These cells start to multiply by way of binary fission. The pro-cyclic cells may turn into meso-cyclic ones, arresting the cycle, or pass into the proventriculus. Those passed to the proventriculus transform into proliferative epimastigotes. These swim to the salivary glands and turn into, either meso-cyclic cells, or metacyclic ones. The metacyclic cells are injected into a mammal by a bite of a fly and the cycle is completed. We, however, have a problem. The titre of stumpy cells in the blood of trypanosomiasis patients is very low minimising the probability of pick-up by a fly. The observed abundance of trypanosomiasis is contradictory to this low probability. It is the so-called transmission paradox. It disappears, if slender cells, which are quite abundant in the blood of patients, are picked-up by tsetse-flies and these are transformed into pro-cyclic cells within the fly, as indicated in figure 3.5. The study of Schuster *et al*. (2021) demonstrated that the slender type of cells may transform inside a tsetse-fly indeed, like the stumpy cells. In addition, a single slender cell was sufficient to infect a tsetse fly under controlled conditions. It is a very high infection efficiency, as we will see in Chapter 4.

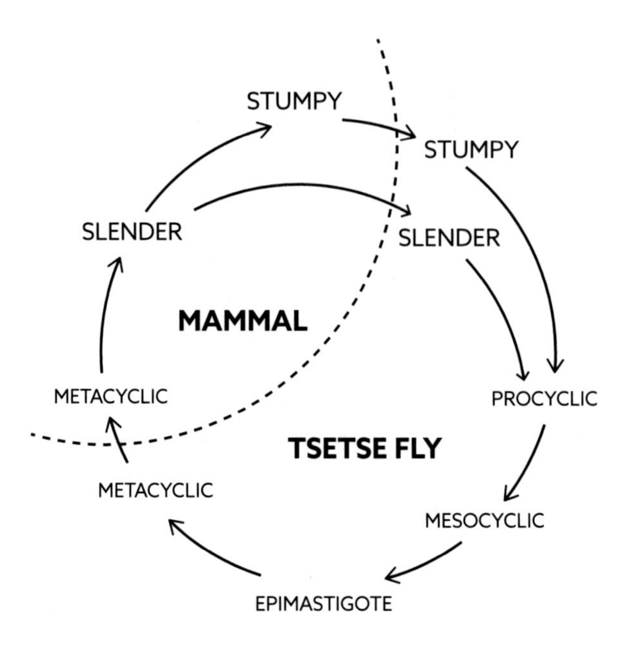

Figure 3.5. Optional life cycle of *T. brucei*. See text for further explanation.

Development of trypanosomiasis upon infection by *T. brucei* is characterised by an early and a late stage (Kennedy and Rogers, 2019). Symptoms in the early stage may be vague and non-specific, like headache and intermittent fever. Diagnosis of the disease may, therefore, be troublesome in an early stage, especially as malaria may both, co-occur and show similar symptoms. We elaborate on malaria below and we continue with trypanosomiasis here. Severe symptoms of affections, like haemolytic anaemia and hepatomegaly, appear upon further spread of the parasite by way of the blood and lymph circulation. Illness aggravates as *T. brucei* passes the blood-brain barrier of a patient entering the central nervous system. It is the late stage of trypanosomiasis, which is also called the encephalitic stage. The characteristic sleeping disorder manifests, in which somnolence and insomnia characterise day and night, respectively. The biological clock reverts completely. The status of a patient will deteriorate, eventually to death, if she, or he, is left untreated. Drugs are relatively efficacious and safe to treat patients in an early stage of trypanosomiasis. In contrast, the only drug available for patients in a late stage, melarsoprol, is very toxic. The case-fatality rate of treatment is estimated at 5-9%. This percentage is quite similar to the one of, untreated, late-stage trypanosomiasis. Less toxic drugs are in the pipeline.

The World Health Organization proposed for sub-Saharan Africa to intervene in the *T. brucei* life cycle in such a way that humans are no longer included in it by 2030. Resolving the transmission paradox is pivotal in achieving this goal. Schuster *et al.* (2021) provided a resolution, as indicated above. Another resolution was provided by Capewell *et al.* (2019) earlier. They argue that people infected by *T. brucei* may be symptomless harbouring *T. brucei* cells extracellularly in the skin. High titres of cells may build up there enabling sufficient up-take by tsetse-flies. If so,

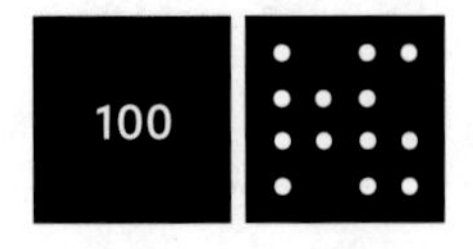

diagnostics and treatment have to be extended to symptomless people. Solving unequivocally the transmission paradox is thus mandatory from a point of view of proper management of trypanosomiasis, anyway.

We turn now from species belonging to the kingdom of Protozoa to those belonging to the Chromista. Malaria is caused by unicellular organisms assigned to species of the genus *Plasmodium*. The dominating one is *P. falciparum* followed by *P. vivax* (Cowman *et al.*, 2016). Four other species are less common among humans. We follow here the life cycle of a *Plasmodium* species infecting man. It starts with a mosquito injecting sporozoites into the skin of a human upon taking a blood meal (Fig. 3.6). These may remain in the skin, or enter a blood vessel. Those remaining in the skin are eliminated by way of the lymphatic system. The ones entering the blood vessel arrive in the liver. The sporozoites change there from a 'migratory mode' into an invasive one, entering hepatocytes. The pathogen subsequently starts to multiply exponentially in the form of merozoites. It may go up to 40,000 merozoites per hepatic cell. The merozoites are released into the blood vessels. These, subsequently, attach to, and enter into, blood cells that are not invaded yet. The erythrocytic stage of the life cycle starts. A merozoite passes the stages of ring and trophozoite erythrocyte and it starts to multiply by way of cell division. It is called schizogony. A single merozoite may produce 16-32 new ones. These mature and destruct the membrane of the erythrocyte. The egressed merozoites enter blood cells, which are uninfected yet, repeating the asexual reproduction. Interestingly, some merozoites develop during schizogony into female and male gametocytes. The stimuli for this change from asexual into sexual reproduction are associated with, a high density of the pathogen, *i.e.*, a high parasitaemia, and drug treatment. The pathogen is able to sense and adapt to the environment. The switch to sexual production is essential to complete the life cycle. Mature gametocytes transfer to a mosquito upon a bite of a patient. The sexual reproduction is, subsequently, completed inside the mosquito. The resulting sporozoites are ready to infect a host. We like to stress here the role of sexual reproduction in creating genetic variation, which is the basis of natural selection, as we will see in Chapter 5.

Malaria in an initial stage is accompanied by non-specific symptoms, like rigours and muscle pains. Efficacious treatment of malaria based on drugs, especially chloroquine and artemisinin, is possible in this stage. Spread of drug resistance among the pathogen, however, reduces drug efficacy, triggering a quest for new treatments. Drug resistance is especially related to *P. falciparum*. No, or partial, treatment of malaria in an initial state may result in progress to a severe stage, and eventually death. Exponential growth of the pathogen in the vascular system results in anaemia and microvascular obstruction with subsequent damage of organs. Severe malaria is more frequent among children than adults in areas where the pathogen is indigenous. Adults seem to have built up sufficient immunity. Advances in malaria control have been achieved by, improved diagnostics, drug combination therapies, preventive drug treatment in pregnancy, and control of mosquitos (Abuga *et al.*, 2021). Vaccines might complete the set of control tools, accomplishing elimination of the disease worldwide. Previous successes in local elimination of malaria, however, show the fragility of keeping the pathogen away (*cf.* Mendis, 2019).

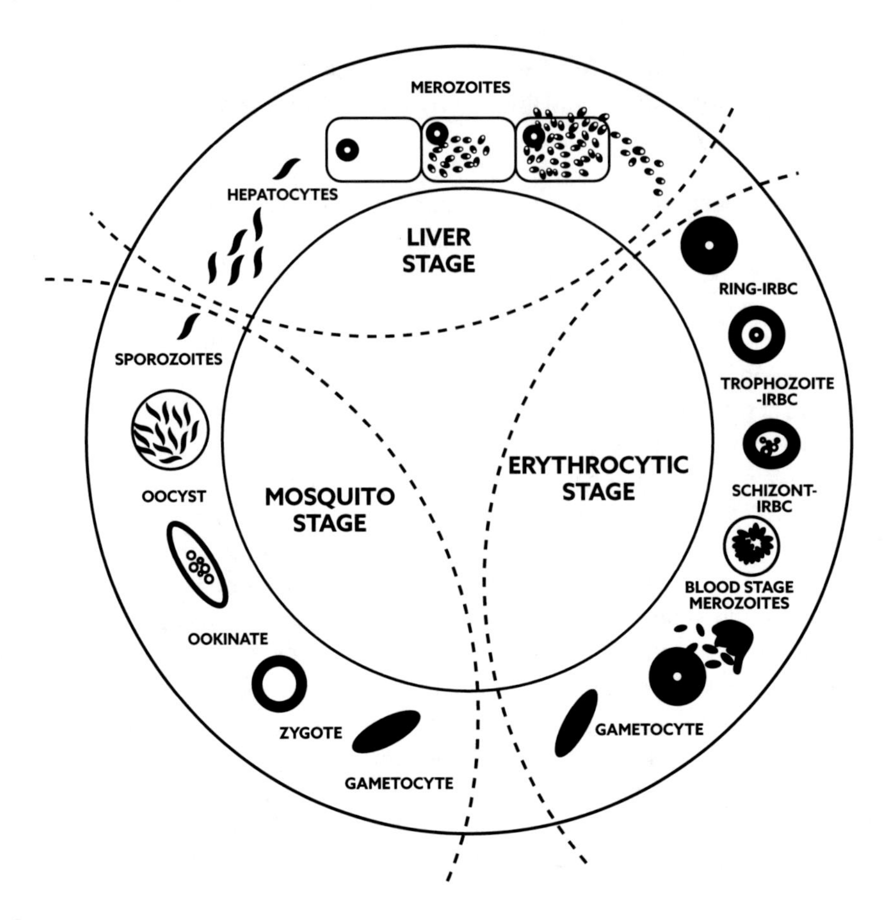

Figure 3.6. Life cycle of *Plasmodium*-species in short, which alternates between species of *Anopheles*, mosquitos, and humans. RBC means red blood cell and I means infected. See text for further explanation.

Toxoplasmosis is caused by the unicellular organism *Toxoplasma gondii*. The life cycle encompasses both, sexual and asexual reproduction (Elsheikha *et al.*, 2021). Sexual reproduction, which is called gametogony, occurs in the intestine of felids, cats, only. A male and female gamete of *T. gondii* fuse to a zygote, which is excreted as an oocyst within the faeces. An oocyst is a real survival unit persisting in soil and water (Shapiro *et al.*, 2019). The excreted oocyst starts a process of internal sporulation producing two sporocysts and each contains four sporozoites. The sporulation adds additional persistence because sporocysts and sporozoites are also enclosed by a cell wall each. The, infectious, sporozoites are, therefore, very well protected against adverse conditions. It enables not survival only, but also dispersal in the environment. Oocysts ingested by mammals and birds pass into the small intestine, in which the sporozoites are released (Elsheika *et al.*, 2021). These invade the epithelial cells, in which differentiation into tachyzoites occurs. Tachyzoites migrate to, and multiply in, various organs, like the eye, brain, and placenta. A parasitophorous vacuole emerges within a tachyzoite upon invading a host cell. It serves as a means of excretion and uptake of compounds. In addition, the vacuole serves as a site of asexual reproduction, in which a tachyzoite divides into two daughter tachyzoites. This process is called endodyogeny. Tachyzoites elicit an immune response of the host. Some tachyzoites encounter the response, or a drug treatment, by turning into bradyzoites. These are enveloped in tissue cysts evading immune responses and cytotoxicity of drugs. The cysts are predominantly located in the brain and muscles. Cysts constitute the symptomless chronic infection. The bradyzoites may turn into tachyzoites again under specific conditions and the infection becomes acute. Cysts, and the enclosed bradyzoites, provide another mode of dispersal of *T. gondii*, besides oocysts including sporozoites. Cysts in the mammalian, or avian, tissue may be ingested by other mammals, or by birds, as these, for example, eat meat contaminated with cysts. If so, the bradyzoites are released from the oocysts and these turn into tachyzoites upon arrival in the intestine. So, all in all, we have three routes of infection of people, (i) ingestion of cysts-contaminated animal tissue, (ii) ingestion of oocysts from water, soil, or food, and (iii) passage of the placenta by tachyzoites from mother to foetus.

The seroprevalence of toxoplasmosis is quite high with about a quarter and a half of people infected in high- and low-income countries, respectively. Most of the people do not show symptoms. The disease may, however, progress, especially in immune-compromised people. It may result in encephalitis, as *T. gondii* is able to pass the blood-brain barrier. Treatment of acute infection is based on drug regimens including pyrimethamine. It, however, fails to clean cysts, the latent infections. Prevention is needed to lower the prevalence of toxoplasmosis, especially among people at risk. Preparing food very well and hand-washing to remove/inactivate cysts and oocysts is pivotal in prevention, as attempts to develop vaccines failed so far.

3.4.2 Helminths

Parasitic worms, the helminths, are assigned to either the phylum Platyhelminthes (flatworms), or Nematoda (roundworms), within the kingdom of Animalia (Parish and Riedel, 2020). Most of these live as ectoparasite in the intestinal tract of a host taking up resources. Some worms inhabit other tissue, like *Wuchereria bancrofti* residing in the lymphatic system of the lower extremities. It causes the disease known as elephantiasis. Another nematode, *Onchocerca volvulus*, inhabits subcutaneous tissue and it may move up to the eyes causing blindness. It is, therefore, called river-blindness.

The life cycle of helminths may include an alternation of hosts, or not. It may also include asexual and sexual reproduction, and worms may be hermaphroditic, or not. Dispersal units may be eggs, or larvae. Dispersal may include abiotic vectors, like water streams, or biotic ones, like mosquitos. Diseases caused by helminths are typical for the (sub-)tropical area, as the sanitary conditions are insufficient often. In addition, the impact of the parasites is higher as people are under-nourished. Children especially suffer from parasitic worms.

We will highlight the disease of human schistosomiasis here, illustrating the specific characteristics of a disease caused by parasitic worms. In addition, control of this disease has priority, because of its devasting effects on man, *i.e.*, 1.4 million disability-adjusted life years in

2017. The disease is caused by species belonging to the genus *Schistosoma*, the schistosomes. These are also known as blood flukes. The genus belongs to the class of Trematoda, which is part of the phylum Nematoda. Human schistosomiasis results especially from infection by, *S. haematobium*, *S. mansoni*, and *S. japonicum* (Colley *et al.*, 2014;). The disease is also known as bilharzia, in reference to the German physician Theodor Maximilian Bilharz (1825-1862). He identified *S. haematobium* as pathogen of schistosomiasis during his employment by the hospital of Cairo. The life cycle of the *Schistosoma*-species encompasses several stages. Man serves as major host and water snails as intermediate host. Occasionally, and depending on *Schistosoma* species, other mammalian hosts may be involved. Active movement in various stages is a hall mark of the pathogen (Linder *et al.*, 2016). Schistosomes can actively search for, and move within, their hosts.

We start looking at the general life cycle of schistosomes at the stage of cercaria (Shiff, 2020). A cercaria consists of a tapering head and a bifurcate tail. The tail is used to swim to man in water. The head attaches to the skin and proteolytic enzymes are secreted. The cercaria bores into the skin and it enters the circulatory system by way of the lymphatic system. The cercaria moves to the lungs and it turns there into the next stage of schistosomulum. The schistosomulum leaves the lung after a few days and it passes to the portal vein in the liver. Schistosomula mature and move in pairs to the vasculature of the end organ. It is the bladder for *S. haematobium* and the intestine for *S. mansoni* and *S. japonicum*. A migrating pair consists of a larger male embracing a slender female. The adult worms ingest red blood cells, which are taken up by the mouth. The intestine has a blind end. It implies that the residues of ingestion, the waste, are transferred back to the mouth to get rid of it. An adult worm has a size of 1 to 1.5cm. It has two suckers, one oral and one ventral. A pair of worms is, in general, together for the entire adulthood. The male folds in such a way that a gynaecophoric canal results, in which the female fits. The canal is used to exchange nutrients and hormones between male and female. It is also used for the sexual reproduction. The female deposits eggs in it, which are, subsequently, fertilised by the sperm of the male. The sperm is produced in the 5-8 testes. It leaves the male by way of a ventral pore. The eggs laid in the fine capillaries of intestine, or bladder, may migrate into the lumen of the intestine, or bladder, secreting proteolytic enzymes. These leave the human body by way of faeces for *S. mansoni* and *S. japonicum*, or urine in the case of *S. haematobium*. Many eggs do, however, not break through the mucosa, remaining in the tissue, or these are flushed away to the liver. The life cycle continues, if eggs are deposited in fresh water, in which a suitable snail host is present (Colley *et al.*, 2014). Fertilised eggs remain vital for 1-2 weeks, irrespectively these are excreted from the human body, or these stay there. An egg in the water hatches, releasing a ciliated miracidium. The miracidium senses for, and actively moves to, a matching snail. The miracidium turns into a sporocyst within the snail and a process of asexual multiplication starts. Daughter sporocysts migrate to the digestive gland of the snail, in which these may produce thousands of cercariae. These are released into the water. The cercariae sense for, and actively move to, a human body in the water, or another mammal. A cercaria needs to establish in a body within 1-3 days, as its reserves deplete within hours, decreasing the infectivity.

The adult worms inhabit the human body, on average, 3 to 10 years. A time span of 40 years has been recorded even. The morbidity does, however, especially result from non-secreted eggs. These become permanently located in the intestine, liver, or urogenital tract. The eggs trigger a granulomatous immune response resulting in inflammation. The clinical manifestation is by way of, a sudden onset of fever, malaise, myalgia, headache, eosinophilia, fatigue and abdominal pain. This acute phase of schistosomiasis wains after 2 to 10 weeks. Acute schistosomiasis is also called the Katayama-syndrome. The immune response is down-regulated at the end of the acute phase and people may progress to the chronic phase of schistosomiasis. It is characterised by intermittent abdominal pain, diarrhoea, and rectal bleeding for intestinal infections, and haematuria, overactive bladder and suprapubic discomfort for urogenital infections. In addition, chronic infection results in the non-specific, disabling, morbidities of, anaemia, malnutrition, and impaired childhood development. Anaemia in schistosomiasis patients is caused by both, inflammation and ingestion of red blood cells by the parasite. It is, therefore, positively associated with parasitaemia. A poor regulation of the immune responses to chronic infections may result ultimately in tissue fibrosis exacerbating the disease.

Schistosomiasis is treated by the drug praziquantel. Its mode of action is not elucidated yet, despite decades of use. The drug seems to paralyse the adult worms by way of interference with the muscles of these. The worms may, subsequently, be removed by the immune system. It explains that the efficacy of praziquantel depends on an effective host response. The drug's efficacy is *Schistosoma* species invariant. It has a low efficacy with respect to the larval stages. The risk of side effects, like headache and dizziness, is positively associated with the parasite load. Evidence of drug resistance of the worms has been reported under controlled conditions. Emergence of drug resistance, therefore, is possible, but it has not been observed in clinical practice yet. The World Health Organization endorsed the preventive use of praziquantel to control schistosomiasis back in 1984. In 2012, it turned even to a strategy of eliminating the disease, in which preventive praziquantel is pivotal. The prevalence of schistosomiasis decreased significantly in sub-Saharan Africa, indeed, comparing the periods of 2000-2010 and 2015-2019 (Kokaliaris *et al.*, 2022). Bayesian restricted, geostatistical, hierarchical models were used to estimate the contribution of preventive praziquantel to the decrease while controlling for potential confounding by various socio-economic and environmental factors. The results indicated the scale-up of preventive drug treatment of school-aged children as a significant, and relevant, factor explaining the observed decrease in schistosomiasis. In addition, results of the study pointed, amongst others, to the need for improved sanitation to achieve sufficient control of schistosomiasis.

3.4.3 Fungi

The kingdom of Fungi encompasses millions of species. The size of the organisms ranges between the micrometres of unicellular yeasts and the square kilometres of the mycelia of filamentous fungi. Fungi are heterotrophic and these need to obtain organic compounds. Fungi, therefore, live as, (i) saprotrophs on dead material, like the familiar mushrooms, (ii) necrotrophs that kill host tissue to feed on it, (iii) biotrophs that live inside a host parasitising it,

(iv) commensals that benefit from a host without damaging it, or (v) mutualists that also provide benefit to the host. Some fungi are facultative with respect to the feeding mode switching from one mode to another, if needed, others have an obligate mode.

Vegetative growth of fungi is based on hyphae. A hypha is a small, cylindrical, structure of one, or more, cells enveloped by a cell wall. The cell wall contains the typical chitin, which is lacking in cell walls of, for example, plants and humans. Cells within a hypha are separated by a septum. The septum has pores enabling transport of compounds between the cells. A hypha enlarges by adding new cells and it branches by way of a new hypha. The resulting structure of hyphae is called the mycelium. So, a fungal individual may become very large as mentioned above. Vegetative growth of fungi may alternate with phases of sexual, or asexual, multiplication. These stages are characterised by a return to unicellular structures, spores, resembling yeasts. Spores are the units of dispersal and survival.

Fungi pathogenic to man are rather the exceptions than the rule. Less than 400 fungal species are known as pathogens (Parish and Riedel, 2020). Fifty of these are responsible for more than 90% of infections among people. In addition, we are dealing predominantly with opportunistic fungi infecting immuno-compromised people. It may explain that fungal diseases, mycoses, receive relatively little attention in medicine.

The dominating incompatibility between fungi and man may be attributed to a lack of adaptation of fungi to the relatively high, endothermic, temperature of the human body (Köhler *et al*, 2015). It might be that parasitic fungi have been especially adapted to poikilotherm organisms, like plants and insects. If so, it may explain that pathogenic fungi of man are especially present on the body surface, *i.e.*, at lower temperature than that of the internal body. In addition, the chitin in the cell walls of fungi is easily recognised as foreign by the human body, inducing immune responses. We do observe more fungal infections indeed among immuno-comprised people than among immuno-competent ones. Finally, a fungus needs to enter the human body, either by way of forcing hyphae into the tissue of interest, or by way of small dispersal units landing in air-filled spaces, like the sinuses, to come into contact with sensitive tissue. Pathogenic fungi, therefore, infect man especially by way of dermal wounds, or small (germinating) spores contacting sensitive tissue.

The fungi pathogenic to immuno-competent people belong to the phyla of, (i) Entomophthoromycota, (ii) Ascomycota, or (iii) Basidiomycota. The soil inhabiting species *Conidiobolus coronatus* is an example of the Entomophthoromycota. It causes a submucosal disease of the nose, sinuses and central face. Tissue swelling may result in impairment of breathing and chronic bacterial sinusitis. Infection is presumed to result from inhalation of fungal ballistospores. Another soil-inhabiting fungus, *Histoplasma capsulatum*, is an example of the Ascomycota. It causes histoplasmosis, which resembles fairly well tuberculosis. The fungus affects especially the lungs, but it may also affect other organs. Histoplasmosis may be fatal, if left untreated. Infection of people results from dispersal of *H. capsulatum* by way of microconidia. These are small enough, *i.e.*, a size of less than five micrometres, to pass into the

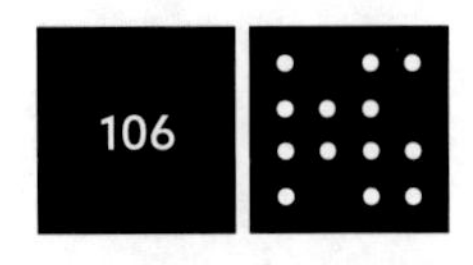

alveoli of mammals, including man. The conidia get absorbed by macrophages, in which the conidia turn into a yeast form. The yeast is able to proliferate inside the host.

The debris-inhabiting *Cryptococcus gattii* belongs to the Basidiomycota. It causes pneumonia and meningoencephalitis among people. Infection by *C. gattii* was known among immuno-compromised people, like AIDS-patients, for a long time. It may, however, infect immuno-competent people, as well. The fungus is inhaled as a dried yeast form, or basidiospores, from which a yeast-form emerges. The yeast may, subsequently, pass into the bloodstream, reaching out to other organs, especially the central nervous system. A *C. gattii* infection is lethal, if left untreated. This fungus indicates that the common division of fungi into primary and opportunistic pathogens, respectively, is not that accurate always. Some fear exists even that *C. gattii* will develop into a major, primary, pathogen of man.

The number of opportunistic fungi infecting immuno-compromised people may be quite large in theory. In practice, the number is quite small, still. Species of the genus *Candida* belong to the known pathogens. These live commonly as commensals on mammalian mucous membranes, especially that of the gastro-intestinal and urogenital tract. These may, however, turn into pathogens, if the host is weakened, or conditionally changed as, for example, during pregnancy. The conditions of the human body, therefore, determine whether the fungus behaves as a commensal, or a pathogen becoming invasive. Another well-known example of such a dual behaviour is *C. albicans*. It may cause opportunistic infections of the skin, intestinal tract, and vagina. Another opportunistic fungus is *Aspergillus fumigatus*. It is a thermophilic saprobe of decayed plants. The conidia disperse to the sinuses and respiratory tract of man. An individual may inhale hundreds of conidia per day. The immune system reacts to the hyphae, which emerge from germinating conidia, eliminating the fungus. The fungus may, however, invade host tissue whenever the immune reaction is limited in neutrophils and macrophages. The fungus may enter the blood circulation, which may, subsequently, result in tissue infarctions.

REFLECTIONS

We may see the complexity of eukaryotic organisms with respect to anatomy and physiology. Complex also are the pathogenic interactions with hosts. The multi-stage life-cycles encompassing multiple hosts pose us for real challenges with respect to management of the resulting diseases. Treatment, or prevention, of a disease is, in general, insufficient to eliminate the causal pathogen. We may, therefore, reflect on proper management of eukaryotic pathogens taking into account all the complexity.

3.5 Outlook

We see a huge variation in pathogens and the human diseases related to these. We have assigned these to the, prions, viruses, bacteria, protists/chromists, helminths, and fungi. We feel the scientific challenges ahead decrease going from prions to the fungi. Prions are rather intractable as these are body-own proteins refolded into 'infectious' ones. The refolded ones propagate inside the body, but these may also be passed to another body. Whereas proteins belong to the area of (bio-)chemistry, prions pass into biology, although these are not organisms. We need to combine methods of chemistry, biology, and medicine, to understand prions really. We have to go a long way still to put prions in the right epidemiological perspective.

Viruses look more biological than prions having RNA, or DNA, besides proteins. We, however, need to consider these as bio-chemical entities still, like we do with the prions. Viruses are not organisms, which explains the difficulties in assigning these to well-defined species. We see a huge challenge ahead, setting up the systematics of viruses properly, as based on the concept of species. We may notice the importance of this concept with regard to investigating host-pathogen interactions, as we will do in Chapter 5.

We face a quite different challenge in dealing with bacteria. Investigating bacteria is relatively easy as we deal with organisms assigned, in general, to well-defined species. We, however, can hardly distinguish between the good and bad ones. Bacteria may, for example, reside in the gut microbiome as the good ones, whereas a gut perforation may result in fatal sepsis caused by the same bacteria. The site and conditions determine whether we may, benefit from, tolerate, or suffer from bacteria. In addition, horizontal gene transfer is common, which may, for example, transfer anti-biotics resistance from one genotype, or even species, to another. We, therefore, need a better understanding of the 'behaviour' of bacteria in various environments.

The protists/chromists are, in general, characterised by rather complex life cycles. The life cycles may include various, stages, vectors, and hosts. Elucidating these and developing efficacious interventions is a real challenge, as demonstrated by the rather consistent failure to control malaria, so far. We may also notice it in the control of helminths. We need to add that socio-cultural factors, in general, are rather important in managing diseases caused by protists/ chromists and helminths, respectively.

Fungi do not really challenge us at the moment. It might change, if the number of immuno-compromised people increases, fungi evolve towards pathogenicity, or environmental conditions change. The increasing urbanisation world-wide might cause such changes in environmental conditions, even as global warming.

We did focus on bio-medical knowledge presenting the wealth of pathogens in this chapter. This knowledge needs to be put into a broader psychological and social perspective to understand the impact of the various pathogens on human health and well-being really. It will be done in Chapter 7, dealing with the impact of epidemics.

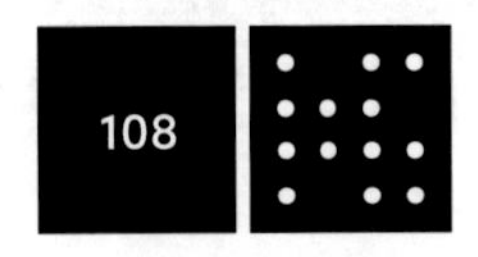

References

Abuga, K.M., Jones-Warner, W. and Hafalla, J.C.R., 2021. Immune responses to malaria pre-erythrocytic stages: implications for vaccine development. Parasite Immunology 43: e12795. DOI: 10.1111/pim.12795.

Arnold, B.J., Huang, I-T. and Hanage, W.P., 2022. Horizontal gene transfer and adaptive evolution in bacteria. Nature Reviews Microbiology 20: 206-218. DOI: 10.1038/s41579-021-00650-4.

Baseler, L. Chertow, D.S., Johnson, K.M., Feldmann, H. and Morens, D.M., 2017. The pathogenesis of Ebola virus disease. Annual Review of Pathology: Mechanisms of Disease 12: 387-418. DOI: 10.1146/annurev-pathol-052016-100506.

Bernander, R., Palm, J.E.D. and Svärd, S.G., 2001. Genome ploidy in different stages of the *Giardia lamblia* life cycle. Cellular Microbiology 3: 55-62. DOI: 10.1046/j.1462-5822.2001.00094.x.

Bertrand, R.L., 2019. Lag phase is a dynamic, organized, adaptive, and evolvable period that prepares bacteria for cell division. Journal of Bacteriology 201: e00697-18. DOI: 10.1128/JB.00697-18.

Blevins, S. M. and Bronze, M.S., 2010. Robert Koch and the 'golden age' of bacteriology. International Journal of Infectious Diseases 14: e744-e751. DOI: 10.1016/j.ijid.2009.12.003.

Breman, J.G., 2021. Smallpox. Journal of Infectious Diseases 224: S379-S386. DOI: 10.1093/infdis/jiaa588.

Breman, J.G. and Henderson, D.A., 2002. Diagnosis and management of smallpox. New England Journal of Medicine 346: 1300-1308. DOI: 10.1056/NEJMra020025.

Capewell, P., Atkins, K., Weir, W., Jamonneau, J., Camara, M., Clucas, C., Swar, N.K., Ngoyi, D. M., Rotureay,B., Garside, P., Galvani, A.P., Bucheton, B. and MacLeod, A. 2019. Resolving the apparent transmission paradox of African sleeping sickness. PloS Biology 17: e3000105. DOI: 10.1371/journal.pbio.3000105.

Carlson, G.A. and Prusiner, S.B., 2021. How an infection of sheep revealed prion mechanisms in Alzheimer's disease and other neurodegenerative disorders. International Journal of Molecular Sciences 22: 4861. DOI: 10.3390/ijms22094861.

Castaneda, D., Gonzalez, A. J., Alomari, M., Tandon, K. and Zervos, X.B., 2021. From hepatitis A to E: a critical review of viral hepatitis. World Journal of Gastroenterology 27: 1691-1715. DOI: 10.3748/wjg.v27.i16.1691.

Castelle, C.J. and Banfield, J.F., 2018. Major new microbial groups expand diversity and alter our understanding of the tree of life. Cell 172: 1181-1197. DOI: 10.1016/j.cell.2018.02.016.

Casadevall, A. and Pirofski, L-A. 2014. Microbiology: ditch the term pathogen. Nature 516: 165–166. DOI: 10.1038/516165a.

Cavalier-Smith, T., 2018. Kingdom Chromista and its eight phyla: a new synthesis emphasising periplastid protein targeting, cytoskeletal and periplastid evolution, and ancient divergences. Protoplasma 255: 297-357. DOI: 10.1007/s00709-017-1147-3.

Cernikova, L., Faso, C. and Hehl, A.B., 2018. Five facts about *Giardia lamblia*. PLoS Pathogens 14: e1007250. DOI: 10.1371/journal.ppat.1007250.

Colley, D.G., Bustinduy, A.L., Secor, W.E. and King, C.H., 2014. Human schistosomiasis. Lancet 383: 2253-2264. DOI: 10.1016/S0140-6736(13)61949-2.

Cowman, A.F., Healer, J., Marapana, D. and Marsh, K. 2016. Malaria: biology and disease. Cell 167: 610-624. DOI: 10.1016/j.cell.2016.07.055.

Dean, K.R., Krauer, F., Walløe, L., Schmid, B.V., Lingjærde, O.C., Bramanti, B. and Stenseth, N.C., 2018. Human ectoparasites and the spread of plague in Europe during the Second pandemic. Proceedings of the National Academy of Sciences USA 115: 1304-1309. DOI: 10.1073/pnas.1715640115.

Elsheikha, H.M., Marra, C.M. and Zhu, X-Q., 2021. Epidemiology, pathophysiology, diagnosis, and management of cerebral toxoplasmosis. Clinical Microbiology Reviews 34: e00115-19. DOI: 10.1128/CMR.00115-19.

Fesharaki-Zadeh, A. 2019. Chronic traumatic encephalopathy: a brief overview. Frontiers in Neurology 10: 713. DOI: 10.3389/fneur.2019.00713.

Fredricks, D.N. and Relman, D.A., 1996. Sequence-based identification of microbial pathogens: a reconsideration of Koch's postulates. Clinical Microbiology Reviews 9: 18-33. DOI: 10.1128/CMR.9.1.18.

Ghaemmaghami, S. 2017. Biology and genetics of PrP prion strains. Cold Spring Harbor Perspectives in Medicine 7: a026922. DOI: 10.1101/cshperspect.a026922.

Grandclément, C., Tannières, M., Moréra, S. Dessaux, Y. and Faure, D., 2016. Quorum quenching: role in nature and applied developments. FEMS Microbiology Reviews 40: 86-116. DOI: 10.1093/femsre/fuv038.

Hill, S.A., Masters T.L. and Wachter, J., 2016. Gonorrhea – and evolving disease of the new millennium. Microbial Cell 3: 371-389. DOI: 10.15698/mic2016.09.524.

Hinnebusch, B.J. and Erickson D.L., 2008. *Yersinia pestis* biofilm in the flea vector and its role in the transmission of plague. Current Topics in Microbiology and Immunology 322: 229-248. DOI: 10.1007/978-3-540-75418-3_11.

Hofmann, J.W., Seeley, W.W. and Huang, E.J., 2019. RNA binding proteins and the pathogenesis of frontotemporal lobar degeneration. Annual Review of Pathology 14: 469-495. DOI: 10.1146/annurev-pathmechdis-012418-012955.

Ioannidis J.P.A., 2021. Infection fatality rate of COVID-19 inferred from seroprevalence data. Bulletin of the World Health Organization 99: 19 - 33F. DOI: 10.2471/BLT.20.265892.

Jagers op Akkerhuis, G.A.J.M., 2010. Towards a hierarchical definition of life, the organism, and death. Foundations of Science 15: 245-262. DOI: 10.1007/s10699-010-9177-8.

Kennedy, P.G.E. and Rodgers, J., 2019. Clinical and neuropathogenic aspects of human African trypanosomiasis. Frontiers in Immunology 10: 39. DOI: 10.3389/fimmu.2019.00039.

Kokaliaris, C., Garba, A., Matuska, M., Bronzan, R.N., Colley, D.G., Dorkenoo, A.M., Ekpo, U.F., Fleming, F.M., French, M.D., Kabore. A., Mbonigaba, J.B., Midzi, N., Mwinzi, P.N.M., N'Goran, E.K., Polo, M.R., Sacko. M., Tchuenté, L-A. T., Tukahebwa, A.M., Uvon, P.A., Yang, G., Wiesner, L., Zhang, Y., Utzinger, J. and Vounatsou, P., 2022. Effect of preventive chemotherapy with praziquantel on schistosomiasis among school-aged children in sub-Saharan Africa: a spatio-temporal modelling study. Lancet Infectious Diseases. DOI: 10.1016/S1473-3099(21)00090-6.

Köhler, J.R., Casadevall, A. and Perfect, J., 2015. The spectrum of fungi that infects humans. Cold Spring Harbor Perspectives in Medicine 5: a019273. DOI: 10.1101/cshperspect.a019273.

Kuhn, J.H., 2021. Virus taxonomy. Encyclopedia of Virology 1: 28-37. DOI: 10.1016/B978-0-12-809633-8.21231-4.

Li, D., Lin, M-H, Rawle D.J., Jin, H., Wu, Z., Wang, L., Lor, M., Hussain, M., Aaskov, J. and Harrich, D., 2021. Dengue virus-free defective interfering particles have potent and broad anti-dengue virus activity. Communications Biology 4: 557. DOI: 10.1038/s42003-021-02064-7.

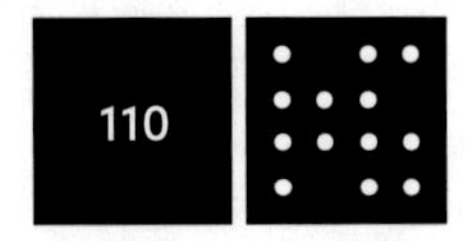

Linder, E., Varjo, S. and Thors, C., 2016. Mobile diagnostics based on motion? A close look at motility patterns in the schistosome life cycle. Diagnostics 6: 24. DOI: 10.3390/diagnostics6020024.

Lipkin, W.I., 2009. Microbe hunting in the 21st century. Proceedings of the National Academy of Sciences USA 106: 6-7. DOI: 10.1073/pnas.0811420106.

Long, J.M. and Holtzman, D.M., 2019. Alzheimer disease: an update on pathobiology and treatment strategies. Cell 179: 312-339. DOI: 10.1016/j.cell.2019.09.001.

Lowe, C., 2021. Basic knowledge on leprosy: a short review. Journal of Biomedical Research & Environmental Sciences 2: 345-349. DOI: 10.37871/jbres1240.

Lurie-Weinberger, M.N. and Gophna, U., 2015. Archaea in and on the human body: health implications and future directions. PloS Pathogens 11: e1004833. DOI: 10.1371/journal.ppat.1004833.

Lutz, C., Erken, M., Noorian, P., Sun, S. and McDougald, D., 2013. Environmental reservoirs and mechanisms of persistence of *Vibrio cholerae*. Frontiers in Microbiology 4: 375. DOI: 10.3389/fmicb.2013.00375.

MacMahon Copas, A.N., McComish, S.F., Fletcher J.M. and Caldwell, M.A., 2021. The pathogenesis of Parkinson's disease: a complex interplay between astrocytes, microglia, and T lymphocytes? Frontiers in Neurology 12: 666737. DOI: 10.3389/fneur.2021.666737.

Mahmoudabadi G., Milo, R. and Philips R., 2017. Energetic cost of building a virus. Proceedings of the National Academy of Sciences USA 114: E4324-E4333. DOI: 10.1073/pnas.1701670114.

Malik, S.S., Azem-e-Zahra, S., Mo Kim, K., Caetano-Anollé, G. and Nasir, A., 2017. Do viruses exchange genes across superkingdoms of life? Frontiers in Microbiology 8: 2110. DOI: 10.3389/fmicb.2017.02110.

Mendis, K., 2019. Eliminating malaria should not be the end of vigilance. Nature 573: 7. DOI: 10.1038/d41586-019-02598-1.

Nelson, K.E., 2020. Transmissible spongiform encephalopathies. In: Nelson, K.E. and Masters Williams, C. (eds.) Infectious Disease Epidemiology: theory and practice. Jones & Bartlett Learning, Burlington MA, United States of America, p. 635-648.

Nelson, C.A., Fleck-Derderian, S., Cooley, K.M. Meaney-Delman, D., Becksted, H.A., Russell, Z., Renaud, B., Bertherat, E. and Mead, P.S., 2020. Antimicrobial treatment of human plaque: a systematic review of the literature on individual cases, 1937-2019. Clinical Infectious Diseases 70: S3-S10. DOI: 10.1093/cid/ciz1226.

Nelson, K.E. and Celentano, D. D., 2020. Human immunodeficiency virus infections and the acquired immunodeficiency syndrome. In: Nelson, K.E. and Masters Williams, C. (eds.) Infectious Disease Epidemiology: theory and practice. Jones & Bartlett Learning, Burlington MA, United States of America.

Parish, N. and Riedel, S., 2020. Microbiology tools for the epidemiologist. In: Nelson, K.E. and Masters Williams, C. (eds.) Infectious Disease Epidemiology: theory and practice. Jones & Bartlett Learning, Burlington MA, United States of America, p. 187-218.

Peng, C., Gathagan, R.J., Covell, D.J., Medellin, C., Stieber, A., Robinson, J.L., Zhang, B., Pitkin, R.M., Olefumi, M.F., Luk, K.C., Trojanowski, J.Q. and Lee, V. M-Y., 2018. Cellular milieu imparts distinct pathological α-synuclein strains in α-synucleinopathies. Nature 557: 558-563. DOI: 10.1038/s41586-018-0104-4.

Prusiner, S.B., 1998. Prions. Proceedings of the National Academy of Sciences USA 95: 13363-13383. DOI: 10.1073/pnas.95.23.13363.

Radolf, J.D., Strle, K., Lemieux, J.E. and Strle, F., 2021. Lyme disease in humans. Current Issues in Molecular Biology 42: 333-384. DOI: 10.21775/cimb.042.333.

Riccardi, N., Giacomelli, A., Antonello, R.M., Gobbi, F. and Angheben, A., 2021. Rabies in Europe: an epidemiological and clinical update. European Journal of Internal Medicine 88: 15-20. DOI: 10.1016/j.ejim.2021.04.010.

Robilotti, E., Deresinski, S., Pinsky, B.A., 2015. Norovirus. Clinical Microbiology Reviews 28: 134-164. DOI: 10.1128/CMR.00075-14.

Rota, P.A., Moss, W.J., Takeda, M., De Swart, R.L., Thompson, K.M. and Goodson, J.L., 2016. Measles. Nature Reviews Disease Primer 2: 16049. DOI: 10.1038/nrdp.2016.49.

Ruggiero, M.A., Gordon, D.P., Orrell, T.M., Bailly, N., Bourgoin, T., Brusca, R.C., Cavalier-Smith, T., Guiry, M.D. and Kirk, P.M., 2015a. A higher level classification of all living organisms. PLoS ONE 10: e0119248. DOI: 10.1371/journal.pone.0119248.

Ruggiero, M.A., Gordon, D.P., Orrell, T.M., Bailly, N., Bourgoin, T., Brusca, R.C., Cavalier-Smith, T., Guiry, M.D. and Kirk, P.M., 2015b. Correction: a higher level classification of all living organisms. PLoS ONE 10: e0130114. DOI: 10.1371/journal.pone.0130114.

Schuster, S., Lisack, J., Subota, I., Zimmerman, H., Reuter, C., Mueller, T., Morriswood, B. and Engstler, M., 2021. Unexpected plasticity in the life cycle of *Trypanosoma brucei*. eLife 10: e66028. DOI: 10.7554/eLife.66028.

Shapiro, K., Bahia-Oliveira, L., Dixon, B., Dumètre, A., De Wit, L.A., VanWormer, E. and Villena, I., 2019. Environmental transmission of *Toxoplasma gondii*: oocysts in water, soil, and food. Food and Waterborne Parasitology 15: e00049. DOI: 10.1016/j.fawpar.2019.e00049.

Shiff, C. J., 2020. Epidemiology of helminth infections. In: Nelson, K.E. and Masters Williams, C. (eds.) Infectious Disease Epidemiology: theory and practice. Jones & Bartlett Learning, Burlington MA, United States of America, p. 917-928.

Singh, P., Benjak, A., Schuenemann, V.J., Herbig, A., Avanzi, C., Busso, P., Nieselt, K., Krause, J., Vera-Cabrera, L. and Cole, S.T., 2015. Insight into the evolution and origin of leprosy bacilli from the genome sequence of *Mycobacterium lepromatosis*. Proceedings of the National Academy of Sciences USA 112: 4459-4464. DOI: 10.1073/pnas.1421504112.

Spyrou, M.A., Tukhbatova, R. I., Wang, C-C., Valtueña, A.A., Lankapalli, A.K., Kondrashin, V.V., Tsybin, V.A., Khokhlov, A., Kühnert, D., Herbig, A., Bos, K.I. and Krause, J., 2018. Analysis of 3800-year-old *Yersinia pestis* genomes suggests Bronze Age origin for bubonic plague. Nature Communications 9: 2234. DOI: 10.1038/s41467-018-04550-9.

Tong, S.Y.C., Davis, J.S., Eichenberger, E., Holland, T.L. and Fowler Jr., V.G., 2015. *Staphylococcus aureus* infections: epidemiology, pathophysiology, clinical manifestations, and management. General Microbiology Reviews 28: 603-661. DOI: 10.1128/CMR.00134-14.

Waites, K.B. and Talkington, D.F., 2004. Mycoplasma pneumoniae and its role as a human pathogen. Clinical Microbiology Reviews 17: 697-728. DOI: 0.1128/CMR.17.4.697–728.2004.

Yin, J. and Redovich, J., 2018. Kinetic modelling of virus growth in cells. Microbiology and Molecular Biology Reviews 82: e00066-17. DOI: 10.1128/MMBR.00066-17.

Zadoks, J.C. and Schein R.D. 1979. Epidemiology and Plant Disease Management. Oxford University Press, Oxford, United Kingdom, 427.

Zhu, Q., Mai, U., Pfeiffer, W., Janssen, S., Asnicar, F., Sanders, J.G., Belda-Ferre, P., Al-Ghalith, G. A., Kopylova, E., McDonald, D., Kosciolek, T., Yin, J.B., Huang, S., Salam, N., Jiao, J-Y., Wu, Z., Xu, Z.Z., Cantrell, K., Yang, Y., Sayyari, E., Rabiee, M., Morton, J.T., Podell, S., Knights, D., Li, W-J., Huttenhower, C., Segata,

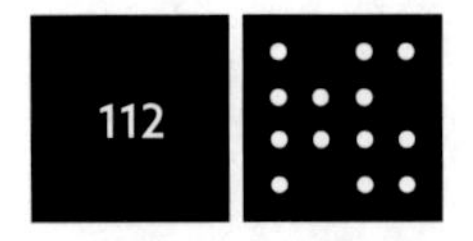

N., Smarr, L., Mirarab, S. and Knight, R., 2019. Phylogenomics of 10,575 genomes reveals evolutionary proximity between domains Bacteria and Archaea. Nature Communications 10: 5477. DOI: 10.1038/s41467-019-13443-4.

EPIDEMIOLOGICAL PARAMETERS

This chapter is an interlude. We finalise the part 'infection at the individual level' and we prepare for the step-up to the part 'infection at the population level'. We will introduce the five key epidemiological parameters. These are, infection efficiency, latent period, infectious period, contact distribution, and the basic reproductive number. This relatively short chapter has neither specific reflections on the content, nor an outlook.

4.1 Infection efficiency

Life cycles of pathogens include a state of dispersal, as we have seen in Chapter 3. Dispersal units serve the passage from one host to another. A unit may be the pathogen itself, like it is for viruses, or it may be a specific form, like a cyst of protists. The dispersal units of a pathogen come into contact with a human body. Not all of the dispersal units are able to establish a pathogenic interaction. These are not in an infectious state, as we may call it. Those in an infectious state may establish on, or within, the body. The dispersal unit turns into an 'infection', as we may call it. Establishment may be relatively simple without any transformation, like it is for *Vibrio cholerae* in the gut, or it may require a complex of processes, in which the pathogen transforms from a dispersal unit into an infection unit, like we saw for *Plasmodium falciparum* causing malaria among man. Establishment means actually that a pathogen is able to take up resources from the host. We call the ratio between dispersal units contacting a host and the number of infection units of the pathogen, the Infection Efficiency (IE). It is one, or 100%, if all dispersal units contacting a host turn into an infection. It is zero, if no dispersal unit turns into an infection unit.

We mentioned in Chapter 3 the determination of IE for *Trypanosoma brucei* among tsetse flies under controlled conditions (Schuster *et al.*, 2021). Inoculation of tsetse flies and a, subsequent, determination of the IE is relatively easy. Determination of the IE of a pathogen among man is a quite different story. How do we, for example, determine the number of virus particles attaching to the epithelial cells in the respiratory tract? We may switch to animal models, but we are limited in ethics sacrificing animals, as well. In addition, the model needs to be representative for the human body. The use of tissue cultures is, in general, limited in the isolation of a specific part of the whole pathogenic process between contact and establishment. We have, for example, seen that the muco-ciliary mechanism prevents attachment of pathogens to the epithelium in the respiratory tract. Inoculation of epithelial cells in culture with a pathogen bypasses the muco-ciliary mechanism and such a model, therefore, provides a flawed determination of the IE. The determination of an IE becomes even more demanding, if a pathogen passes from one organ to another, or it transforms from one stage to another. We will describe some approaches of determining an IE below using the various categories of pathogens described in Chapter 3. We start with prions.

Prions

A proper determination of an IE for prions seems rather impossible at the moment. We may inject homogenates of prion-infected tissue in animals and we may, subsequently, score the proportion of animals diseased. We arrive then at estimates of the IE in the sense of 'a not-specified number of prions' results in a certain percentage of animals diseased (*cf.* Eckland *et al.*, 2018). Such experiments are, of course, of value to demonstrate a prion as pathogen. In addition, we may compare groups of strains of a prion, or conditions, arriving at a relative expression of the IE. We may, for example, set the IE at 1 for the most pathogenic strain and rank other strains relative to this strain. We, however, need to have an indication of doses of inoculum to design appropriate interventions for management of prion diseases. In addition, results of animal models need to be adapted to be representative of the human body. We feel an integrated approach based on, non-interventional, clinical data and modelling might be the way ahead in determination of an IE of prions, as it has been done for the Herpes simplex virus-2 already. We look at the determination of IE of viruses now.

Viruses

The Herpes simplex virus-2 causes genital herpes. Adequate sampling of the virus at dispersal from an infected to an uninfected partner is limited in a relatively high temporal variation in viral load in the genital tract (Schiffer *et al.*, 2014). It may vary by a factor of 10. Sampling just prior to coitus is rather impossible. An integrated modelling approach to the determination of IE was, therefore, used. It included, a simulation model of virus shedding, a mathematical model of transmission, and a statical model applied to available data of genital swabs and diaries of intercourse. The IE for Herpes simplex virus-2 was estimated at 10^{-4}, or less. It means that at least 10,000 virus particles need to disperse from an infected to an uninfected partner during coitus for settlement of the virus at the uninfected partner.

The collection of specimens may be easier for other viruses than Herpes simplex virus-2. If so, we may quantify the inoculum, which is usually based on a standard (*cf.* Karimzadeh *et al.*, 2021). The Tissue Culture Infectious Dose$_{50}$ ($TCID_{50}$) is such a standard. It indicates the inoculum that kills, or causes a cytopathic effect, of 50% of a tissue culture. The choice of the tissue culture thus affects the resulting $TCID_{50}$. The choice of the tissue culture also affects another standard that is commonly used, the Plague-Forming Unit (PFU). It expresses the inoculum that causes a (light) microscopical visible plague, lesion, in a tissue culture. The $TCID_{50}$ and PFU may, subsequently, be translated into a number of virus particles, using electron microscopy. It is a laborious exercise.

Determination of IE of a pathogen directly in man is limited in ethics, as we mentioned above. It was a bit different in the past and we turn to a clinical study directed to the determination of the IE of Adeno virus type 4 in man (Couch *et al.*, 1969). We call it the human mastadenovirus nowadays (https://ictv.global/). It causes various diseases among mammals, including man. The study of Couch et al. (1969) was directed to an acute respiratory disease that was caused by the virus among military recruits. Volunteers were inoculated by way of, either nasal inoculation, or inhalation of virus-containing aerosols. Various doses of inoculum were used. The doses were in the range 3 to 400 $TCID_{50}$ for nasal inoculation and 0.1 to 171 $TCID_{50}$ for inhalation of aerosols. The number of subjects inoculated per dose and method of inoculation was between two and nine. Subjects were checked for disease. Controls were not included. The lowest dose of virus that caused disease was between 10 to 14 $TCID_{50}$ and 0.1 to 1 $TCID_{50}$ for nasal inoculation and aerosol inhalation, respectively. The IE of the virus was thus higher for inhalation. The $TCID_{50}$ was equal to about 13 virus particles, as determined by way of electron microscopy. So, the IE would be between 0.008 and 0.08 for the virus, if we adopt the most likely route of dispersal of the virus by way of inhalation of aerosols.

We may state that the study of Couch *et al.* (1969) is quite exceptional. People were inoculated with non-attenuated virus and the inoculum was expressed in number of virus particles. It resulted in an absolute value of the IE of the virus. The number of subjects included was relatively low, impairing the statistical validity of the study. But, the inclusion of subjects remains exceptional from our current view on ethics. We cannot imagine similar studies with, for example, SARS-CoV-2 nowadays. We also have ethical limitations in determining the IE of bacteria, to which we will turn now.

Bacteria

In contrast to prions and viruses, we may express relatively easily inoculum of bacteria in numbers. Various methods may be used to determine the number of bacterial cells in inoculum. Optical density measurement is the most common one nowadays, as it is a high-throughput method (Beal, *et al.* 2020). The association between optical density and number of cells, however, is not straightforward. In addition, the measurement depends on the configuration of the instrument used. A standard calibration protocol is required to arrive at a validated number of bacterial cells independent of instrument and laboratory of measurement. The traditional cell count method of plating out may be used to set up such a protocol. A bacterial suspension is then plated out on an artificial medium, enabling growth of bacterial cells. The suspension is eventually diluted. Individual cells will start to multiply establishing a colony. The colonies on the plate are, subsequently, counted. The amount of inoculum is expressed by the number of Colony-Forming Units (CFU). This method is valid for bacteria that may grow as an individual cell on artificial media. Viable cells are, therefore, counted only. Another standard calibration protocol is based on dilution series of silica microspheres. The calibration accounts for viable and non-viable bacterial cells and plating on media is not required. This calibration protocol turned out to be superior in a comparative study, in which 244 laboratories were included.

The IE of bacteria seems to span a range from 0.1 down to infinitesimally small. We may arrange the effective dose of bacteria, *i.e.*, the lowest dose needed to interact with the host, in such a way that we may decern a pattern (Fig. 4.1). The hypothesis was put forward that the distance of interaction between bacterial and host cell would determine the effective dose (Schmid-Hempel and Frank, 2007). In short, a bacterial cell excretes bio-active molecules interacting with a host cell. The host cell may be in the close vicinity of the excreting bacterial cell, or further away. If further away, a dilution effect may occur. If so, the bacterial cell has to excrete higher numbers of bio-active molecules to pass a critical threshold of infection at the targeted host cell. It is depicted in figure 4.1 as the distant action of, for example, *Vibrio cholerae*, whereas *Mycobacterium tuberculosis* exerts a local action. Bacteria of the genus *Yersinia* may be classified as either acting locally, or distantly.

The hypothesis of Schmid-Hempel and Frank (2007) was investigated further by Leggett *et al.* (2012). Data of 43 human pathogens were analysed including, various viruses and bacteria, two chromists, and one fungus. These were assigned to local and distant action, respectively. The results of the study supported the hypothesis. The authors, however, posed a cautionary note

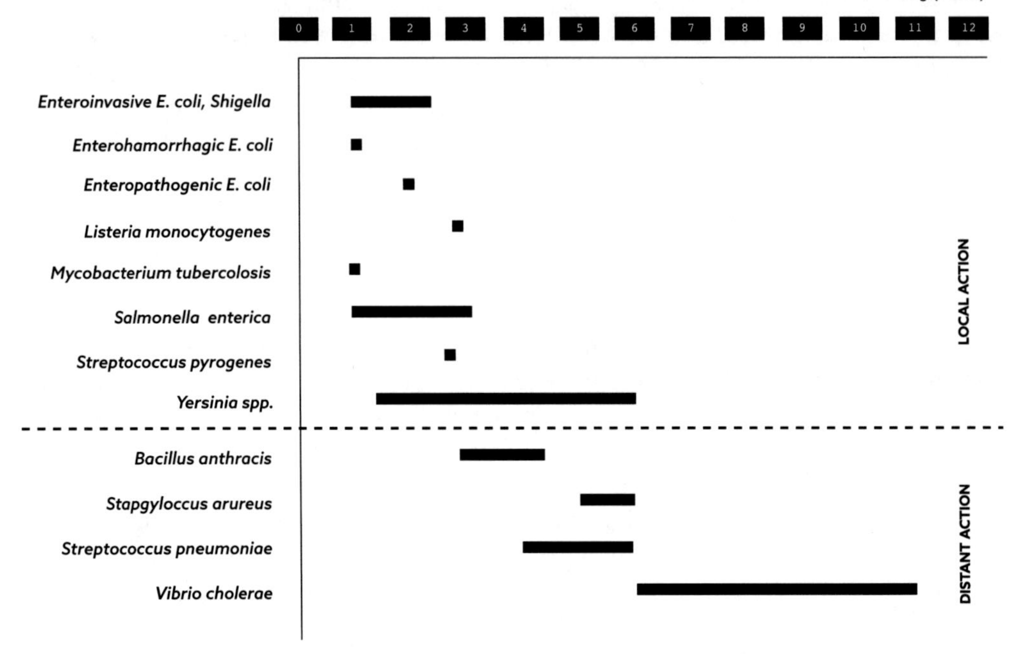

Figure 4.1. Infective dose of various bacteria on a logarithmic scale. Entries of upper panel are bacterial species of which bio-active molecules interact with host cells at very short distance, whereas those indicated in the lower panel do with host cells at larger distance. See text for further explanation.

with respect to the data used. Data resulted from, feeding experiments, from observations of epidemics, or the method of data generation was unknown. The latter was actually the case for most of the pathogens included in the study. Anyway, we may pick up here the message that bacteria do not need a direct contact with a host cell to interact, *i.e.*, to establish, and we may pass now to the category of protists and chromists. A category of pathogens characterised by multi-stage life cycles, as we have seen in Chapter 3.

Protists and chromists

The protist *Giardia lamblia* causes a diarrheal disease among mammals including man. We may use the results of a study with Mongolian gerbils (Schaefer *et al.*, 1991) to estimate the IE of the protist. The gerbils were inoculated orally with cysts, or not. The number of cysts inoculated was in the range of zero to 10,000. The cysts were derived from one original cyst by way of multiplication in gerbils. So, the inoculum was genetically quite uniform among the doses. Eleven doses were used for inoculation, including the control. Ten gerbils were inoculated per dose. A gerbil was declared as infected, if trophozoites were detected in the small intestine, or cysts were present in the faecal material. A linear relationship was determined between the inoculum dose and the proportion of gerbils infected. The regression coefficient was nearly one. We may derive from this relationship that about 25 cysts were needed to infect at least one gerbil. It indicates an IE of about 0.04. The authors also referred to an older study, in which less than 10 cysts of the protist were needed to infect a volunteer prisoner. If so, the IE of *G. lamblia* would be 0.1, at least. A relatively high IE. It is not surprising as each cyst releases 4 mature trophozoites that may infect a host, as we have seen in Chapter 3.

The chromist *Plasmodium falciparum* causes malaria. A dose-finding trial was executed in volunteers, of which we may get an estimation of the IE of this chromist (Mordmüller *et al.*, 2015). We focus on the volunteers who were inoculated intravenously with sporozoites. No control was included. An inoculum dose of 50 sporozoites was sufficient to generate infection of one out of the three volunteers inoculated. It suggests an IE of the chromist of 0.02. A lower dose than 50 was not included in this study. So, the IE might have been higher. In addition, the number of three persons inoculated was rather low from a statistical point of view. We may conclude in general that the biological validity of studies involving animals, and certainly man, is high, whereas the statical one is rather low. Anyway, we turn to helminths now. We will encounter a quite different approach to the estimation of IE, looking at pathogenic worms.

Helminths

The entry and establishment of the parasitic worm *Schistosoma mansoni* in man was investigated using human skin in so-called Franz cells (Bartlett *et al.*, 2000). A Franz cell is composed of an upper well, the donor, and a lower well, the receptor. The skin was clamped in between. The skin was obtained from five women undergoing cosmetic surgery. The upper well was filled with a suspension that contained about 1000 cercariae of the worm. About half of the number of cercariae attached irreversibly to the skin within one minute. About 15% didn't attach even after one hour. Eight to eleven cercariae were able to pass the skin fully being recovered as schistosomula. We may see this as an establishment of the worm in the human body, as we know from Chapter 3. Let us say that 10 out of the 1000 cercariae were able to establish, which implies an IE of 0.01. It is lower than we may infer from another study, in which a mouse model was used (Lindner *et al.*, 2020). Mouses became infected at the lowest inoculum dose of 25 cercariae. It implies an IE of 0.04, at least. We notice that, on the one hand, the mouse model included the complete infection route of the worm, but, on the other hand, the skin of mice is not similar to the one of man. Anyway, we have some idea about the IE of *S. mansoni*, colonising the human body, and we look finally at fungi. Fungi pathogenic for man are relatively rare, as we have recognised in Chapter 3. It may explain that we know relatively few about IE of fungi.

Fungi

We could trace a single estimate of IE of fungi only (Leggett *et al.*, 2012). A value of 0.1 was mentioned for *Histoplasma capsulatum*. The fungus causes histoplasmosis, as we have seen in Chapter 3. A disease that resembles tuberculosis. We do not have any idea about the approach to the determination of the IE. If the estimation is valid, the relatively high IE may result from the small size of the microconidia of *H. capsulatum* that enter relatively easily the human alveoli.

Fungi conclude this section about IE. We may have some idea of it in various categories of human pathogens now. We continue with the presentation of two other epidemiological parameters, the latent and infectious period, respectively. These express the dimension of time of epidemics.

4.2 Latent and infectious period

The latent period, which is symbolised by *p*, is the time between establishment of a pathogen on a host and the take-off from the host by way of dispersal units. The period includes growth, if applicable, and multiplication of a pathogen. The latent period may be as short as one day, like it is for the bacterium *Yersinia pestis* causing plague, and it may go up to fifty years, as observed for the PrP prion in Kuru disease. The latent period may be similar to the incubation period of a disease, or not. The incubation period is the time between establishment of a pathogen and the appearance of clinical symptoms of the disease on the host. If similar, or shorter, clinical symptoms may serve as markers of the end of latency. Markers that may be easily recognised, and these, thus, may indicate infectious cases, enabling timely interventions to inhibit an epidemic. The latent period of the coronavirus SARS-CoV-1, for example, is equal to the incubation period of the Severe Acute Respiratory Syndrome (SARS). We observed an epidemic of SARS back in 2003. An epidemic that was quite well managed indeed. A latent period longer than the incubation period may have facilitated the eradication of the Variola virus, which caused smallpox, by way of massive vaccination campaigns (*cf.* Breman, 2021).

Virologists assumed that the latent period of SARS-CoV-2 on man was similar to the incubation period of COVID-19, as it was determined previously for the SARS-CoV-1 virus and SARS. The assumption turned out to be wrong (He *et al.*, 2020). Throat swabs of 94 laboratory-confirmed COVID-19 cases were used to estimate the dynamics of viral shedding. Analysis of the data suggested pre-symptomatic shedding already. It was confirmed by an analysis of data of serial intervals of 77 transmission pairs. A serial interval is the period of onset of symptoms between two successive cases. A subsequent study indicated that the latent period of SARS-CoV-2 is, on average, 1.5 days shorter than the incubation period of COVID-19 (Xin *et al.*, 2022). Larger data sets were used in that study. The difference of 1.5 days seems relatively small, but it has marked consequences, as we will see later on. A pre-symptomatic end of the latent period may even go up to years before onset of disease symptoms of the host. A striking example is the HIV-virus and the resulting AIDS. We see the devastating effects of such a difference in years still. The epidemic, pandemic, had been progressed considerably before it got noticed. It leaves us with millions of infected people. People who cannot be cured completely yet. We may inhibit disease progression only. The epidemic has to pass away literally, except we may develop more efficacious therapies of AIDS.

Dispersal of a pathogen from a host starts thus after the latent period. It is the start of the infectious period, symbolised by *i*. This period continues as long as dispersal units of a pathogen leave a host. The infectious period may be as short as one day, *e.g.*, the Norwalk virus causing gastroenteritis, and as long as 40 years, *e.g.*, parasitic worms of the genus *Schistosoma* causing schistosomiasis. The number of dispersal units liberated during the infectious period may vary. The number, in general, is low at the start, it has a peak and gets to zero again. The distribution may be positively, or just negatively, skewed, or it is a normal one. The distribution of, for example, SARS-CoV-2 may fit quite well a normal one (He *et al.*, 2020). The peak value, therefore, is at about the middle of the infectious period, which corresponds the more, or the less, with the onset of clinical symptoms of cases. We may notice that the virus may be infectious even on people that do not get diseased. The incubation period is actually lacking for those people.

The take-off from a host by dispersal units of a pathogen has a dimension of time, as expressed in the latent and infectious period. It also has a dimension of space, as expressed in a contact distribution. We will elaborate the spatial dimension in the next section.

4.3 Contact distribution

Dispersal units of a pathogen take off from a host and settle at some distance from it. The distance between the point of origin and its point of settlement, is called the 'dispersal distance'. Settlement may be on a host, a non-host, or any other site, of which a dispersal unit cannot displace anymore. The probability density function describing the distances of dispersal of all units from an individual host is called the 'contact distribution'. It is symbolised by *D*. The general pattern of a contact distribution is one of a decreasing function of number of dispersal units settled versus distance from the infected host (Fig. 4.2). We may notice that probability functions are characterised by approaching the abscissa and ordinate only. This has some practical meaning. One, we cannot determine all the dispersal units that take off from a host, even under controlled conditions. So, we may state that an unknown number of units 'disappears' at distance zero, at the host. A wind burst may, for example, uplift SARS-CoV-2 viruses immediately after exhalation by a subject. Two, we are limited in the distance, at which we may detect a pathogen. We cannot trap a pathogen along kilometres. It is a challenge to catch it within meters from an infected host even. We thus see, in general, a pattern of settling of dispersal units in a relatively large proportion in the vicinity of the infected host and a decreasing proportion further away. A proportion that may become infinitely small at larger distance. We encounter a statistical hurdle in fitting a curve, a contact distribution, to such a data of dispersal. The probability density is highest in the vicinity of the origin, as dispersal units settle predominantly there, at least with respect to the dispersal units we detect. This data, therefore, determines largely the fit of a probability curve, *i.e.*, it has the largest statistical weight. In contrast, the units settling further away have a rather low statistical weight. We, therefore, cannot determine the tail properly by way of fitting simply probability functions to data of pathogen dispersal.

The common management of epidemics focuses on the high-density part of contact distributions. We saw it, for example, in keeping an interpersonal distance of one to two metre to control dispersal of SARS-CoV-2 (*e.g.*, Welsch *et al.*, 2021). Such an intervention may be efficacious in protecting individuals, if we assume a rather low infection efficiency (IE) of a pathogen. It may be, in general, a valid assumption for the majority of people, who are relatively healthy, indeed. It may be questionable for the most vulnerable part of a community. A relatively low number of dispersal units, say the tail of a contact distribution, may be detrimental for those people still.

A contact distribution may be based on the dispersal units of a pathogen. We may extend it to one based on infections, cases. We then determine the number of infected individuals, cases, around an originally infected individual, case, taking into account the distance from this one. It is a quite common method in botanical epidemiology (*cf.* Frantzen and van den Bosch, 2000; Severns *et al.*, 2019). Interference by other sources of pathogen dispersal, however, is likely out-of-doors blurring a contact distribution. We may arrive at a proper contact distribution by way of a stepwise procedure nevertheless (*cf.* Frantzen and van den Bosch, 2000). This procedure also enables us to tackle the statistical problem mentioned above with respect to determining the tail of a contact distribution. We proceed as follows. Step one, we fit various types of probability density functions to the data of infections, cases, observed. Step two, we use the parameters of the various probability density functions to model epidemics. Step three, we compare the predictions with an epidemic observed, preferably with one under semi-controlled conditions. The best fitting model incorporates then the proper contact distribution. If so, we will see that a non-exponentially decreasing contact distribution, the dotted line in figure 4.2, results in a relatively fast-expanding epidemic, whereas an exponentially decreasing one, the continuous

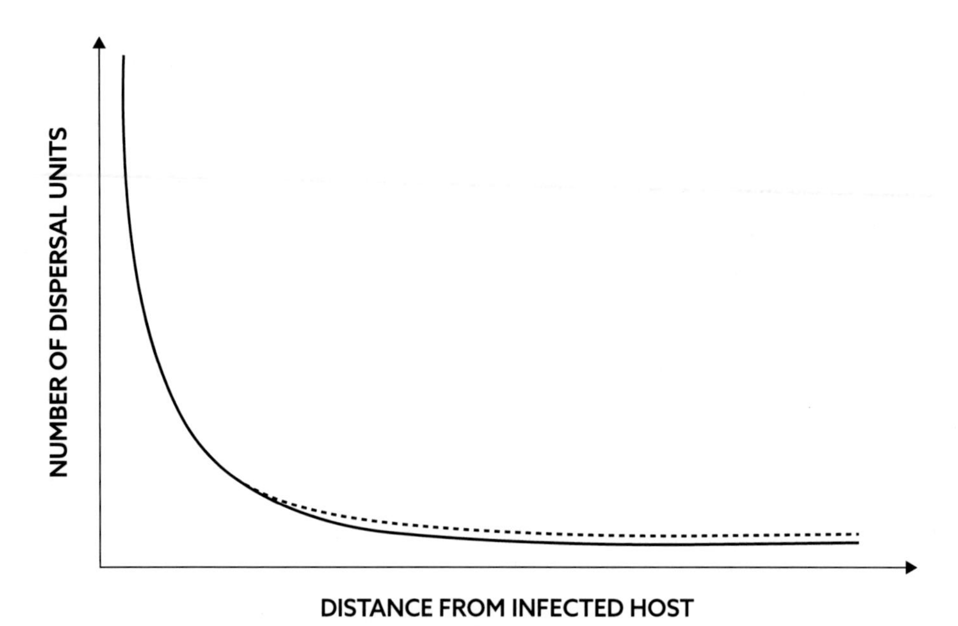

Figure 4.2. Hypothetical dispersal gradients according to, a power law distribution, as indicated by the dotted line, and an exponential distribution, as indicated by the continuous line. The distance from an, immotile, infected host is indicated on the abscissa and the number of dispersal units on the ordinate. See text for further explanation.

line in figure 4.2, in a relatively slow expanding epidemic (*cf.* Kot *et al.*, 1996). A relatively small difference in the tail of the contact distribution is amplified considerably at the population level. The devil is in the (de)tails indeed. We will elaborate on this issue in Chapter 6

We have considered contact distributions from a point of view of dispersal of a pathogen only, so far. It is certainly a valid view for sessile hosts, like plants. It also is appropriate for motile hosts with a relatively short extent, like most animals. We may not stick to this view only with respect to man. We have to add another view, in which distance of dispersal of a pathogen is measured relatively to the home of an infected subject, rather than the subject itself. Such a view is needed, if the following applies, (i) dispersal of a pathogen depends relatively strongly on movement of people, (ii) people travel abroad of their neighbourhood, and (iii) a travelling person should carry the pathogen in an infectious state. If so, a contact distribution may have a tail that extends thousands of kilometres. Having indicated this, we turn to another epidemiological parameter, the basic reproductive number. It has neither a dimension of time, nor one of space. It is a dimensionless parameter, like the IE.

4.4 Basic reproductive number

The basic reproductive number, symbolised by R_0, expresses the ability of an organism to multiply, assuming suitable conditions for reproduction. It is the average number of offspring per mother. We need to redefine R_0 with respect to pathogens, as new infections constitute the offspring of a pathogen. It, therefore, is the number of new infections resulting from one infection, or in terms of infection of a host, the number of newly infected individuals related to an originally infected one. We may see R_0 as the net result of, (1) multiplication of a pathogen within a host, (2) dispersal of the pathogen to (a) new host(s), and (3) the infection efficiency. In addition, the R_0 is defined with respect to a completely susceptible, human, population. We call it the effective reproductive number, R_t, if the assumption of a completely susceptible population is not met, like it is in an ongoing epidemic.

Estimation of R_0, or R_t, of pathogens is impossible at the individual level of a host. We have to step up to the population level. We may arrive then at estimates based on observations in combination with, or without, mathematical modelling. The former means inevitably accepting bias due to all the biotic and abiotic factors affecting the, multiplication, dispersal, and infectivity of pathogens. The latter implies accepting the suite of assumptions underlying the models. Values of R_0 have been reported ranging from values just above 1, *e.g.*, Influenza A virus that causes, erratically, flu among man, up to 203 for the Measles Morbillivirus that provokes measles, and all values in between. The estimation, and use, of R_0 is not that straightforward (Delamater *et al.*, 2019; Bauch, 2020). In addition, we feel the assumption that the R_0 needs to be estimated from a completely susceptible population is rather impossible to meet. It indicates that R_0 is a rather theoretical parameter. A parameter of value and flaws, especially in the theory of evolutionary dynamics, as we will see in the next chapter. In addition, we feel the warnings regarding the estimation of R_0 may apply as well to R_t. If so, we may question the use of R_t instead of R_0.

We have presented all the key epidemiological parameters now. We are ready to step up to the population level. It is the level at which we start to talk about epidemics. We will, however, address firstly another phenomenon at the population level, the one of evolutionary dynamics of pathogens.

References

Bartlett, A., Brown, M., Marriott, C. and Whitfield, P.J., 2000. The infection of human skin by schistosome cercariae: studies using Franz cells. Parasitology 121: 49-54. DOI: 10.1017/s0031182099006034.

Bauch, C.T., 2020. Estimating the COVID-19 R number: a bargain with the devil? Lancet Infectious Disease 21: 151-153. DOI: 10.1016/S1473-3099(20)30840-9.

Beal, J., Farny, N.G., Haddock-Angelli, T., Selvarajah, V., Baldwin, G.S., Buckley-Taylor, R., Gershater, M., Kiga, D., Marken, J., Sanchania, V., Sison, A., Workman, C.T. and iGEM Interlab Study Contributors, 2020. Robust estimation of bacterial cell count from optical density. Communications Biology 3: 512. DOI: 0.1038/s42003-020-01127-5.

Breman, J.G., 2021. Smallpox. Journal of Infectious Diseases 224: S379-S386. DOI: 10.1093/infdis/jiaa588.

Couch, R.B., Knight, V., Douglas, R.G., Black, S.H. and Hamory, B.H., 1969. The minimal infectious dose of adenovirus type 4; the case for natural transmission by viral aerosol. Transactions of the American Clinical and Climatological Association 80: 205-211. DOI: not applicable.

Delamater, P.L., Street, E.J., Leslie, T.F., Yang, Y.T. and Jacobsen, K.H., 2019. Complexity of the basic reproduction number. Emerging Infectious Diseases 25: 1-4. DOI: 10.3201/eid2501.171901.

Eckland, T.E., Shikiya, R.A. and Bartz, J.C., 2018. Independent amplification of co-infected long incubation period low conversion efficiency prion strains. PLoS Pathogens 14: e1007323. DOI: 10.1371/journal.ppat.1007323.

Frantzen, J. and van den Bosch, F., 2000. Spread of organisms: can travelling and dispersive waves be distinguished? Basic and Applied Ecology 1: 83-92. DOI: 10.1078/1439-1791-00010.

He, X., Lau, E.H.Y., Wu, P., Deng, X., Wang, J., Hao, X., Lau, Y.C., Wong, J.Y., Guan, Y., Tan, X., Mo, X., Chen, Y., Liao, B., Chen, W., Hu, F., Zhang, Q., Zhong, M., Wu, Y., Zhao, L., Zhang, F., Cowling, B.J., Li, F. and Leung, G.M., 2020. Temporal dynamics in viral shedding and transmissibility of COVID-19. Nature Medicine 26: 672-675. DOI: 10.1038/s41591-020-0869-5.

Karimzadeh, S., Bhopal, R. and Nguyen Tien, H., 2021. Review of infective dose, routes of transmission and outcome of COVID-19 caused by the SARS-COV-2: comparison with other respiratory viruses. Epidemiology and Infection 149: e96. DOI: 10.1017/S0950268821000790.

Kot, M., Lewis, M.A. and Van den Driessche, P., 1996. Dispersal data and the spread of invading organisms. Ecology 77: 2027-2042. DOI: 10.2307/2265698.

Leggett, H.C., Cornwallis, C.K. and West, S.A., 2012. Mechanisms of pathogenesis, infective dose, and virulence in human parasites. PloS Pathogens 8: e1002512. DOI: 10.1371/journal.ppat.1002512.

Lindner, T., Stenzel, J., Koslowski, N., Hohn, A., Glass, Ä., Schwarzenböck, S.M., Krause, B.J., Vollmar, B., Reisinger, E.C. and Sombetzki, M., 2020. Anatomical MRI and [^{18}F]FDG PET/CT imaging of *Schistosoma mansoni* in a NMRI mouse model. Scientific Reports 10; 17343. DOI: 10.1038/s41598-020-74226-2.

Mordmüller, B., Supan, C., Lee Sim, K., Gómez-Pérez, G.P., Salazar, C.L.O., Held, J., Bolte, S., Esen, M., Tschan, S., Joanny, F., Calle, C.L., Löhr, S.J.z., Lalremruata, A., Gunasekera, A., James, E.R., Billingsley, P.F., Richman, A., Chakravarty, S., Legarda, A., Muñoz, J., Antonijan, R.M., Ballester, M.R., Hoffman, S.L.,

Alonso, P.L. and Kremsner, P.G., 2015. Direct venous inoculation of *Plasmodium falciparum* sporozoites for controlled human malaria infection: a dose-finding trial in two centres. Malarai Journal 14: 117. DOI: 10.1186/s12936-015-0628-0.

Schaefer, F.W., Johnson, C.H., Hsu, C.H. and Rice, E.W., 1991. Determination of *Giardia lamblia* cyst infective dose for the Mongolian gerbil (*Meriones unguiculatus*). Applied and Environmental Microbiology 57: 2408-2409. DOI: 10.1128/aem.57.8.2408-2409.1991.

Schiffer, J.T., Mayer, B.T., Fong, Y., Swan, D.A. and Wald, A., 2014. Herpes simplex virus-2 transmission probability estimates based on quantity of viral shedding. Journal of the Royal Society, Interface 11: 20140160. DOI: 0.1098/rsif.2014.0160.

Schmid-Hempel, P. and Frank, S.A., 2007. Pathogenesis, virulence, and infective dose. PloS Pathogens 3: 1372-1373. DOI: 10.1371/journal.ppat.0030147.

Schuster, S., Lisack, J., Subota, I., Zimmerman, H., Reuter, C., Mueller, T., Morriswood, B. and Engstler, M., 2021. Unexpected plasticity in the life cycle of *Trypanosoma brucei.* eLife 10: e66028. DOI: 10.7554/eLife.66028.

Severns, P.M., Sackett, K.E., Farber, D.H. and Mundt, C.C., 2019. Consequences of long-distance dispersal for epidemic spread: patterns, scaling, and mitigation. Plant Disease 103: 177-191. DOI: 10.1094/pdis-03-18-0505-fe.

Welsch, R., Wessels, M., Bernhard, C., Thönes, S. and Von Castell, C., 2021. Physical distancing and the perception of interpersonal distance in the COVID-19 crisis. Scientific Reports 11: 11485. DOI: 10.1038/s41598-021-90714-5.

Xin, H., Li, Y., Wu, P., Li, Z., Lau, E.H.Y., Qin, Y., Wang. L., Cowling, B.J., Tsang, T.K. and Li, Z., 2022. Estimating the latent period of Coronavirus disease 2019 (COVID-19). Clinical Infectious Diseases 74: 1678-1681. DOI: 10.1093/cid/ciab746.

INFECTION AT THE POPULATION LEVEL

5

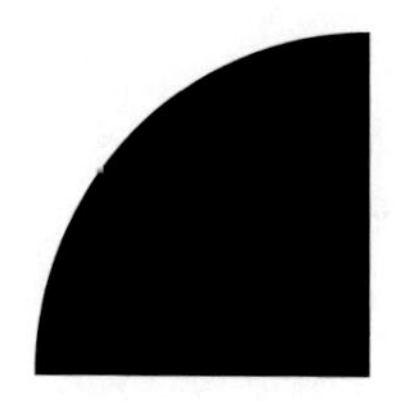

EVOLUTIONARY DYNAMICS OF PATHOGENS

We step up to the population level in this chapter, entering the popular, but tricky, topic of evolution. The concept, model, of evolution is based, ultimately, on evidence rather than empirical demonstration. We will, therefore, focus on the key elements of the theory of evolution, *i.e.*, genetic variation and the mechanism of natural selection that can be supported by empirical data. We will elaborate on theory and add empirical data of the various categories of pathogens presented in Chapter 3, so far available. We will explain the mechanism of host tolerance, besides that of host resistance and its counterpart pathogenicity. Tolerance is a frequently overlooked mechanism in host-pathogen interactions. We will also highlight the topic of pathogen competition, which is at the heart of evolutionary dynamics of pathogens.

5.1 Genetic variation in pathogenicity and resistance

A human is, like any organism, a potential source of food for parasites. The human body, therefore, has various defence mechanisms to protect it against such parasites, like the skin and the immune system. The great majority of potential parasites is not able to break this human defence at all. We call this basic incompatibility. In contrast, parasites that master the defence show basic compatibility. We call these thus the pathogens of humans, of which we provided a glimpse in Chapter 3.

We do not have to wonder about the high frequency of basic incompatibility between parasites and humans. Viruses, bacteria, fungi, and most helminths are not able to penetrate the human skin, except a vector is involved. The malaria causing *Plasmodium falciparum*, for example, is able to penetrate the human skin by way of a blood-taking mosquito only. It, however, implies that *P. falciparum* is adapted to two, completely different, hosts, *i.e.*,, a mammal and an insect. Mammals have a relatively high and continuous body temperature, whereas insects do not have an own body temperature. These are rather cold. So, a parasite like *P. falciparum* has to cope with a completely different body temperature of the two hosts, besides all other differences in morphology and physiology. The probability of a parasite adapting to two completely different hosts is rather low explaining the small proportion of parasites being pathogens of humans. We saw a high body temperature of humans as reason for incompatibility with most fungi in Chapter 3 already (*cf.* Köhler *et al.*, 2015). The epithelium also is a reason of basic incompatibility, as a parasite needs to have specific proteins to dock onto and enter it. The epithelium may, therefore, explain, amongst others, the very small proportion of viruses being pathogenic to humans, although these may disperse very easily among man, contacting epithelium in the respiratory tract, or intestine.

Basic compatibility is in operation as soon as one genotype/phenotype of a parasite at least is able to feed on one genotype/phenotype of a host at least, under conditions suitable for a parasitic interaction. We may see from this definition already that we are dealing with genotype-genotype interactions between a pathogen and a host. These interactions have been, and are, studied extensively for plant diseases, whereas we are limited in ethics and methodology investigating infectious diseases in humans. We, therefore, adopt the framework of pathogenicity and resistance of pathogen-plant interactions here (*cf.* Frantzen, 2000), and we, subsequently, consider the likelihood of pathogen-human interactions to fit in that frame as well. We, however, turn first of all to a typical study directed to pathogen-plant interactions.

Investigating pathogen-plant interactions starts with a culture of various genotypes of both, a pathogen and its host of interest. We can, of course, not reproduce a completely identical genotype, as we will have some mutations always, even in case of completely asexual reproduction. We have, therefore, to deal with (near) isogenic lines. We differentiate the lines of plant and pathogen by using the term 'variety' and 'strain', respectively. Culturing plant varieties and pathogen strains is, in general, feasible for most of the agricultural crops and their pathogens. It has also been done for wild plants and some of their pathogens. The various

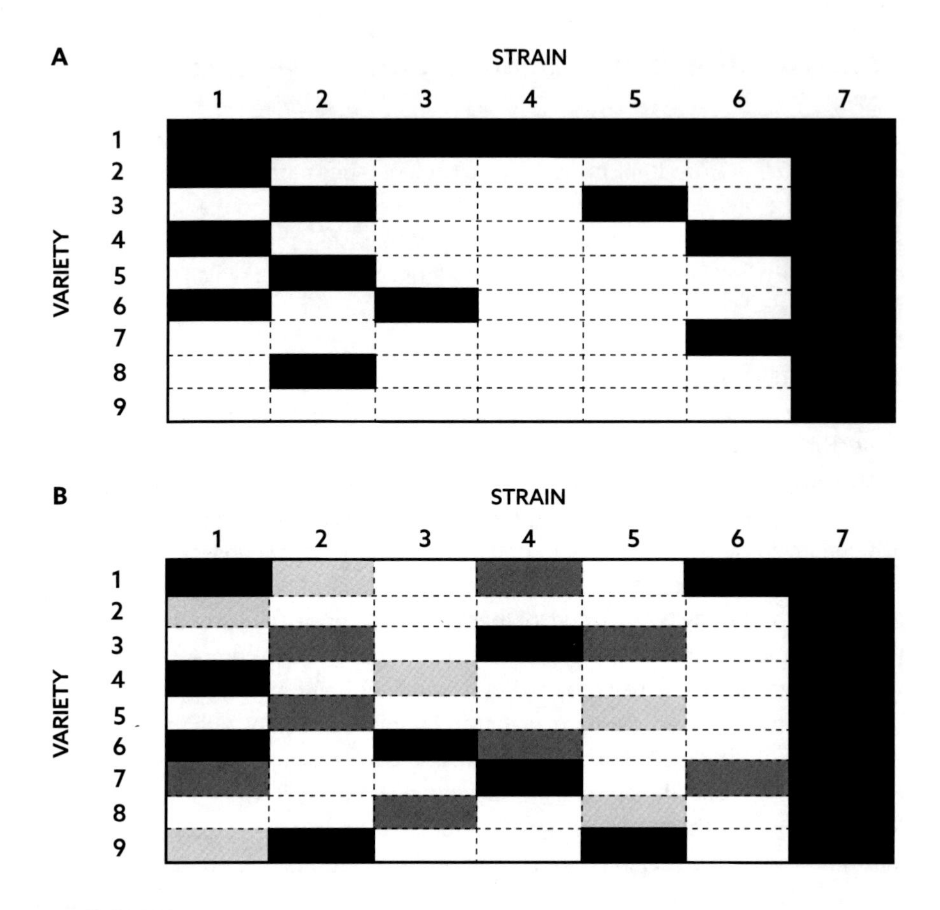

Figure 5. 1. Hypothetical testing of 7 strains of a pathogen on 9 varieties of a host, in which the outcome is disease, or not (panel A), or a degree of disease (panel B). Black indicates completely diseased, white indicates no disease at all, and grey colours intermediate levels of disease. See text for further explanation.

plant varieties and pathogen strains, respectively, need to be cultured under similar conditions to avoid bias by way of 'maternal' effects. So, culturing of the biological test material under similar conditions is a demanding, primary, task to study pathogen-host interactions. The subsequent testing needs also to be done under similar conditions to determine really genetic effects. It means actually that replicates of pathogen strains and plant varieties, respectively, are randomly assigned to a place within an experimental area. So, the use of a fully randomised design in combination with sufficient replicates constitutes the basis of testing pathogen-host interactions. In addition, the pathogen should be in an infectious state at the time of inoculation even as the host plant should be in a susceptible state. The abiotic conditions need to be suitable for infection. All in all, the study set-up should demonstrate basic compatibility, *i.e.*, at least one pathogen strain should cause infection, disease, in at least one plant variety. If not, the set-up of the study is not valid. We, thus, have a check on the validity of the study conditions.

Plants are checked on symptoms of infection, or disease, after some time, *e.g.*, one week after inoculation. The scoring is, in general, possible by eye. We put, subsequently, the average, scores in a matrix of plant variety by pathogen strain resulting in a checkerboard pattern (Fig. 5.1). The pattern may be really black and white indicating infected/diseased, or not. We are dealing then with qualitative resistance of the host plant and virulence of the pathogen. We see in panel A of figure 5.1 that plant variety 1 has no defence at all against all the pathogen strains, or in other words, all strains are virulent regarding this variety. In contrast, variety 9 is resistant to all pathogen strains, except strain 7. The strains 1-6 are avirulent with respect to variety 9. We also see that strain 7 is virulent to all varieties. Variety 1 is thus susceptible to all strains, or, in other words, these strains are all virulent to variety 1. We may also detect a checkerboard pattern with various shades of grey as shown in panel B of figure 5.1. We are dealing then with so-called quantitative, or partial, resistance. Infection is in a range between very low infection/disease and completely infected/diseased. The counterpart of the pathogen is called aggressiveness. So, a pathogen strain shows a degree of aggressiveness. We see strain 7 as the most aggressive strain to all plant varieties. No variety is resistant to all strains. So, we may conclude that plant resistance and pathogen virulence are not absolute phenomena. These are relative to the varieties and strains tested in a study. In addition, the outcome of the testing may depend on the experimental conditions. We may notice that determining quantitative resistance and aggressiveness is even more sensitive to environmental conditions than determination of qualitative resistance and virulence, as minor genes of host and pathogen may be involved, besides major genes. We need to do the testing under various conditions to arrive at a real understanding of the genetics of pathogen-plant interactions.

The mechanisms underlying qualitative and quantitative resistance of plants may be constitutive, or inducible. Constitutive mechanisms are in operation permanently. Thickness of the cuticula and hairiness of a plant are examples of constitutive mechanisms. A plant invests in such a defence, irrespectively the presence of a pathogen, or not. If not, it is waste of investment. The general hypothesis, therefore, is that investment in constitutive resistance occurs in environments characterised by a high likelihood of pathogenic attack. In contrast, inducible mechanisms turn on upon contact with a pathogen. The hypersensitive response is a typical example of such an induced defence. The entering pathogen is isolated by deposition of callose and the surrounding, infected, host cells are killed by the host itself to deprive the pathogen from resources. Such induced resistance is worthwhile in an environment, in which epidemics occur rather infrequently. The plant invests only when needed by an attack of a pathogen. It may be a risky strategy, as the defence may turn on too slowly to inhibit the pathogen on time. A compromise is to do a small investment in pre-disposition, immunisation, to turn the inducible defence mechanisms on faster upon contact with a pathogen. It is like a vaccination. Various factors may serve such a pre-disposing, immunising.

Genetic variation in resistance and pathogenicity is omnipresent in plants and their pathogens, irrespectively we are dealing with, viruses, bacteria, fungi, or nematodes. This variation provides the basis for the process of natural selection. We will return to that process in section 5.3 below. We need to notice here that some caution is needed with respect to viruses, of which we do not have a proper species concept yet. The genetic variation of viruses is, therefore, expressed at the level of viral 'species' rather than 'strains'. We, for example, compare then the Tobacco mosaic virus with the Tobacco ringspot virus, which both infect tobacco plants.

We turn now to the question, whether genetic variation in resistance, which is expressed qualitatively or quantitatively, and pathogenicity, which is expressed in virulence or aggressiveness, is present in pathogen- human interactions. It is clear that we cannot execute experiments on man, using randomised designs under controlled conditions, like we do in investigating pathogen-plant interactions. We need to adopt an indirect approach to pathogen-human interactions, in which Genome-Wide Association Studies (GWAS) are essential. In a GWAS, the phenotypic expression of a trait, like resistance and pathogenicity, is associated with genetic variants of the expressing organism by way of the use of population-based data. We like to stress here the term 'association', as we are dealing with non-controlled, observational, studies. Insight into the underlying mechanisms of the trait expression is, subsequently, needed to infer causal relations between genotype and phenotype. We may also notice that a GWAS is unidirectional. We investigate either the side of the host, or the pathogen, when we investigate pathogen-human interactions.

The execution of a GWAS starts by selecting the study population (Uffelmann *et al.*, 2021). The trait of interest may be dichotomous, like it is for qualitative resistance and virulence, or quantitative, like it is for quantitative resistance and aggressiveness. A dichotomous trait provides, in general, a more straightforward analysis. We may then, for example, include cases and matching controls in the study. We feel, however, that quantitatively expressed interactions are dominant with respect to human infectious diseases. We need, thus, to quantify disease/infection caused by a pathogen. Confounding, collider bias, may occur in selecting populations, irrespectively the data is obtained from repositories, or collected at new. The sampling of a catchment population should be representative, valid, from a point of view of quality and quantity. It concerns both, the phenotypic and genotypic sampling, as well as the inclusion of data of potential confounding factors. Sample-size calculations are mandatory to have sufficient statistical power in the testing.

Genotyping in GWAS is commonly directed to Single-Nucleotide Polymorphisms (SNPs). It results in blocks of associated SNPs that are associated with the trait of interest. We call these 'genomic risk loci' often. These may hint to underling mechanisms and the genes governing these. The use of whole-genome sequencing for the genotyping is superior to the microarray-based genotyping of SNPs and the whole-exome sequencing, but it is quite expensive still. Whole-genome sequencing is superior in detecting relatively rare genetic variants that may be of considerable effect size. We may, therefore, expect that whole-genome sequencing becomes the primary method of genotyping in short, assuming a considerable cost reduction.

Quality control is a next step in a GWAS. The collected data needs to be checked on missing and fault entries. Ancestry and relatedness in the genetic data need especially attention to avoid population stratification, which may result in bias of test statistics. The subsequent testing on associations is based on linear regression in the case of a quantitative trait, or logistic regression for a qualitative one. The linear regression model is of the type of:

(1) $Y \sim W\alpha + X_s\beta_s + g + e$

in which Y is a vector of phenotype-values, W is a matrix of covariates, α a corresponding vector of effect sizes, X is a vector of genotype values for all individuals at a specific locus s, β_s the corresponding fixed effect size of genetic variant s, *g* is a random effect that captures the polygenic effect of other loci, and *e* is the random effect of residual errors of the model. The choice of a regression model is likely, if we take into account the complexity of the data and the output we like to have. It is, however, not correct from a purely statistical point of view. In regression, the variation in the dependent variable, which is Y here, is regressed on one, or more, independent variables on the right-hand side of equation (1). These are fixed, *i.e.*, these do not show random variation. Application of a medical intervention in a randomised clinical trial is, for example, such a fixed factor without random variation. We have an intervention, or not. We, however, have no fixed variable at all in equation (1), violating a basic assumption of regression. We cannot estimate the significance of the bias that may result. In contrast, the multiple testing of all the potential associations in the data, which may go up to millions, inevitably introduces bias. It is corrected by elevating the threshold of calling an association significant, *i.e.*, we reduce the level of the probability P of declaring an association significant, whereas it is not. We refer to the Italian mathematician Carlo Emilio Bonferroni (1892-1960) often, who introduced the statistical correction for multiple testing. It is one of the available methods to correct for bias of multiple-testing. These do all, finally, the same by reducing the P-value to declare an outcome of testing significant.

The analysis of GWAS-data results in many loci that may guide us to key pathways and genes involved in these. We are, in general, dealing with polygenic phenomena, in which many genes may contribute very little to the resulting phenotypic effect. In addition, these may be obscured by correlations with parts of the genome that are not involved in the phenotypic expression of a trait. It is really puzzling to detect the clinically relevant genetic variation by way of a GWAS, especially the ones based on SNPs only. Clearly, we need to combine the statistical outcome of a GWAS with insight in molecular processes to determine the heritability of a trait that expresses resistance, or pathogenicity. In addition, a GWAS needs to be replicated in another, independent, study population to confirm the results. A GWAS may, subsequently, be followed up by studies zooming in on candidate genes (*e.g.*, Gómez *et al.*, 2020). We may, therefore, conclude that a single GWAS does not provide unequivocal indications of relevant genetic variation in resistance, or pathogenicity, although the power of a GWAS may be increased in various ways. We may, for example, execute multipoint imputation in GWAS-samples using a known human resistance/susceptibility locus, as it was done for malaria (Jallow *et al.* 2009), or including a stratification for pathogen strain, like executed for tuberculosis (Omae *et al.* 2017).

Imputation is the addition of rare and low-frequency variants from reference, genomic, panels to GWAS-samples using specific algorithms (Quick *et al.*, 2020). It is employed when whole-genome sequencing is not feasible, as it is common still. The quality of the imputation hinges on the representation of various populations in the reference panel. Sequencing a subset of participants in a GWAS and, subsequently, using these as a reference panel for imputation, is an alternative for populations that are under-represented in reference panels.

A nice example of a compilation of evidence of various GWAS and candidate genes studies, identifying genetic variation in host resistance, is the one directed to COVID-19 (Velavan *et al.*, 2021). The genes identified ranged from, Mucin 5b that is involved in a first line of human defence, *i.e.*, the muco-ciliary clearance of viruses in the respiratory tract, ACE1 that is involved in the cell entry of SARS-CoV-2 particles, to IFN1-genes involved in immunity. We may notice that the phenotypic trait of interest varied among studies. It was, for example, hospitalisation, or not, in one study, and a positive, or negative, score on a diagnostic test in another study. Interestingly, all traits were considered as binary ones in the sequence of studies, which seems not to represent properly the pathogenesis of COVID-19.

We may turn to the bacterium *Streptococcus agalactiae* for a GWAS directed to genetic variation in pathogenicity (Gori *et al.*, 2020). The bacterium is an opportunistic pathogen of humans invading the gastrointestinal and urogenital tract. Isolates of *S. agalactiae* infecting humans were assigned to six major clonal complexes, using multi-locus sequence typing. These clonal complexes differ in invasiveness of man, as indicated by clinical data. The GWAS was directed to identification of potential genes that cause the difference in invasiveness, say virulence, among and within the clonal complexes. Various cluster complex-specific genes could be attributed to invasiveness, virulence, of *S. agalactiae*. These fit pre-existing knowledge of the significance of, bio-film formation and the composition of the bacterial capsule and pili, respectively. We indicated the significant role of bio-film formation in survival of bacteria already in Chapter 3. The GWAS was not replicated by another, independent, study population, impairing the validity of the study. A similar drawback had another GWAS, which was directed to *Streptococcus pneumoniae* (Chaguza *et al.*, 2020). In addition, the effect sizes that were determined in this study were rather small and the heritability indicated was negligible. It forced the authors to conclude that multiple genetic, and other, factors are involved in invasiveness, aggressiveness, of *S. pneumoniae* of the central nervous system.

A GWAS directed to the pathogenicity of *Escherichia coli* did neither include an independent study population to validate the results (Galardini *et al.*, 2020). The validation was sought in a different way, as we will outline here. The genus *Escherichia* encompasses varies species, of which *E. coli* is one. It is a commensal colonising the gut of vertebrates, including man, and an opportunistic pathogen causing diarrheic and extra-intestinal diseases, like sepsis. Pathogenicity in the study presented here was defined as the ability of *E. coli* to cause extra-intestinal disease. The power of the GWAS was increased by both, a reduction of the host variation and a concomitant increase of the variation of the bacterium. Host variation was reduced by using a rather, genetically, uniform, mouse model. Females of the mice OF1-variety were used. This variety resulted originally from the CF-1 variety that was inbred firstly for 20 generations. A pair of mouses was selected and multiplied by outbreeding. The variety was introduced in Charles River Laboratories France in 1967 and it was re-named in Oncins France 1 (OF1). Pathogen variation was increased by the inclusion of 326 strains of *Escherichia coli*, seven strains of *E. albertii*, five of *E. fergusonii* and 32 of cryptic *E.* clades. An *E. coli* strain of known aggressiveness was included as positive control, killing all mice, and, similarly, a strain was included as negative control, *i.e.*, killing none of the mice. Ten mice were inoculated with each of the 370 strains in the study. Pathogenicity of *E. coli* strains, and other *Escherichia* species, was expressed by the mortality rate of mice at seven days after inoculation. The results of the GWAS confirmed the so-called 'high-pathogenicity island' as a major genomic region of pathogenicity with respect to extra-intestinal disease. It encodes the bacterial siderophore yersiniabactin. The pathogen needs the siderophore to take up sufficient iron within a host. The siderophore also impairs the efficacy of the human immune system. So, the results of the GWAS fit the known mechanisms.. In addition, the GWAS was validated by way of gene knockout experiments.

We may conclude that the identification of relevant genetic variation in human resistance to pathogens, and vice versa in pathogenicity of human pathogens, is a real challenge accompanied by a myriad of pitfalls. In contrast, the worldwide and open collaboration that emerged quickly during the COVID-19 pandemic was very encouraging, as it enables facing such challenges. We may remember that scientific collaboration is a major in the outlook for epidemiology we provided in Chapter 1. Anyway, we identified genetic variation in human resistance to pathogens, and similarly genetic variation in pathogenicity of human pathogens. The next question is, from a point of view of evolutionary dynamics, whether variation in human resistance is directly related to variation in pathogenicity. We know from plant diseases that such relationships exist, indicating gene-for-gene interactions between a host and its pathogen. If so, we have a basis of co-evolutionary dynamics of host and pathogen. We cannot infer it from a, unidirectional, GWAS straightforward, as mentioned above.

We turn to a study of Gagneux *et al.* (2006). It was directed to the adaptation of *Mycobacterium tuberculosis* to humans. The focus was on genetic variation in virulence of *M. tuberculosis*, but genetic variation in human resistance was actually included as well, although not defined in terms of genomics. The bacterium was isolated from *c.* 2100 patients with tuberculosis in San Francisco (USA). Demographic, and other epidemiological, data was collected including the place of birth and self-defined ethnicity. The isolates were genotyped using Restriction Fragment Length Polymorphisms (RFLP), even as it was done for representative samples of *M. tuberculosis* from 80 other countries worldwide. The analysis of this global sampling indicated six major lineages of *M. tuberculosis*, which were highly geographically structured, *e. g.*, the West-African 1 lineage. The *M. tuberculosis* isolates of five sub-populations of subjects in San Francisco were assigned to these major lineages. Almost all of the isolates belonged to three *M. tuberculosis* lineages, *i.e.*, the Euro-American (*c.* 48%), Indo-Oceanic (*c.* 26%), and East-Asian (*c.* 26%). The five sub-populations were defined by place of birth, *i.e.*, (i) USA, (ii) China, (iii) Philippines, (iv) Vietnam and (v) Central America. Tuberculosis patients were also sub-divided with respect to the genotyping of the *M. tuberculosis* isolates, which were either clustered, or unique. The clustered patients were considered being part of a recent chain of transmission and the unique cases as re-activations of latent infections. In addition, cases were considered as primary, or secondary. The hypothesis was that a rare lineage in a specific host population would hardly transmit and cause secondary cases in that population. Indeed, the secondary case rate ratio of the Euro-American lineage of *M. tuberculosis* was by far highest in the USA-borne sub-population of San-Francisco and the Central America-borne sub-population, respectively, whereas it was the Indo-Oceanic lineage in the Philippines- and Vietnam-borne sub-populations, and the East-Asian lineage in the sub-population of the China-borne sub-population. It all pointed to a strong pathogen-host gene interaction and, thus, co-evolutionary dynamics. If so, it may have consequences regarding the control of tuberculosis. We may, however, not rule out that the San-Francisco findings resulted from a socially stratified dispersal of *M. tuberculosis*, or other confounding factors.

The study of Gagneux *et al.* (2006) suggested strongly a genetic interaction between *M. tuberculosis* and *Homo sapiens*. We do, however, need insight in the process of natural selection, which is based on the concept of fitness, to infer whether genetic variation in *M. tuberculosis* pathogenicity resulted from *H. sapiens* resistance and, *mutatis mutandis*, genetic variation in resistance of *H. sapiens* resulted from *M. tuberculosis* pathogenicity. We will address the topic of natural selection and fitness in section 5.3. Firstly, we, however, need to explore the topic of genetic variation in disease tolerance, as it may influence the process of natural selection of pathogens.

REFLECTIONS

The gold standard of detecting gene-for-gene interactions in pathogen-host interactions has been set in Plant Pathology. We cannot use the same methodology for pathogen-human interactions. We may, however, reflect about the use of elements of it, either in an adapted form, or not. And, in general, we may reflect about the use of knowledge about pathogen-plant interactions in our research on pathogen-human interactions.

5.2 Genetic variation in tolerance

We dealt with human resistance to pathogens in the previous section. In this section, we address the topic of tolerance of pathogens. The concept of tolerance was proposed for plant crops back in the seventies (*cf.* Frantzen, 2007). It is the ability of a host variety to suffer less from an (infectious) disease compared with another variety, if both have a similar level of infection, a similar outgrowth of pathogen. The mechanisms underlying tolerance do not interfere with the pathogen itself, like resistance mechanisms do by way of reducing the establishment and multiplication of a pathogen, but these enable a minimisation of adverse effects exerted by the establishment of a pathogen. And here, we face the challenge in the concept of tolerance. We need to compare two varieties at a similar level of pathogen load. How to achieve a similar level of pathogen load? A pathogen lesion on a stem of a plant may, for example, kill the whole plant, whereas a single lesion on the leaf may have nearly any impact. So, determining simply the number of lesions, virus particles, or whatever, is not sufficient to determine the pathogen load precisely. Let us, however, assume that we can do it for various varieties of a host. We may then delineate the difference between resistance and tolerance in a single figure (Fig. 5. 2). If we also assume quantitative resistance of the host and aggressiveness of the pathogen, we arrive at a linear relation between pathogen load and impact on the host expressing these on a scale from zero to 100%. The pathogen load decreases, the stronger the resistance of the host variety. The impact of the pathogen, therefore, decreases as well. The impact of the pathogen may also decrease, if tolerance of the host increases keeping a similar level of resistance. If we assume qualitative resistance of the host and virulence of the pathogen, respectively, the pathogen load can take two values only, zero or 100%, reducing the relationship to two points on the abscissa. Tolerance may operate still, reducing the impact of the pathogen on the host from 100% to zero at minimum. Whereas resistance is defined with respect to pathogenicity, as we have seen in section 5.1, tolerance is independent of pathogen strain, independent of pathogenicity. This pathogenicity-independency of tolerance may affect the evolutionary dynamics of pathogens as we will elaborate in the next section (5.3). Determining tolerance, like outlined above, is a real challenge for plants, although we are able to culture rather isogenic lines of hosts and pathogens. In addition, we may manage relatively well the pathogen load. Determining tolerance of pathogens in humans in this, rather quantitative, way is rather impossible. We may, however, determine human tolerance of pathogens diving in the mechanisms behind both, resistance and tolerance, as we will outline here.

Tolerance was investigated as a means to cope with chronic infection of *Plasmodium falciparum* causing malaria (Nahrendorf, *et al.*, 2021). The researchers were triggered by the observations that children, who trapped the pathogen for the first time, suffered severely from malaria, whereas it is less so at re-infection, or in the phase of chronic infection, if children survive the first, acute, phase of infection at all. They hypothesised that a dampening of inflammation would serve tolerance of malaria by the children upon re-infection, or a flare-up. The focus of the study, therefore, was on the differentiation of monocytes in macrophages, which stimulates inflammation, as we outlined in Chapter 2. A mouse-model and two strains of *Plasmodium chabaudi* were used. The model mimics very well humans infected by *Plasmodium* species, although *P. chabaudi* is not compatible with man. The two *P. chabaudi* strains differ significantly in aggressiveness. The pathogen was cultured within mosquitos of *Anopheles stephensi* and the resulting sporozoites were collected from the salivary glands. We refer to section 3.4.1 for a more detailed description of the life cycle of *Plasmodium* species, explaining the method of sporozoites production and the subsequent use of these in the study. Mice were inoculated using 200 sporozoites per individual, a number that is well above the minimum effective dose of 50 (*cf.* section 4.1).

Inoculation of naïve mice with *P. chabaudi* resulted in a strong, inflammatory, immune reaction indicating an attempt to eliminate the pathogen, *i.e.*, a resistance response. A subsequent inoculation at a similar dose, which caused a similar level of parasitaemia, did result in a silenced inflammation response only, even in mice that were treated with chloroquine to remove *P. chabaudi* that resulted from the first inoculation. The deviant inflammatory response could be attributed to a functionally re-programming of monocytes that prevented differentiation of these into inflammatory macrophages. The functional re-programming seemed to be based on imprinting of the monocytes in the, re-modelled, spleen rather than an epigenetic re-programming of bone marrow progenitors.

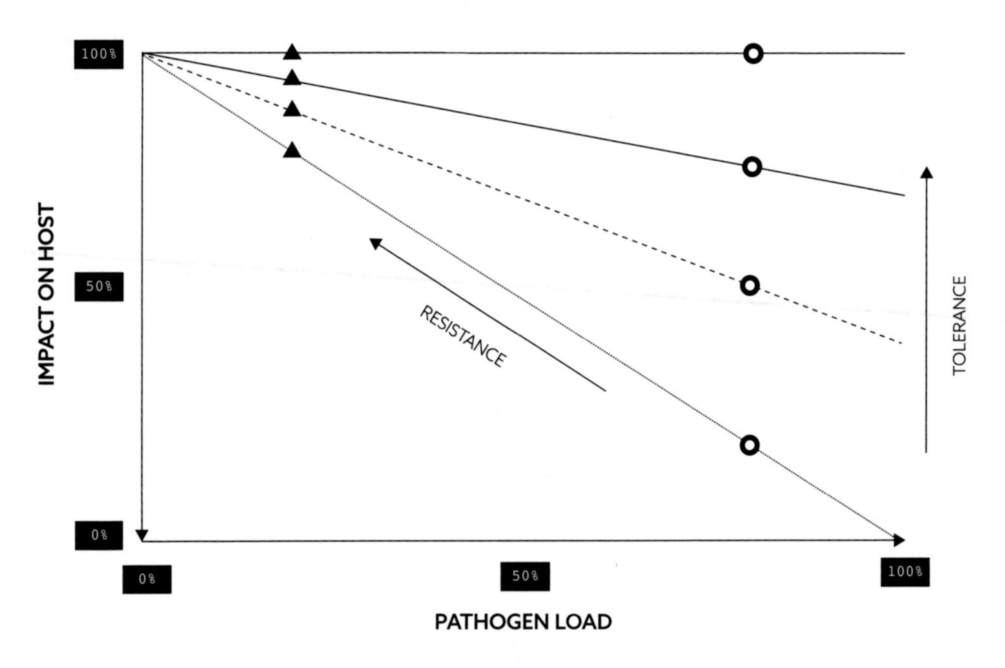

Figure 5.2. Hypothetical relation between pathogen load and impact on a host, both on a scale from zero to 100%. The effects of increase of host resistance and tolerance are indicated. The open circles indicate four host varieties with a different level of tolerance and all showing a relatively low level of resistance. The black triangles indicate four host varieties with a different level of tolerance and all at a relatively high level of resistance. See also text.

The study of Nahrendorf *et al.* (2021) presented here, triggers some intriguing questions with respect to tolerance. One, is demonstrating tolerance of a pathogen in an animal model representative for human tolerance of pathogens? We may answer 'yes' regarding *Plasmodium spec.* tolerance, if we adopt a biological view in a narrow sense, a purely physical view. We tend to 'no' broadening our view and adopting the trinity of human life presented in Chapter 2. If so, we need to take the mental status into account with respect to immune responses. We may assume the mental status affects the immune response positively, or just negatively, and thus tolerance, as well. Our mental status is determined by both, the condition of our body and our social environment. And these two, of course, interact as well. We feel the way ahead with respect to research on human tolerance may be a combination of well-designed animal studies, like the one of Nahrendorf *et al.* (2021), with observational studies of humans, which are directed to mental and social aspects.

Two, do we focus too much on the function of resistance of the immune system overlooking, or ignoring, tolerance? The answer is 'yes', looking at the answer on the preceding question, but it is also 'yes', if we take a physical view on tolerance, as Schneider (2021) did. He pointed to the conditions during the logistic phase of the SARS-CoV-2 pandemic, in which we lacked efficacious anti-viral treatments. Oxygen-suppletion and anti-inflammatory drugs were used to support tolerance of COVID-19 by critically ill patients. Schneider also indicated, implicitly, to the fact that tolerance, like resistance, may not be a matter of genetics, or not only. In this case, tolerance was mediated by oxygen-suppletion and anti-inflammatory drugs. Did the response of patients on the medication also depend on their genetic constitution? We do not know.

Three, is the human immune system the mechanism behind tolerance, or do we have other mechanisms? The role of the immune system in tolerance is likely, as the negative side effects of the immune response, which is directed against the establishment and growth of pathogens, cause the major burden of infectious diseases often. We, however, know that other mechanisms of tolerance exist in plants. One is, for example, an increase of photosynthesis in cells not invaded by a pathogen to compensate for the loss of photosynthetic capacity of the invaded cells. This mechanism may even result in a net benefit of the plant in terms of photosynthesis. Likely, humans also have such compensatory mechanisms. We need to investigate the whole physiological response of a body towards a pathogen to detect such mechanisms (*cf.* Schneider, 2021).

Four, may we better focus on tolerance of a pathogen than on resistance to it in coping with epidemics? We can, of course, not provide a simple 'yes', or 'no', as an answer to this question. We, however, feel that the focus is far too much on treatments directed towards inhibition of, establishment, growth, and multiplication of pathogens, by way of vaccination, drugs, quarantine of cases, and so on, rather than stimulating tolerance. We will return to this topic in the Chapters 9 and 10.

Five, how do we position tolerance in the overall human reaction to a pathogen? Some researchers position it as a means of host defence (*cf.* King and Li, 2018), which we certainly disagree, as we outlined above. And do we tolerate a disease, or is it tolerance of a pathogen? The term 'disease tolerance' has appeal, as tolerance may also apply to non-communicable diseases. In contrast, defining and quantifying 'disease', or 'illness' with regard to humans, is rather impossible making the term 'disease tolerance' less tangible. We, therefore, have some preference of the term 'tolerance of a pathogen', even if it is not the pathogen itself, but the negative effects of the invasion of the pathogen that we tolerate.

Finally, do we have genetic variation in tolerance among people, *i.e.*, do we have innate tolerance of pathogens, infectious diseases? It is likely, but it has not been demonstrated yet, so far known. Genetic variation in tolerance may enable completely different evolutionary dynamics of pathogens than genetic variation in human resistance to pathogens, as we will see in the next section.

REFLECTIONS

Unequivocal demonstration of tolerance of pathogens, or diseases, is a real challenge from a scientific point of view. Is it worthwhile to invest in more research on tolerance, or do we have enough theoretical and clinical evidence already to exploit it further in management of infectious diseases?

5.3 Natural selection, the concept of fitness

Background

Charles Darwin (1809-1882) wrote his well-known book 'The origin of species' in 1859 (Darwin, 1859). He proposed the process of natural selection an as explanation of the huge variety of species observed. Darwin set a next step in evolutionary thinking that started a century before (Mayr, 1982). Interestingly, the scientific work of the Swedish taxonomist Carl Linnaeus (1707-1778) was pivotal in the conceptualisation of the process of natural selection by Darwin. Linnaeus, and other taxonomists, had assigned the natural variation in plants and animals to genera and species. Species were thus well-defined and fixed, triggering the scientific question about the origin of all these species. The question that was addressed in Darwin's book. A question ignored by Linnaeus, as he considered the variety of species as given, as a static fact.

The French naturalist Jean-Baptiste Lamarck (1744-1829) introduced the notion of adaptation, which he called the perfection of an organism. So, Lamarck identified the need, and ability, of an organism to adapt to its environment and, therefore, to evolve. Darwin arrived at the same insight later on. Darwin and Lamarck, however, had a different view on the mechanism behind evolution, as we call it now. Lamarck's view prioritised changes in the environment as the trigger of adaptation by an organism. In short, the variation comes from outside an organism. In contrast, Darwin assumed random variation in an organism that is ordered by the environment, natural selection. In short, the variation comes from inside. A view shared by the British naturalist Alfred Russell Wallace (1823-1913), who developed the concept of natural selection independently of Darwin. A publication of Wallace on this topic in 1858 actually triggered Darwin to publish 'The Origin of Species' in 1859.

Lamarck's view fits 'vertical evolution', as we may call it (Mayr, 1982). He took the attitude of a palaeontologist to explain the origin of species. Fossils were known at his time, even as the changes in the earth's environment in time, which are reflected in the geologic stratification of the soil. Interestingly, Darwin was educated in geology, but he took, like Wallace, a different view, which fits 'horizontal evolution', as we may call it. Darwin was triggered especially by the relatively large differences in soil and vegetation among the isles of Galapagos. He saw this variation reflected in the beaks of finches, which differ among the isles. So, we need not to dig into the soil to start thinking about evolution, but we may see it also above the ground, since Darwin and Wallace. Darwin, also, added the notion that evolution may proceed along relatively small changes rather than disruptive events that cause immediately a change from one species into another one. We may notice that the terms 'vertical' and 'horizontal' do not have a mechanistic meaning with respect to evolution. These just reflect the quite opposite way of thinking about evolution, as described by Mayer (1982).

The idea of natural selection was picked-up by scientists and worked out in scientific theory, as supported by empirical evidence. We follow here John Endler (1986) to outline the theory of natural selection. A process of natural selection is, according to him, in operation if,

1. Individuals of a population of an organism differ in a (phenotypic) trait/attribute, *i.e.*, variation of it exists within a population;
2. A consistent relationship between the trait and the ability of survival and reproduction exists, *i.e.*, the individuals of a population differ in fitness;
3. A consistent relationship exists between parents and offspring regarding the trait, which is at least partially independent of environmental effects, *i.e.*, inheritance of the trait.

The frequency distribution of a trait will differ between offspring and parents in a population, if these three conditions apply. The phenotypic variation indicated under (1) may result from recombination and mutation in the genome, environmental factors, or both. Such a trait may, for example, be resistance to a pathogen. The subsequent selection of expressions of the trait, which is indicated under (2), is at the phenotypic level always. In addition, the expression of the trait should be related to the fitness of an organism. A young adult, for example, who resists an infection by the Ebola-virus, or tolerates it, is able to survive and reproduce, whereas a susceptible, or non-tolerant, one is not. The trait of resistance, or tolerance, may, subsequently, inherited to her, or his, children, which is indicated under (3).

Fitness in the sense of having offspring does not depend simply on the response to the Ebola-virus, as indicated in the example here. Whether we mate, or not, and give birth to children depends on a variety of physical, mental, and social factors. We may, therefore, define fitness in an absolute sense, as the expected time to extinction of a gene, or an allele. It is rather impractical. In contrast, it becomes tractable defining it in a relative sense. We may compare individuals, who differ in a specific trait, regarding their offspring over one, or more, generations. We may then determine a significant change in frequency of the trait in a population, or not. If so, a consistent, major, selecting factor should be in operation.

We have observed the process of natural selection in relatively simple ecosystems, like the high-input agriculture. For example, a single crop, like corn, is grown on a relatively large area under relatively uniform and resource rich conditions. The conditions do, however, not facilitate growth of the crop only. It also enables the growth of un-wanted plants, which we call weeds. These capture resources, affecting the growth of the crop. Herbicides are then used to eliminate the weeds. Herbicides are, in general, used year after year. Several weeds, however, turned out to be resistant, or tolerant, regarding the application of herbicides (*cf.* Radosevich *et al.*, 1997). Resistance to the active compound triazine of herbicides was, for example, detected in the annual plant *Senecio vulgaris*, which is considered as a weed, back in the seventies. The resistance is based on a point mutation in a single-protein binding site in photosystem II of the plant. This single mutation determines whether a *S. vulgaris* plant succumbs the application of a triazine-based herbicide, or not. The frequency of herbicide-resistant *S. vulgaris* plants, therefore, increased at sites of regular herbicide-use. Triazine-resistant plants have a fitness advantage at those sites. In contrast, these plants have a lower fitness, compared with triazine-susceptible plants, at sites not treated with triazine-based herbicides. Triazine-resistant plants are out-competed at those sites and the resistance trait vanishes from the local *S. vulgaris* population.

We may abstract the example of triazine-resistant *S. vulgaris* in the following points that are specific of the process of natural selection:

1. A mutation in the genome that was expressed in the phenotype; mutations are common whether these result from, errors in the process of recombination, other internal factors, or external factors like irradiation.
2a. The mutation had a relatively large impact on the fitness under conditions of both, presence and absence of the selecting factor, *i.e.*, application of triazine, or not. We feel such a large impact of a single mutation is rather uncommon, but it exists.

2b. A dominant selecting factor, in the sense of a relatively large effect in a relatively short term, that was present at agricultural sites over several generations of the plant, or not at all at (semi-)natural sites. We may regard presence of such a factor as uncommon, except in relatively simple ecosystems, like agriculture.
3. The mutation was heritable, enabling an increase of the frequency in populations under selection and a decrease, or elimination, of it in populations not exposed to triazine-based herbicides. Inheritance of a mutation, however, is not necessary. The mutation needs to be in the genome that is transferred to the off-spring.

We may thus demonstrate unequivocally the operation of the process of natural selection, although in relatively simple biological systems. The result of it is a shift in the frequency of a specific trait over generations. Is this evolution? The answer is no, if we take the perspective of Darwin, who intended to explain the variety in species by way of natural selection. And no, we couldn't see any evidence of (incipient) species formation in the processes of natural selection observed among plants and animals, so far. We may, of course, be limited in time of observing species formation, which may outreach the relatively short life span of man. This would fit the reasoning of Darwin. The limitation in time does, however, not apply to bacteria, which have relatively short generation times. We are then confronted with horizontal gene transfer, obscuring the concept of species, as outlined in section 3.3.2. We may also adopt, and adapt, the reasoning of Lamarck that species formation in plants and animals may need extreme conditions, enforcing radical changes of organisms resulting in new species. Anyway, we cannot demonstrate evolution of species, as based on the process of natural selection, so far.

We may answer 'yes' on the question whether evolution is the result of the process of natural selection, if we define evolution as "any net directional change in the characteristics of organisms, or populations, over many generations" (Endler, 1986). This definition is rather subjective regarding the phrases 'any net directional change' and 'many generations'. We may then see everywhere evolution, as it happens commonly indeed. We have an inflation of the term 'evolution'. In addition, we exclude prions and viruses by using this definition, as these are non-organisms (*cf.* Chapter 3). We do, therefore, not use the term 'evolution' further in this book to avoid any confusion about the meaning. Instead, we focus on processes that can be observed and quantified and we put these in a framework of evolutionary dynamics, of which the result might be evolution of species.

Pathogen-host interactions are fascinating as pathogens may affect hosts, but hosts may also affect pathogens. We have read in section 5.1 about the existence of genes underlying phenotypic variation in both, pathogenicity and resistance of organismal pathogens and humans, respectively. The conditions (1) and (3) set for a process of natural selection, which are indicated above, are fulfilled from the side of the pathogen as well as that of people. In addition, we may, in general, state that pathogenicity and resistance are traits that may provide potentially advantages in terms of fitness, which fits condition (2) above.

We turn now to a commonly used proxy of fitness, the basic reproduction number R_0. We introduced it in section 4.4 as a key epidemiological parameter already.

Maximisation of R_0

R_0 is a major fitness proxy in the theory on evolutionary dynamics of host-pathogen interactions and life-history theory in general (Lion and Metz, 2018). It is then generally defined as the "the average lifetime off-spring number in a given environment". The R_0 for man is, for example, in the range of about 0-4 children in high-income countries, whereas it is in the range of 5-10 children in low-income countries, expressing clearly a difference in environment. But, can we, therefore, conclude that the fitness of people in low-income countries is higher than it is for man in high-income countries?! This simple statement about the differences in the human R_0 fits the major conclusion of a theoretical study of Lion and Metz (2018) directed to pathogens: we have to go beyond R_0 maximisation. They arrived at this conclusion as follows.

The traditional view on maximisation of R_0 results basically from the following equation,

$$(1) \quad R_0 = \frac{\beta S}{\mu + \alpha + \gamma}$$

in which, S is the density of a susceptible host population, β the rate at which a pathogen multiplies and disperses from one infected host individual to other (uninfected) individuals, μ the mortality rate of uninfected host individuals, α the pathogen-induced mortality rate among the host, and γ the recovery rate of infected individuals. Parameter α is set equally to pathogenicity. We may see from equation (1) that an increase of α results in a decrease of R_0, except β increases concomitantly, or γ decreases. But, yes, an increase of the number of new infections (β) may also result from an increase of pathogenicity (α), even as a decrease of the number of recovered host individuals (γ). The overall conclusion from equation (1), therefore, is that maximisation of R_0 results in an intermediate level of pathogenicity. This conclusion became a paradigm in the study of evolutionary dynamics of pathogens, as it is derived from the commonly used Susceptible Infected Recovered (SIR-) model type.

Writing simply R_0 masks that it abstracts both, a suite of traits of the pathogen and all the variables of the environment, including the host. In equation (1), the variables of the environment are reduced to those determining the dynamics of the host population, in which a pathogen settles. If so, a pathogen will spread inside a host population, if R_0 exceeds one. In equation,

$$(2) \quad R_0 (X|E_0) > 1$$

in which X represents the traits of the pathogen given an environment E_0, which is the environment produced by the dynamics of the host population in the absence of the pathogen, as indicated by the subscript zero. It implies that increasing, or maximising, R_0 is beneficial for a pathogen, if we consider R_0 as an absolute parameter of fitness. We may notice that R_0 is defined for an environment, E_0, characterised by a susceptible, uninfected, host population. The process of natural selection is, however, based on differences in fitness among phenotypes/

genotypes, say strains, of a pathogen. So, we need to consider situations, in which a pathogen enters a host population, in which another strain is present already. The new, mutated, strain may invade indeed if,

(3) $R(Y|\hat{E}) > 1$

in which R is the basic reproductive number in an evolutionary sense, Y represents the traits of the invading strain given the host environment Ê, which is already invaded by the pathogen. We may notice that we have as well a $R_0 (Y \mid E_0)$ of the new strain. The fitness, the invasiveness, of the strain characterised by traits Y may be expressed by,

(4) $R(Y|\hat{E}) = [R_0(Y)\varphi(\hat{E})]^{q(Y,X)}$

in which $\varphi(\hat{E})$ is a function summarising the effects of the environment and q (Y, X) a function of the traits of both strains. Setting the function *q* at 1, the fitness of the strain is a multiplicative function of R_0, the epidemiological one, and φ. If so, the condition expressed in equation 3 is full-filled when $R_0(Y) > R_0 (X)$ and maximisation of R_0 is subject of the process of natural selection indeed. The epidemiological R_0 functions as an optimisation principle. It means that we can find potential Evolutionary Stable Strategies (ESSs) of the pathogen (*cf.* Dieckmann *et al.*, 2002). An ESS is a strategy that creates an environment in which no other strategy can be employed. We will come back to ESSs in the next section, which is directed to pathogen competition (5.4). A model with an optimisation principle also bears inevitably a so-called 'pessimisation principle'. It means here that maximisation of R_0 results in lowering the density of susceptible individuals of the host to such a low a level that a mutant of the pathogen cannot invade the host population anymore. The density is at a minimum with respect to survival of the pathogen. In other words, the maximisation of R_0 has a ceiling and the pathogenicity is assumed to settle at an intermediate level, as mentioned above.

The optimisation principle of equation (4), and thus maximisation of R_0, holds if the effect of the environment on the fitness of the pathogen is summed up by a single number such that increasing function φ can change the sign of R-1 from negative to positive only. We call such a form of environmental feedback one-dimensional. We may state it in another way, the environment becomes relatively simple, as it is determined by a single variable that provides feedback. The density of susceptible individuals of a host is such a variable in the SIR-model presented by equation (1). We may pose the question whether man serves as a relatively simple, one-dimensional, environment of pathogens, supporting the paradigm of R_0 maximisation. The answer is certainly 'no' taking into account the complexity of the life cycles of pathogens (*cf.* Chapter 3). The human body may just be a side-path, or an intermediate host, and a pathogen has to cope with various environments in- and outside the human body. We have to turn to ESS in multi-dimensional environments to get more grip on the evolutionary dynamics of human pathogens, as we will do in the next section (5.4). Here, we will address first the question whether pathogens served as selecting factors with respect to the human evolutionary dynamics.

Selection of humans by pathogens

The traditional way of investigating natural selection is top-down. We start from observations of phenotypes of organisms, like Darwin did, analyse these using statistics, and put these in a framework of genetics. The availability of advanced molecular tools also enabled a bottom-up approach. We may look into the genome of humans, or other organisms, to trace signs of selection by pathogens, as we will outline in the following.

A mutation in an allele providing defence to a pathogen will increase in frequency, until it may get fixed in a host population, *i.e.*, the frequency equals one (Booker *et al.*, 2017). The mutation may be accompanied by linked, selection neutral, polymorphisms in the vicinity of it. So, we may not see an increase of the mutation in a population only, but also an increase in frequency of the surrounding genomic region of the mutation. The consequence is a decrease of the genomic heterogeneity around such a mutation, as it becomes fixed in a population. The classical model is one of hard sweeps, which means that the part of the haplotype dragged with the mutation is similar among all individuals with the mutation. Other types of sweeps may, however, occur when, the same mutation in an allele occurs at different haplotypes (soft sweep), a beneficial allele results from various mutations (soft sweep), or the mutation becomes effectively neutral arriving at a certain frequency in the population, *i.e.*, it did not get to fixation (partial sweep). The latter seems common regarding polygenic traits, *e.g.*, those governing quantitative resistance to pathogens. Evidence is accumulating that soft and partial sweeps are more common than previously assumed. We may not rely simply on the hard sweep model anymore, challenging us the more in the exploration of natural selection at the level of the genome.

We may look up now the haplotype homozygosity around a target Single Nucleotide Polymorphism (SNP), using sequencing data of individuals in a population (Voight *et al.*, 2006). We use the Extended Haplotype Homozygosity (EHH) statistic. It will be 1 at the target SNP itself, *i.e.*, haplotypes of all individuals have, of course, this SNP. The statistics decreases to zero at some distance of the SNP, *i.e.*, haplotypes of all individuals become unique at that distance from the target SNP. We arrive at the integrated EHH (iHH) by plotting the EHH versus the distance from the target SNP and we compute the integral of the area under the curve. We may then compare an original SNP with its mutation using the integrated Haplotype Score (iHS), which is defined by,

$$iHS = ln\left(\frac{iHH_o}{iHH_m}\right)$$

in which the iHS approaches zero if, the iHH of the mutant SNP, indicated by 'm', equals that of the original one, which is indicated by 'o'. Large negative values hint to long haplotypes carrying the mutated SNP and large positive values to long haplotypes with the original SNP. The iHS was standardised to have an average of zero and a variance of one, making it independent of the frequency of the target SNP in a population. The iHS, however, remains less reliable for SNPs of which the minor variant has a frequency of less than 5%. Such SNPs are commonly skipped from an analysis based on the iHS.

The iHS was developed and used to detect positive selection in the human genome. Positive means that SNPs, the genes associated, increase in frequency, as these are beneficial, adaptive, for us. A step further is to allocate positive selection to specific pathogens that attack, or have attacked, people. The SNP-data of the 650,000 samples of the Human Genome Diversity Panel was used (Corona *et al*, 2018). The samples were drawn from 53 populations worldwide. In addition, interactome data of the, IntAct Molecular Interaction Database (https://www.ebi.ac.uk/intact/home), Biological General Repository for Interaction Data sets (https://thebiogrid.org/), and VirusMint (no more accessible online) were used. Data was retrieved by January 2014. The interactome data was used to identify human genes interacting with a specific pathogen. Data of interactions between man and 26 pathogens were available, of which 22 were viruses and four bacteria. The availability of interactome data of viruses thus surpassed that of bacteria. Data was not available for pathogenic eukaryotic organisms. The results of the study do, therefore, certainly not allow conclusions about selection exerted by the various categories of human pathogens described in Chapter 3. Genes, or their products, interacting with more than one pathogen were removed from the analysis. SNPs within a certain range of a target gene were selected and the iHs-score of each SNP was scored ignoring the sign of the score, *i.e.*, selection of both, an original and mutant SNP was included. SNPs in linkage disequilibrium were removed from the analysis to guarantee independence of the iHS-scores. The sum of all iHS-scores of the remaining SNPs, which belong to genes interacting with a specific pathogen, were calculated for a population. A bootstrapping approach was used to infer whether the sum was significantly higher than would be expected from randomness. The significance values resulting for each pathogen – human population pair were finally used to determine significant positive selection over the 53 populations, as indicated by the overall level of significance. The analysis was replicated using the SNP-data of three human populations, *i.e.*, an African, European, and East Asian, of the HapMap phase-II cohort (https://www.genome.gov/10001688/international-hapmap-project).

Nine out of the 26 pathogens in the study were identified as exerting, or had exerted, positive selection on the human genome. Two out of the nine were bacteria. A positive selection of mutations in the human genome could be detected for *Bacillus anthracis* and *Yersinia pestis*, whereas it could not be detected regarding *Staphylococcus aureus* and *Escherichia coli*. The number of human genes involved in the positive selection exerted by *B. anthracis* and *Y. pestis* was 493 and 730, respectively. The number of human genes involved in the seven viruses that exerted positive selection was lower. It was in the range of 6 to 261, suggesting defence against, or tolerance of, viruses is less complex than it is for bacteria. We may notice that the results could be replicated for *Y. pestis* and the Measles Morbillivirus only. The replication was, also, limited in the number of populations available in the HapMap-data, *i.e.*, three, which is low compared with the 53 in the Biological General Repository for Interaction Data used in the main analysis.

We may conclude cautiously from the study presented here (Corona *et al.*, 2018) that pathogens may have impact on human evolutionary dynamics. The use of data of the interactome suggests that the underlying mechanism is that of resistance. We may, however, not rule out the mechanism of tolerance, as confounding may have influenced the results. We may remember from section 5.2 that tolerance differs significantly from resistance. Whereas the efficacy of human resistance against pathogens depends on the pathogenicity of these, tolerance of pathogens does not at all per definition (*cf.* Figure 5.2). This implies that pathogens may not 'break' tolerance, as tolerance does not act as a selecting factor for pathogens, whereas resistance does. We may, therefore, conclude that investing in tolerance is a real sustainable approach of a host, in general, to cope with pathogens. In contrast, investing in resistance, and more specifically immunisation, includes always a risk of breaching it by pathogens in the long, or short, term.

We may look now the other way around posing the question whether humans may act as a selecting factor for pathogens. The answer to that question is inevitably related to the topic of competition among pathogens. We had a glimpse on it already tackling the subject of R_0-maximisation above. We will elaborate pathogen competition in the next section.

REFLECTIONS

The mechanisms of resistance and tolerance of humans differ significantly in impact on the evolutionary dynamics of pathogens. The distinction of these, therefore, is of utmost importance. We may reflect about approaches to distinguish these in investigating natural selection exerted by pathogens on man. In addition, we may reflect about the strong focus we have on resistance, including immunisation. Do we overestimate the meaning of it?

5.4 Pathogen competition

Concept of species

We need first of all to take a closer look on the concept of species to deal with competition of pathogens. Recognition of individuals of a species is relatively easy using morphological characters as long as it concerns animals and plants. We have added physiological aspects, *e.g.*, growth on a specific substrate by classifying micro-organisms like fungi and bacteria. The ability to exchange genetic information, sexually and a-sexually, among individuals of a species while keeping the major morphological and physiological characteristics of a species is pivotal in the concept of species. You may, for example, be pretty sure that the offspring of a dog is a dog, although it may show differences in various features, like the size and the fur.

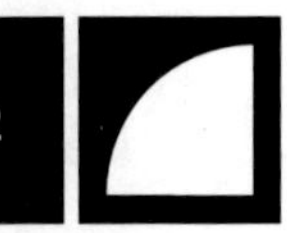

And yes, we have the, exceptional, case of mating of horses and donkeys, which passes the species border. The resulting offspring of mules is, however, infertile. We will see an offspring of horses only, if a male and female horse mate and, similarly, donkeys need to mate with donkeys to generate an off-spring of donkeys. We see the maintenance of major features of a species even among bacteria, despite the relatively high frequency of horizontal gene transfer. We may remember from Chapter 3 that horizontal gene transfer is the transfer of genes between individuals belonging to a different species. We may, therefore, be pretty sure that, for example, sampling a colony of *Escherichia coli*, and plating it onto agar, results in a new *E. coli* colony. And here, we arrive at a crucial point of the concept of species. It needs to be predictable, besides providing an efficient and stable organisation of the observed variation (*cf.* Peterson, 2014).

Predictability is the Achilles heel of the species concept in virology, so far. It hinges largely on sequence differentiation, and therefore, statistics. Just to make the point clear. Humans and apes are easy to distinguish, joint offspring is not possible, but the similarity in genome is relatively high nevertheless. Thus, we might expect mating of apes and man, if we look at the similarity of their genomes only, but it doesn't happy in the real world. In addition, the genome of viruses, RNA or DNA, is prone to frequent mutations and re-assortments, which are, in general, insignificant (*cf.* Grubaugh *et al.* 2020). In addition, mutations and re-assortments are relatively easily expressed in the phenotype of a virus, which is relatively simple anyway, as compared with organisms. The phenotype on which the process of natural selection acts, as we have seen in the preceding section. So, we have neither a robust genotype, nor a robust phenotype, to apply actually the, biological, concept of species to viruses. We feel the term 'species', as used by the International Committee on Taxonomy of Viruses (https://ictv.global/), has a different meaning from that in biology. Regarding prions, we do not even talk about species, as the 'infectiousness' of these is not determined by DNA, or RNA (*cf.* Chapter 3).

The concept of species is significant in defining 'pathogenicity' as well. We may repeat here from section 5.1 that basic compatibility is in operation as soon as, at least, one genotype/phenotype of a pathogen (species) is able to feed on, at least, one genotype/phenotype of a host (species) under conditions suitable for a parasitic interaction. Resistance of a host, and pathogenicity of a pathogen operate when basic compatibility exists between a host species and a pathogen species. We have further sub-divided resistance in 'qualitative' and 'quantitative', which have the counterparts in 'virulence' and 'aggressiveness' of the pathogen. Such terms are hardly to apply to non-organisms that do not fit the, biological, concept of species. And indeed, we see in the literature of viruses that basic compatibility (pathogenicity), aggressiveness, and virulence are commonly put together under the term 'virulence' (*e.g.*, Geoghegan and Holmes, 2018).

The term 'population' is also related significantly to the, biological, concept of species. A term that is inherent to the process of natural selection, in general, and pathogen competition specifically. A population is defined as a local group of individuals of a species. The exchange of genes within a population is significantly higher than the exchange with individuals of another population. So, the rate of gene exchange determines the borders of a population. If the rate of exchange between two populations of a species becomes zero, it may be the start of a separate evolutionary trajectory resulting in two different species. Distinguishing populations is relatively easy regarding organisms that are sedentary during a large proportion of life-time, like plants. It is more difficult concerning animals that move around. Fortunately, these can be tracked, in general, relatively easily enabling both, identification of individuals that aggregate in groups and determination of the gene flows within, and among, the groups. We run in troubles delineating populations of micro-organisms, viruses, and prions, which are quite invisible. In addition, these may disperse relatively easily over large distances, especially the ones dispersing by air. The increasing availability of sequencing tools, however, enables us to detect spatial structures of micro-organisms and viruses. We may, therefore, define populations for micro-organisms at least, whereas it remains limited in the lack of a, biological, species concept for viruses. We do not have tools to detect spatial structures of prions yet.

Prions and viruses are enforced by the relatively simple structure to survive and multiply exclusively in hosts, except dispersal during a relatively short period of time under rather specific conditions. Survival units of both are unknown. We may, therefore, wonder whether prions and viruses obey the principles of competition, and thus the process of natural selection, as it is outlined for organisms. We need to keep this in mind in dealing further with competition among pathogens. We turn to theory now.

Theory

We pick up the Susceptible Infected Recovered (SIR) host model mentioned in section 5.3 above. The number of susceptible individuals in a host population increases with a birth rate, b, it decreases with a mortality rate, μ, it decreases with a transmission rate, β, and it increases with a rate of loss of immunity, v. It reads in a differential equation, as follows,

$$(1)\quad \frac{dS}{dt} = b(S,I,R) - \mu S - \beta SI + vR$$

in which we assume that offspring of infected individuals is not infected.

The number of infected individuals in a host population increases with the transmission rate, β, and it decreases with, the mortality rate, μ, a rate of mortality induced by a pathogen, α, and a rate of recovery from the infected state, γ. It reads in a differential equation, as follows,

$$(2)\quad \frac{dI}{dt} = \beta SI - (\mu + \alpha + \gamma)I$$

in which α is, in general, considered to express the pathogenicity.

The number of individuals in a host population that recovered from infection increases with the rate of recovery from the infected state, γ, and it decreases with the mortality rate, μ, and the rate of loss of immunity, ν. It reads in a differential equation, as follows,

$$(3)\quad \frac{dR}{dt} = \gamma I - (\mu + \nu)R.$$

We focus here on the dynamics of the pathogen. It may show a stable equilibrium, which means in mathematical terms that the trait vector of the pathogen has a value that is not invadable by other values of that vector. We call this an Evolutionary Stable Strategy (ESS). It is a strategy that creates an environment, in which no other strategy can be employed. We may think of a strain of a pathogen species in a host population that does not allow the invasion of another strain of the same species. It implies that the resident strain has the highest fitness of all strains, at least for that specific (host) environment. We may state it in another way, the strain has the highest invasion fitness regarding that (host) environment. The invasion fitness is determined by the parameters β and α in the SIR-models, which are linked. An increase of the pathogenicity, α, may result in a higher multiplication of the pathogen increasing the transmission rate, β, as well. In contrast, a higher α may imply a reduction of the host density decreasing β. If so, the pathogen faces a trade-off between α and β. The consequences of such a trade-off, as related to the invasion fitness, may be visualised in a so-called 'pairwise invasibility plot' (Fig. 5.3). Such a plot shows that the pathogenicity will decrease, if we assume a relatively weak trade-off as it is shown in the left panel. The loss of pathogenicity would result ultimately in a loss of basic compatibility between pathogen and host. In contrast, a relatively strong trade-off would generate an ESS that results in an intermediate level of pathogenicity, as it is shown in the right panel. A similar result, the more or the less, as the one of R_0 maximisation in a relatively simple, one-dimensional, environment, as we outlined in section 5.3 above.

The SIR-model provides a starting point to explore ESSs of pathogens. An adaptation of it is needed to include the various environments that pathogens may encounter. The inclusion of a pathogen reservoir is such an extension. The bacterium *Vibrio cholerae* is a typical example of a pathogen that resides outside the human body for a prolonged period of time, *i.e.*, in a reservoir, as we may call it (*cf.* Chapter 3). The following three differential equations (4)-(6) describe an extended model that includes a reservoir for the pathogen (Pandey *et al.*, 2022),

$$(4)\quad \frac{dS}{dt} = (b - \mu)S - \beta SI - \beta_2 S\,\frac{P}{P+1},$$

in which we see the net growth of the number of susceptible individuals of the host, as unaffected by the pathogen, in the first term. The second term expresses the number of susceptible individuals turning into an infected one, due to dispersal of, and infection by, the pathogen residing within the host population. We have seen the parameters of the first and second terms in the equations (1)-(3) already. The third term indicates the number of susceptible individuals turning into an infected one due to dispersal of, and infection by, the pathogen coming from the reservoir, P. We added the subscript of 2 to β to distinguish both transmission rates in the equation. Equation (2) is extended by adding a third term,

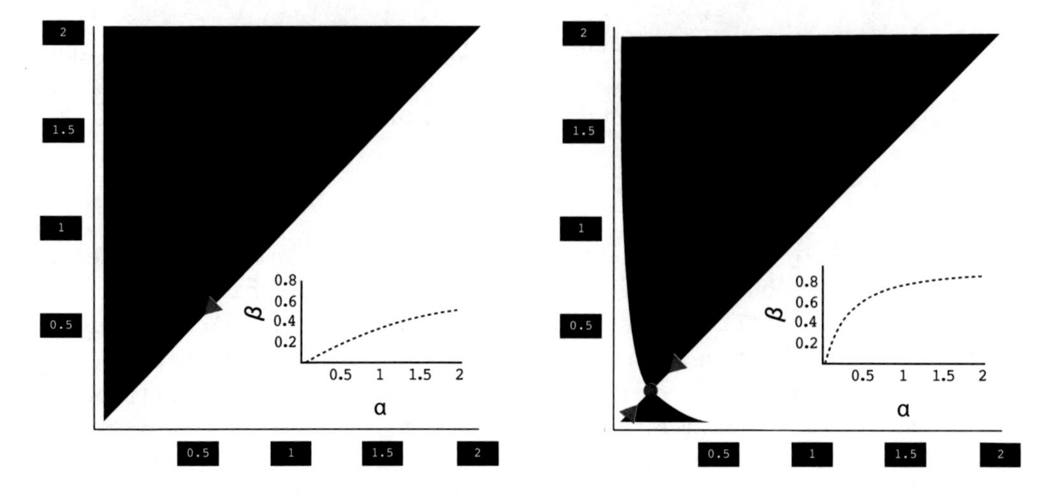

Figure 5.3. Pairwise-invasibility plots of an invading and resident strain, as affected by pathogenicity of each. Pathogenicity of the invader is expressed on the ordinate and that of the resident on the abscissa. A higher value indicates a stronger pathogenicity. The dark area indicates the values of pathogenicity of resident and invader, respectively, that does not enable invasion. The left-hand panel shows the invasibility by assuming a relatively weak trade-off between pathogenicity (α) and transmission rate (β), see inserted graph, whereas the right-hand panel shows the invasibility by assuming a strong trade-off as shown in the inserted graph. Arrows show the direction of natural selection, which may direct to a stable point (ESS), as indicated by a purple dot.

$$(5)\quad \frac{dI}{dt} = \beta SI - (\mu + \gamma + \alpha)I + \beta_2 S \frac{P}{P+1},$$

in which the first term describes the increase of the number of susceptible individuals turning into an infected one, due to dispersal of, and infection by, the pathogen residing within the host population. The second term expresses the decrease of the number of infected individuals due to mortality that results from the pathogen, α, and other factors, μ, respectively, and the recovery of individuals from infection, γ. The third term indicates the number of individuals that become infected by the pathogen coming from the reservoir. And we have a third equation of the dynamics of the pathogen with respect to the reservoir,

$$(6)\quad \frac{dI}{dt} = \theta I + rP\left(1 - \frac{P}{K}\right) - \mu_p P,$$

in which θ is the rate of pathogen reproduction on infected host individuals that, subsequently, enter the reservoir, *r* is the rate of pathogen reproduction in the reservoir that is limited in the carrying capacity K of the reservoir, and μ_p is the mortality rate of the pathogen in the reservoir.

The ESSs that correspond with various trade-offs may, subsequently, be explored, having this basic model at hand (Pandey *et al.*, 2022). We look first at the trade-off between pathogenicity, α, and the mortality rate μ_p of the pathogen in the reservoir (Fig. 5.4). If the trade-off implies

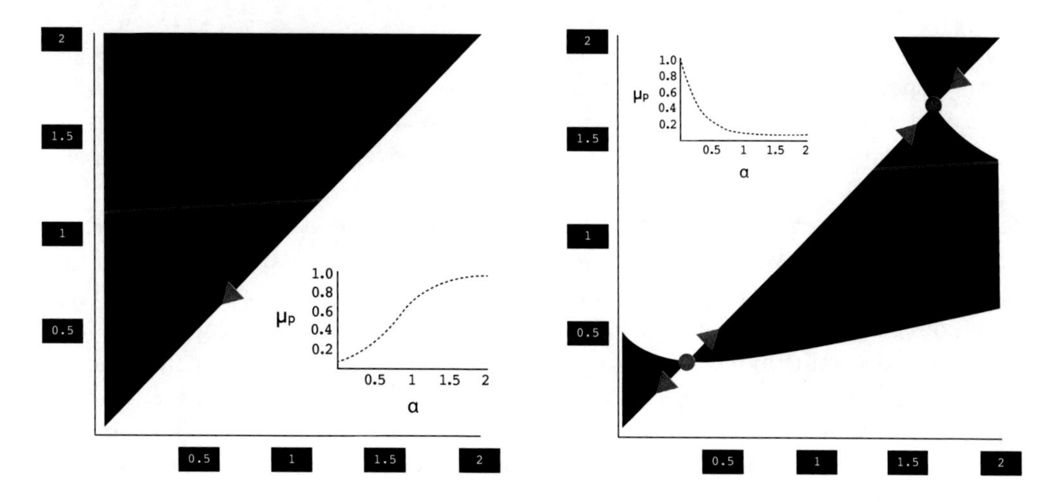

Figure 5.4. Pairwise-invasibility plots of an invading and resident strain, as affected by pathogenicity of each. Pathogenicity of the invader is expressed on the ordinate and that of the resident on the abscissa. A higher value indicates a stronger pathogenicity. The dark area indicates the values of pathogenicity of resident and invader, respectively, that does not enable invasion. The left panel shows the 'invasibility', if we assume a negative trade-off between pathogenicity (α) and pathogen mortality rate in the reservoir (μ_p), see inserted graph, whereas the right panel shows the 'invasibility', if we assume a positive trade-off as shown in the inserted graph. Arrows show the direction of natural selection, which may direct to a stable point (ESS), as indicated by a purple dot, or from an 'evolutionary repellor', as indicated by a pink one.

that an increase of pathogenicity towards the host results in an increase of the mortality of the pathogen in the reservoir, we will observe a minimisation of the pathogenicity towards loss of basic compatibility, as indicated in the left-hand panel of figure 5.4. If the trade-off implies that an increase of the pathogenicity towards the host decreases the mortality in the reservoir, we see bi-stability in the evolutionary dynamics, as indicated in the right-hand panel. We see a so-called 'evolutionary repellor', a point that repels, at relatively low levels of pathogenicity of both, the resident and invading strain of the pathogen. The 'repellor' results either in minimising pathogenicity towards loss of basic compatibility, or an increase of pathogenicity towards a stable point, an 'evolutionary attractor', at a relatively high level of pathogenicity. In short, an ESS that corresponds to an intermediate level of pathogenicity (*cf.* Fig. 5.3, right hand panel) is not possible on the assumption of a trade-off between pathogenicity towards a host and persistence of the pathogen in the reservoir. We feel it also to be remarkable that stability in pathogenicity may be reached in opposite directions, depending on the initial levels of pathogenicity of resident and invading strain, respectively.

We explore the model further by including multiple trade-offs of pathogenicity versus, infection rate, mortality in the reservoir, and reproduction in the reservoir. One of the possible configurations is presented in figure 5.5, left-hand panel. We see two stable points and an 'evolutionary repellor'. The evolutionary path of pathogenicity may also be affected by the carrying capacity, K, of the reservoir, although no trade-off exists between both. We assume here a relatively low value of K and we see an evolutionary branching point, besides an 'evolutionary attractor' and an 'evolutionary repellor', respectively (Fig. 5.5, right-hand panel).

We may notice the general applicability of the model provided by Pandey *et al.* (2022), replacing the compartment 'reservoir' by, a side host, a host needed to complete the life-cycle of the pathogen, or simply another organ of the host. We are, for example, not aware of any pathogen that resides permanently in the same organ of humans. So, we need to extent the standard SIR-model, in general, to model the ESSs of pathogens, while taking into account intraspecific competition. In addition, we may infer from the study that predictions about evolutionary trajectories of pathogenicity, as affected by intraspecific pathogen competition, are not straightforward, at least from a theoretical point of view. An inference that fits Lion and Metz (2018), as we mentioned above.

The preceding was directed to intraspecific competition of pathogens. We may use a similar approach to interspecific competition by way of including pathogen species instead of strains of a pathogen. So far, so good. We may, however, need a different modelling approach to pathogen competition, if we take into account the spatial dimension of evolutionary dynamics.

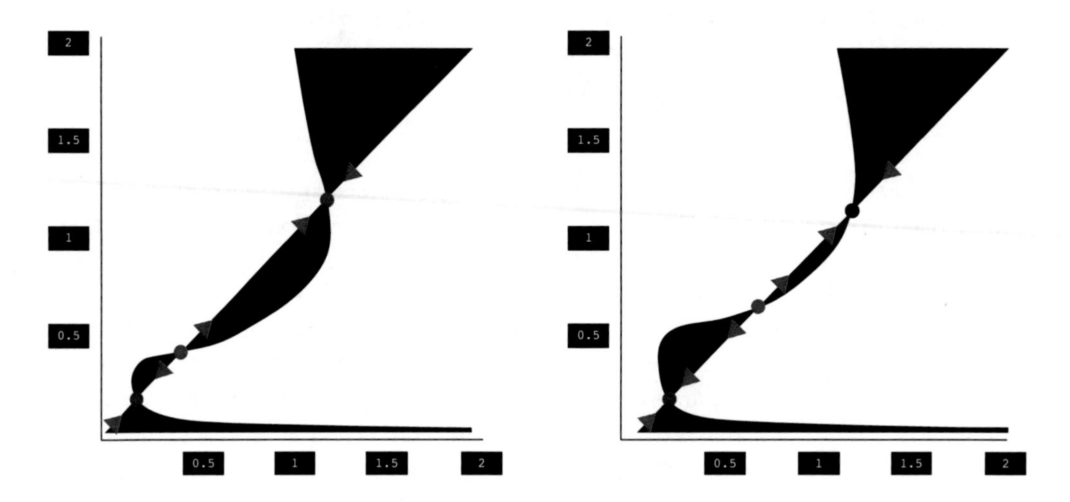

Figure 5.5. Pairwise-invasibility plots of an invading and resident strain, as affected by pathogenicity of each. Pathogenicity of the invader is expressed on the ordinate and that of the resident on the abscissa. A higher value indicates a stronger pathogenicity. The dark area indicates the values of pathogenicity of resident and invader, respectively, that does not enable invasion. The left-hand panel shows invasibility, if we assume multiple trade-offs of pathogenicity. The right-hand panel shows the evolutionary dynamics at a relatively low carrying capacity of the reservoir. Arrows show the direction of natural selection, which may direct to a stable point (ESS), as indicated by a purple dot, from an 'evolutionary repellor', as indicated by a pink one, or to a branching point, as indicated by a black dot.

We turn to a study that models competition between two pathogens, as influenced by host mobility (Poletto *et al.*, 2015). A study that included cross-resistance as well. Cross-resistance means here that a host, which has built up adaptive immunity to one pathogen, may also have it with respect to another species. We may summarise the study set-up in three major elements.

One, the study was based on the concept of a meta-population. A meta-population is a set of populations, of the same species, that are connected. Individuals may move from one population to another with a certain probability, p_m, at a specific interval. The probability may vary among specific pairs of populations. The result may be that some populations are more isolated than others and that the densities of the populations also vary. We see spatial dynamics of the host. The intensity of movement depends on the type of host considered. Clearly, the intensity may be, for example, much lower for elderly people included in a setting of a meta-population than for young adults. The concept of meta-population allows much flexibility regarding the sizes of, and movements between, the populations, as well as the type of host included.

Two, the two pathogens included in the study differed in the basic reproductive number, R_0, which is expressed in the ratio between both values, R_0^r. The two pathogens also differed in the velocity of development on the host. The difference in development on the host causes a difference in recovery of the host from infection. The recovery is delayed for a host individual infected by the 'slow' pathogen compared with one infected by the 'fast' pathogen. In addition, infection of a host individual by one pathogen may induce resistance, immunity, to the other pathogen, which arrives later on. The level of cross-resistance, expressed by *cr*, may vary between zero and one, expressing complete resistance down to no resistance at all.

Three, the SIR-like simulation model, including the parameters p_m, R_0^r and *cr*, was run in a Monte Carlo setting to arrive at the outcome of the modelling, the proportion of populations occupied by the 'slow' and 'fast' pathogen, respectively.

We see that most of the populations remain free of the pathogens, if the probability of host movement is rather low (Fig. 5.6). A higher R_0 of the 'fast' pathogen, compared with the 'slow' one, results in a very narrow bandwidth of occurrence of the 'slow' pathogen (left panel). In contrast, a higher R_0 of the slow one results in a relatively large bandwidth, including co-occurrence of both pathogens (right panel). The 'invasibility' of populations by the pathogens was further explored regarding the level of cross-resistance at a relatively high probability of host movement, *i.e.*, 0.01. We note than, the stronger the cross-resistance, the higher the fraction of populations invaded by the 'fast' pathogen solely, except the R_0 of the 'slow' pathogen exceeds substantially R_0 of the 'fast' one. Invasion of populations by the 'slow' pathogen requires that cross-reference is nearly absent, independent of the value of R_0^r.

We may, overall, conclude that we have sufficient tools at hand to explore theoretically competition of pathogens in various settings. Modelling is inevitable to get grip on the evolutionary dynamics of pathogens, which may be fuelled by competition for host resources

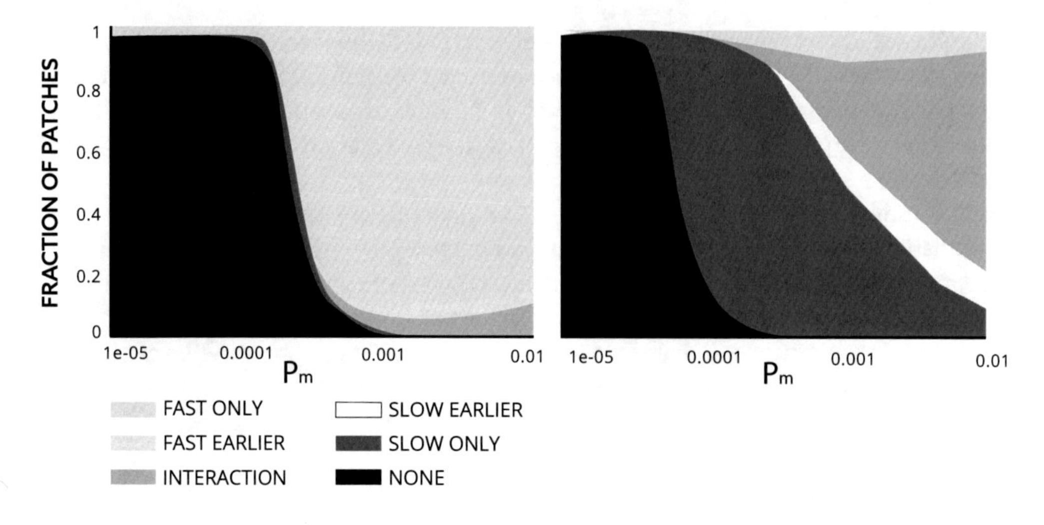

Figure 5.6. Fraction of host populations occupied by a 'slow' and 'fast' developing pathogen, respectively, as related to the average probability of movement of host individuals between populations, p_m, taking into account the velocity of development of a pathogen on the host. Various scenarios of abundance, or absence, of the two pathogens are indicated, including the time of arrival compared with each other (see legend). The left panel shows the scenarios assuming a higher R_0 of the 'fast' pathogen, *i.e.*, a value of $R_0{}^r$ of 0.8, and the right panel a higher one of the 'slow' pathogen, *i.e.*, a value of $R_0{}^r$ of 1.6. See also text.

at relatively large temporal and spatial scales. Insight in evolutionary dynamics generated by modelling may, subsequently, guide us in obtaining empirical evidence (*cf.* Lion and Metz, 2018). And so, which empirical evidence of pathogen competition do we have at hand at the moment?

Empirical evidence

We turn first to bacteria, which are relatively easy to investigate. We distinguish two types of competition. One is driven by the exploitation of a resource, in which growth is limited. The other is direct interference by the exudation of anti-microbial substances, anti-biotics, that limit the growth of individuals of another strain, or species. Such substances may even cause mortality among individuals of the interfered strain. Quorum sensing is pivotal in both types of bacterial competition. Direct bacterial interference was investigated in a study directed to the non-pathogenic food bacterium *Carnobacterium maltaromaticum* (Ramia *et al.*, 2020). The study included 73 strains of *C. maltaromaticum* collected from various habitats. The number of strains per habitat varied between 2 and 37. One strain could not be attributed to a specific habitat. Strains were grown individually, *i.e.*, free of competition, and the resulting culture was centrifuged to trap potential anti-biotics in the Cell Free Supernatant (CFF). In addition, the time needed to achieve a certain cell-density, *i.e.*, an OD_{595nm} of 0.2, was determined for each of the strains in absence of interference. The 73 strains were tested pairwise for interactions. Two roles were assigned to each strain, that of 'sender' and that of 'receiver', respectively. If 'sender', the corresponding CFF was added to a culture of the 'receiver' in a well. The Growth Inhibition Indicator (GII) of the strain serving as receiver was calculated by subtracting the time needed to arrive at the specified cell density without addition of the CFF from the time needed with the addition. A threshold of 300 minutes was used to define an interaction as inhibitory. The testing of 219 pairs, out of the 5329, were repeated, resulting in a reproducibility of 98%. All the pairs were, subsequently, subject of a network analysis (Fig. 5.7).

Strains could be assigned to one out of the four categories of interactive profiles determined:

(i) Producer-Sensitive-Resistant (PSR); producer means that a strain is able to produce anti-biotics that do inhibit the growth of at least one other strain. Sensitive implies that the growth of the strain is inhibited by the anti-biotics of at least one other strain and resistant indicates that a strain resists, or tolerates, the anti-biotics of at least one other strain. So, PSR-strains included all the possible types of interaction and the specific type of interaction depended on the interacting strain. Thirty-six strains showed this PSR-profile.
(ii) Sensitive-Resistant (SR); strains that are sensitive, or just resistant/tolerant, to anti-biotics whereas these do not excrete anti-biotics interfering with any strain included in the study. Twenty-four strains showed this SR-profile.
(iii) Producer-Resistant (PR); strains producing anti-biotics inhibiting growth of at least one other strain and showing resistance/tolerance to anti-biotics of all strains producing these. Five strains showed this PR-profile.
(iv) Resistant (R); strains resistant/tolerant to anti-biotics of all other strains, but unable to produce anti-biotics interfering with any strain in the study. Eight strains showed this R-profile.

We may notice that four strains could not be attributed to one of these categories. In addition, nearly 80% of the 5329 pairs of strains included could be characterised as 'reciprocal neutral'. So, both strains of a pair did not interfere with each other. About 20% of the pairs showed non-reciprocal inhibition. It means that one strain was interfering another strain, but the reverse did not happen. Reciprocal inhibition was rather rare, *i.e.*, about 1% of all the 5329 pairs of strains tested. The network could be characterised, overall, as 'sender-determined'. Intraspecific competition was thus demonstrated in *C. maltaromaticum* and it turned out to be a multi-faceted phenomenon. We may add that the anti-biotics of some of the strains may also be involved in interspecific competition. We see such an interspecific interference, for example, with respect to the bacterium *Listeria monocytogenes* a food-borne pathogen that causes listeriosis in man.

We stay on bacteria and turn to the impact of interspecific competition on the host. The target organism is *Vibrio cholerae*. We know from Chapter 3.3.2. that pathogenic strains of *V. cholerae* excrete a toxin to cope with the severe interspecific competition in the gut of a host. A side-effect of the toxin is damage to the gut epithelium and the resulting severe diarrhoea. The effect of the toxin on interspecific competition, including its side-effect, was investigated using flies of *Drosophila melanogaster* (Fast *et al.*, 2018). The focus was on the type-six secretion system (T6SS) of *V. cholerae*. The T6SS mediates toxicity to other Gram-negative bacteria, whereas Gram-positive bacteria are not affected. Three strains of *V. cholerae* were used, (1) a strain showing a relatively low level of aggressiveness, as expressed in the mortality rate of flies, (2) a strain showing a relatively high level of aggressiveness, and (3) the aggressive strain modified in such a way that the T6SS did not function. The microbiome of the fly gut was manipulated, or not. If not, the microbiome was dominated by the Gram-negative bacteria of *Acetobacter pasteurianus* and the Gram-positive bacteria of *Lactobacillus brevis* and *L. plantarum*. These three bacterial species are considered as commensals. If manipulated, the gut was inhabited solely by, either *A. pasteurianus*, or *L. brevis*. Fly guts free of microbiome served as controls. The description of the study design did not include statistical testing. We, therefore, cannot judge the validity of the study completely.

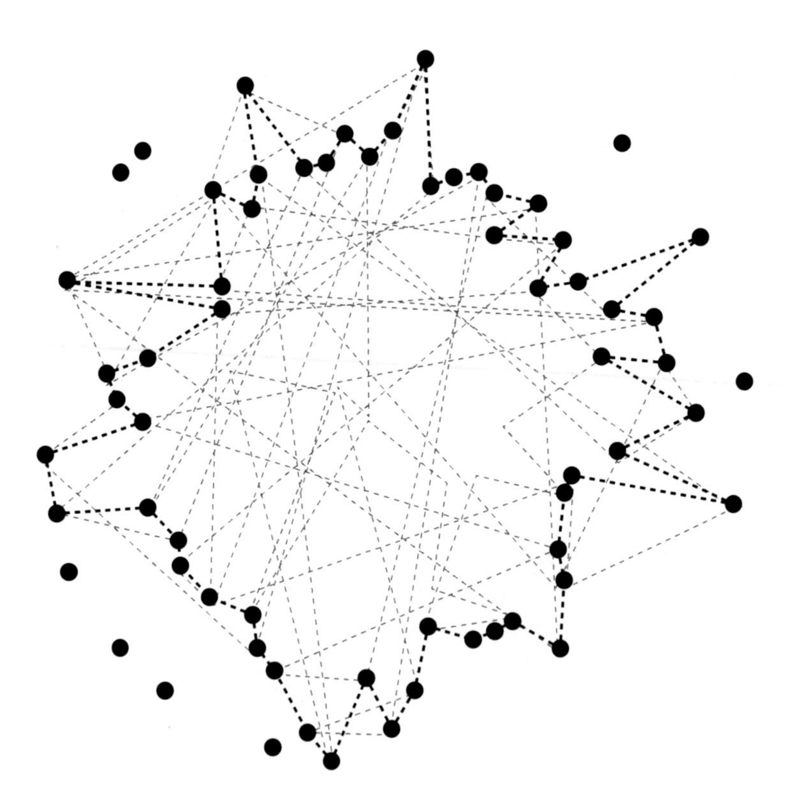

Figure 5.7. Snapshot of network of intraspecific competition among strains of *Carnobacterium maltaromaticum*. Dots indicate strains and lines indicate interference between strains. Eight strains are insensitive to interference, as indicated by the dots outside the network. Sensitivity of interference increases going from outside to the centre of the network (graphically not shown).

The results indicated that abundance of the Gram-negative bacteria of *A. pasteurianus* is essential for the toxicity mediated by the T6SS of *V. cholerae* towards the host, whereas an effect of the Gram-positive bacteria *L. brevis* and *L. plantarum* could not be detected. So, a tool of *V. cholerae* to compete with Gram-negative bacteria of other species, which are competing for similar resources in the gut, destroys as well resources by impacting on the host. The impact is either by causing diarrhoea, or killing the host even. We should be careful, of course, to translate the findings of a relatively simple *D. melanogaster* – *V. cholerae* system to humans, but we may state that cholera results from collateral side-damage of intense bacterial, interspecific, competition in the microbiome rather than targeted parasitism of man. This statement is contradictory to the distant action on the host that is mentioned in Chapter 4. Pathogens like *V. cholerae* keep challenging us.

We turn now from bacteria to eukaryotic pathogens, and more specifically chromists. We also turn from direct interference of competing organisms to resource competition among these. Resource competition among strains of *Plasmodium chabaudi* in rodents was investigated, serving as a model for malaria in man (Wale *et al.*, 2017). Two strains of *P. chabaudi* were used in the study, a pyrimidine-resistant strain and a sensitive one. Drugs with the active ingredient pyrimidine were prescribed for treatment of malaria in the past. The prescription is not valid anymore, due to the high frequency of pyrimidine resistance among *Plasmodium* species. Results of competition experiments in mice had indicated that the pyrimidine-sensitive strain is the superior competitor of the two strains. The competition is strongly a-symmetric, which means that the pyrimidine-resistant strain suffers from interference by the sensitive one and the reverse doesn't hold hardly. Mice were inoculated solely with the pyrimidine-resistant strain, or inoculated with both strains, using an additive competition design. It implies that the inoculum dose overall was twice as high in mixture than in mono-culture, which may have caused bias in the results of the study. Mice were supplemented with para-aminobenzoic acid, or not. If so, the supplementation was at a relatively, high, medium, or low concentration. The para-aminobenzoic acid is required in the pyrimidine synthesis and methionine metabolism of *P. chabaudi*. Five mice were inoculated in each of the eight treatment categories, which resulted from, inoculation or not, mono or mixture, and the (level of) supplementation of para-aminobenzoic acid, or not. Infections of the mice were monitored daily by way of sampling blood in the period between three and 21 days after inoculation. The pathogen density was determined by way of quantitative Polymerase Chain Reaction. The density of red blood cells was determined using flow cytometry, as the pathogen feeds on these. We may notice that the number of replicates, the number of mice per category, was relatively low from a point of view of statistics. In addition, some mice passed away before the end of the monitoring, causing an imbalance in the design. Data were presented in graphs only. So, the set-up of the study was, overall, rather poor from a point of view of statistics.

The general dynamics of *P. chabaudi*, as expressed in the density of merozoites, was one of a steep increase between the third and seventh days after inoculation. It was followed by a decrease that plateaued at a certain level, or not. The decrease continued when the pyrimidine-resistant strain grew in mono-culture on mice that were not supplemented with para-aminobenzoic acid, or a suppletion at a low concentration only. Suppletion at higher concentrations resulted in *P. chabaudi* plateauing at a higher number of merozoites. The plateau was the higher, the higher the concentration of para-aminobenzoic acid. Growth of the pyrimidine-resistant strain in competition showed no plateau at a medium concentration of para-aminobenzoic acid. Data of growth at high concentration were missing due to mortality of mice. The impact of *P. chabaudi* on mice, as expressed in the density of red blood cells, was relatively low until *c.* 5 days after inoculation, becoming high between 7 and 12 days, and the impact became low again afterwards. The impact tended to be a bit larger on mice inoculated with both strains. We may state that the results of this study are not very convincing on their own. It is not surprising as, (i) determining resource competition is, in general, less easy than direct interference, (ii) studying competition of pathogens inside a host, and certainly within a mammal, is a real challenge, and (iii) the number of animals included in the study, and so the power of it, is limited in ethics. Anyway, the study provided some evidence of intraspecific resource competition in pathogenic chromists and the results fit those of preceding studies.

We stay on *P. chabaudi*, but we turn to interspecific competition of it with a completely different organism, the helminth *Nippostrongylus brasiliensis* (Wait *et al.*, 2021). Mice were inoculated with doses of, 10^3, 10^4, 10^5 or 10^6 sporozoites of a strain of *P. chabaudi*. Sixteen mice were inoculated per dose. Another series of mice was inoculated at similar doses, but it was in combination with administering 200 third stage larvae of *N. brasiliensis*. This hookworm parasitises red blood cells of the host, like *P. chabaudi* does. So, the chromist and worm compete for the same host resource. The number of mice per dose also was 16. The density of red blood cells was monitored daily upon 11 days after inoculation. In addition, the proportion of red blood cells infected by merozoites of *P. chabaudi* was determined. Seven mice were excluded from the analysis due to failure of the inoculation, or 'substantial deviations' of the expected infection dynamics. We feel the latter is a bit subjective. Anyway, the study design showed some imbalance, which might have affected the statistical analysis. The following statistical model was applied to the data,

$$I_{t+1} = P_{e,max} S_t I_t S_{dt},$$

in which, I represents the density of red blood cells infected with merozoites at a specific day t, or t + 1 day, $P_{e,\,max}$ is the effective propagation (e) of merozoites among red blood cells at maximum (max), S is the number of susceptible red blood cells not infected by merozoites yet, and S_d a term indicating deviations of the *P. chabaudi* population dynamics, which cannot be attributed to the availability of susceptible red blood cells on the preceding day. The immunity response of mice may be a likely factor of causing such deviations. The assumption, therefore, was that S_d expresses the immune response of mice only.

The average of the *P. chabaudi* infection peak, as expressed in $P_{e,\,max}$, was highest at the lowest inoculum dose, irrespectively *P. chabaudi* grew alone, or together with *N. brasiliensis*. The infection peak, however, occurred earlier at higher inoculum doses, *i.e.*, at day 3 post inoculation at the highest dose and day 7 at the lowest dose. The immune response of mice, as expressed in S_d, was also activated earlier at the highest dose compared with the lowest dose. Co-occurrence of the hookworm delayed the peak about one day at all inoculum doses, except the highest one. An effect of co-occurrence of both parasites on the strength of the immunity response could not be detected. A mechanistic model was, subsequently, used to interpret the results in terms of the underlying mechanisms. The results suggested that the growth of the *P. chabaudi* population was limited in the availability of red blood cells in an early stage due to the abundance of *N. brasiliensis*, which, therefore, caused a delay of the *P. chabaudi* infection peak. The study thus indicated that interspecific competition may occur even between two pathogens belonging to completely different categories of pathogens, *i.e.*, chromists and helminths. An intriguing finding, of which the relevance needs to be determined in terms of co-occurrence of pathogens in humans yet.

We like now to highlight a quite different type of competition between pathogens. It is based on differences in the velocity of colonising a host. We may call it 'the first come, the first service'-type of competition. We touched on this type of competition in the part 'theory' above already. We focus here on a study directed to 'social immunity' in the Argentine ant and two of its pathogens, the fungi *Metarhizium brunneum* and *M. robertsii* (Milutinović *et al.*, 2020). We should not confuse social immunity with bodily immunity. Social immunity refers to cleaning up, and structuring, of an insects' community to minimise dispersal and establishment of pathogens. In the study here, grooming of ants was investigated as an expression of social immunity. Grooming turned out as pivotal in the competition between the two *Metarhizium* species. Conidiospores of the fungus settle on the cuticle of an ant, germinate and, subsequently, penetrate and establish within the ant. The ant is killed and the fungus starts to multiply on the cadaver by shedding conidiospores. In the study, 30 individual ants were inoculated with the fungus, or a group of 50 ants together, except for two ants that were not inoculated to keep these alive for grooming. The fraction of ants killed per 30, or 50, was determined and spores were collected from the cadavers. Three strains of each of *M. brunneum* and *M. robertsii* were used. Ants were inoculated with, either two strains of the same species, or one strain of *M. brunneum* and one of *M. robertsii*. All possible combinations of strains were applied. The spores collected were identified and quantified for each species and strain. In addition, germination tests were executed for each of the six strains used.

The results of the study indicated *M. robertsii* as the superior competitor, if co-inoculated with *M. brunneum* on individual ants. This superiority could not be detected when co-inoculating a group of ants. The lack of superiority in a group of ants may be explained by a combination of grooming and germination of conidiospores. Grooming of spores is more complete in a group, as both, auto- and allo-grooming occurs, whereas it is auto-grooming only in the individual setting. In addition, conidiospores of *M. robertsii* germinate, on average, slower than those of *M. brunneum*. Those of *M. robertsii* are, therefore, exposed longer to grooming. The combination of longer exposure, which also occurs in the individual setting, and allo-grooming, may explain the loss of the competitive superiority of *M. robertsii* in the group setting. A conclusion that was supported by way of executing a stochastic competition model. We see thus a very subtle interplay between social interactions of the host and biology of the pathogens. Subtility seems the hallmark of pathogen competition anyway, if we look at the examples presented here.

REFLECTIONS

The concept of species is at the centre of competition of pathogens. We may use it relatively easily with all categories of pathogens, except viruses and prions. We do not distinguish species in prions anyway, but you may reflect about improving the concept of viral species to fit the concept of pathogen competition, as outlined here. Alternatively, you may reflect about another scientific view on competition of pathogens.

5.5 Outlook

We see four major challenges in advancing our knowledge about evolutionary dynamics of pathogens. First of all, considerable effort is required to distinguish human resistance against pathogens, including immunity, from tolerance of these. The COVID-19 pandemic has triggered awareness of tolerance in managing the disease in an early stage of the epidemic, in which vaccines were missing still and the immune response needed to be silenced to tolerate COVID-19 by relatively frail people. We like to stress here the role of host tolerance in the evolutionary dynamics of pathogens. Tolerance is independent of pathogenicity and it, therefore, does not exert selection on pathogens. Actually, tolerance provides a mechanism to the host that might trigger a change from pathogenicity to commensalism. We, therefore, foresee that existing theory on evolutionary dynamics of pathogens will be expanded with respect to tolerance of pathogens by hosts. In addition, mechanisms underlying human tolerance will be investigated thoroughly, as we cannot distinguish resistance and tolerance without having sufficient insight in the mechanisms. Finally, we will see the emergence of appropriate methods to determine, and quantify, tolerance unequivocally with respect to man.

We may indicate a challenge as well regarding the establishment of a direct link between genetic variation of a specific host and pathogenicity of a specific pathogen to determine the likelihood of natural selection. It will become possible by testing theoretical predictions of Evolutionary Stable Strategies of the target pathogen by using advances methods in, sequencing, imaging, metabolomics, and so on, adding as well mental aspects, if humans serve as host.

A third challenge ahead is the up-grading of the concept of viral species towards a biological meaningful one. It is a tough job for such a non-organism. We will have to go beyond sequencing to arrive at a, biologically, proper concept of species. Alternatively, we may revise the concept of natural selection, as it has been developed the last two centuries. We feel it less promising to invest in the latter.

The last challenge is to deal with the evolutionary dynamics of prions. A tremendous one. We feel an intensive collaboration between chemists, (evolutionary) biologists, and physicians will arise. It will be like the collaboration in bio-mimetics, in which chemists and engineers learn from biologists. In contrast to bio-mimetics, biologists need to learn from chemists to come up with appropriate systematics of prions. If so, we may start to explore the evolutionary dynamics of these non-organisms, of these proteins showing an infectious character.

References

Booker, T.R., Jackson, B.C. and Keightley, P.D., 2017. Detecting positive selection in the genome. BMC Biology 15: 98. DOI: 10.1186/s12915-017-0434-y.

Chaguza, C., Yang, M., Cornick, J.E., Du Plessis, M., Gladstone, R.A., Kwambana-Adams, B.A., Lo, S.W., Ebruke, C., Tonkin-Hill, G., Peno, C., Senghore, M., Obaro, S.K., Ousmane, S., Pluschke, G., Collard, J-M., Sigaùque, B., French, N., Klugman, K.P., Heyderman, R.S., McGee, L., Antonio, M., Breiman, R.F., Von Gottberg, A., Everett, D.B., Kodioglu, A. and Bentley, S.D., 2020. Bacterial genome-wide association study of hyper-virulent pneumococcal serotype 1 identifies genetic variation associated with neurotropism. Communications Biology 3: 559. DOI: 10.1038/s42003-020-01290-9.

Corona, E., Wang, L., Ko, D. and Patel, C.J., 2018. Systematic detection of positive selection in the human-pathogen interactome and lasting effects on infectious disease susceptibility. PloS One 13: e0196676. DOI: 10.1371/journal.pone.0196676.

Darwin, C., 1859. On the Origin of Species, as republished in 1996 by Oxford University Press, Oxford, United Kingdom, 439 pp.

Dieckmann, U., Metz, J.A.J., Sabelis, M.W. and Sigmund, K., 2002. Adaptive Dynamics of Infectious Diseases: in pursuit of virulence management. Cambridge University Press, Cambridge, United Kingdom, 532 pp.

Endler, J. A., 1986. Natural Selection in the Wild. Princeton University Press, Princeton, New Jersey, United States of America, 336 pp.

Fast, D. Kostiuk, B., Foley, E. and Pukatzki, S., 2018. Commensal pathogen competition impacts host viability. Proceedings of the National Academy of Sciences USA 115: 7099-7104. DOI: 10.1073/pnas.1802165115.

Frantzen, J., 2000. Resistance in populations. In: Slusarenko, A.J., Fraser, R.S.S. and van Loon, L.C. (eds.) Mechanisms of Resistance to Plant Diseases. Kluwer Academic Publishers, Dordrecht/Boston/London, the Netherlands, pp. 161-187

Frantzen, J., 2007. Epidemiology and Plant Ecology, Principles and Applications. World Scientific Publishing, Singapore, 172 pp.

Gagneux, S., DeRiemer, K., Van, T., Kato-Maeda, M., De Jong, B.C., Narayanan, S., Nicol, M., Niemann, S., Kremer, K., Gutierrez, M.C., Hilty, M., Hopewell, P.C. and Small, P.M., 2006. Variable host-pathogen compatibility in *Mycobacterium tuberculosis*. Proceedings of the National Academy of Sciences USA 103: 2869-2873. DOI: 10.1073/pnas.0511240103.

Galardini, M., Clermont, O., Baron, A., Busby, B., Dion, S., Schubert, S., Beltrao, P. and Denamur, E., 2020. Major role of iron uptake systems in the intrinsic extra-intestinal virulence of the genus *Escherichia* revealed by a genome-wide association study. PloS Genetics 16: e1009065. DOI: 10.1371/journal.pgen.1009065.

Geoghegan, J.L. and Holmes, E.C., 2018. The phylogenomics of evolving virus virulence. Nature Reviews Genetics 19: 756-769. DOI: 10.1038/s41576-018-0055-5.

Gómez, J., Albaiceta, G.M., García-Clemente, M., López-Larrea, C., Amado-Rodríguez, L., Lopez-Alonso, I., Hermida, T., Enriquez, A.I., Herrero, P., Melón, S., Alvarez-Argüelles, M.E., Boga, J.A., Rojo-Alba, S., Cuesta-Llavona, E., Alvarez, V., Lorca, R. and Coto, E., 2020. Angiotensin-converting enzymes (ACE, ACE2) gene variants and COVID-19 outcome. Gene 762: 145102. DOI: 10.1016/j.gene.2020.145102.

Gori, A., Harrison, O.B., Mlia, E., Nishihara, Y., Mun Chan, J., Msefula, J., Mallewa, M., Dube, Q., Swarthout, T.D., Nobbs, A.H., Maiden, M.C.J., French, N. and Heyderman, R.S., 2020. Pan-GWAS of *Streptococcus agalactiae* highlights lineage-specific genes associated with virulence and niche adaptation. mBio 11: e 00728-20. DOI: 10.1128/mBio.00728-20.

Grubaugh, N.D., Petrone, M.E. and Holmes, E.C., 2020. We shouldn't worry when a virus mutates during disease outbreaks. Nature Microbiology 5: 529-530. DOI: 10.1038/s41564-020-0690-4.

Jallow, M., Ying Teo, Y., Small, K.S., Rockett, K.A., Deloukas, P., Clark, T.G., Kivinen, K., Bojang, K.A., Conway, D.J., Pinder, M., Sirugo, G., Sisay-Joof, F., Usen, S., Auburn, S., Bumpstead, S.J., Campino, S., Coffey, A., Dunham, A., Fry, A.E., Green, A., Gwilliam, R., Hunt, S.E., Inouye, M., Jeffreys, A.E., Mendy, A., Palotie, A., Potter, S., Ragoussis, J., Rogers, J., Rowlands, K., Somaskantharajah, E., Whittaker, P., Widden, C., Donnelly, P., Howie, B., Marchini, J., Morris, A., SanJoaquin, M., Achidi, E.A., Agbenyega, T., Allen, A., Amodu, O., Corran, P., Djimde, A., Dolo, A., Doumbo, O.K., Drakeley, C., Dunstan, S., Evans, J., Farrar, J., Fernando, D., Hien, T.T., Horstmann, R.D., Ibrahim, J., Marsh, K., Michon, P., Modiano, D., Molyneux, M.E., Mueller, I., Parker, M., Peshu, N., Plowe, C.V., Pujialon, O., Reeder, J., Reyburn, H., Riley, E.M., Sakuntabhai, A., Singhasivanon, P., Sirima, S., Tall, A., Taylor, T.E., Thera, M., Troye-Blomberg, M., Williams, T.N., Wilson, M., Kwiatkowski, D.P., Wellcome Trust Case Control Consortium and Malaria Genomic Epidemiology Network, 2009. Genome-wide and fine-resolution association analysis of malaria in West Africa. Nature Genetics 41: 657-665. DOI: 10.1038/ng.388.

King, I.L. and Li, Y., 2018. Host-parasite interactions promote disease tolerance to intestinal helminth infection. Frontiers in Immunology 9: 2128. DOI: 10.3389/fimmu.2018.02128.

Köhler, J.R., Casadevall, A. and Perfect, J., 2015. The spectrum of fungi that infects humans. Cold Spring Harbor Perspectives in Medicine 5: a019273. DOI: 10.1101/cshperspect.a019273.

Lion, S. and Metz, J.A.J., 2018. Beyond R_0 maximisation: on pathogen evolution and environmental dimensions. Trends in Ecology and Evolution 33: 458-473. DOI: 10.1016/j.tree.2018.02.004.

Mayr, E., 1982. The Growth of Biological Thought: diversity, evolution, and inheritance. The Belknap Press of Harvard University Press, Cambridge, United States of America, 974 pp.

Milutinović, B., Stock, M., Grasse, A.V., Naderlinger, E. and Hilbe, C., 2020. Social immunity modulates competition between coinfecting pathogens. Ecology Letters 23: 565-574. DOI: 10.1111/ele.13458.

Nahrendorf, W., Ivens, A. and Spence, P. J., 2021. Inducible mechanisms of disease tolerance provide an alternative strategy of acquired immunity to malaria. eLife 10: e63838. DOI: 10.7554/eLife.63838.

Omae, Y., Toyo-oka, L., Yanai, H., Nedsuwan, S., Wattanapokayakit, S., Satproedprai, N., Smittipat, N., Palittapongarnpim, P., Sawanpanyalert, P., Inunchot, W., Pasomsub, E., Wichukchinda, N., Mushiroda, T., Kubo, M., Tokunaga, K. and Mahasirimongkol, S., 2017. Pathogen-lineage-based genome-wide association study identified CD53 as susceptible locus in tuberculosis. Journal of Human Genetics 62: 1015-1022. DOI: 10.1038/jhg.2017.82.

Pandey, A., Mideo N. and Platt, T. G., 2022. Virulence evolution of pathogens that can grow in reservoir environments. The American Naturalist 199: 141-158. DOI: 10.1086/717177.

Peterson, A. T., 2014. Defining viral species: making taxonomy useful. Virology Journal 11:131. DOI: 10.1186/1743-422X-11-131.

Poletto, C., Meloni, S., Van Metre, A., Colizza, V., Moreno, Y. and Vespignani, A., 2015. Characterising two-pathogen competition in spatially structured environments. Scientific Reports 5: 7895. DOI: 10.1038/srep07895.

Quick, C., Anugu, P., Musani, S., Weiss, S.C., Burchard, E.G., White, M.J., Keys, K.L., Cucca, F., Sidore, C., Boehnke, M. and Fuchsberger, C., 2020. Sequencing and imputation in GWAS: cost-effective strategies to increase power and genomic coverage across diverse populations. Genetic Epidemiology 44: 537-549. DOI: 10.1002/gepi.22326.

Radosevich, S., Holt, J. and Gherse, C., 1997. Weed Ecology: implications for management. John Wiley and Sons, New York, USA, 589 pp.

Ramia, N. E., Mangavel, C., Gaiana, C., Muller-Gueudin, A., Taha, S., revol-Junelles, M. and Borges, F., 2020. Nested structure of intraspecific competition network in *Carnobacterium maltaromaticum*. Scientific Reports 10: 7335. DOI: 10.1038/s41598-020-63844-5.

Schneider, D. S., 2021. Immunology's intolerance of disease tolerance. Nature Reviews Immunology 21: 624-625. DOI: 10.1038/s41577-021-00619-7.

Uffelmann, E., Huang, Q.Q., Munung, N.S., De Vries, J., Okada, Y., Martin, A.R., Martin, H.C., Lappalainen, T. and Posthuma, D., 2021. Genome-wide association studies. Nature Reviews Methods Primers 1: 59. DOI: 10.1038/s43586-021-00056-9.

Velavan, T.P., Pallerla, S.R., Rüter, J., Augustin, Y., Kremsner, P.G., Krishna, S. and Meyer, C.G., 2021. Host genetic factors determining COVID-19 susceptibility and severity. EBioMedicine 72: 103629. DOI: 10.1016/j.ebiom.2021.103629.

Voight, B.F., Kudaravalli, S., Wen, X. and Pritchard, J.K., 2006. A map of recent positive selection in the human genome. PLoS Biology 4: e72. DOI: 10.1371/journal.pbio.0040072.

Wait, L.F., Kamiya, T., Fairlie-Clarke, K.J., Metcalf, C.J.E., Graham, A.L. and Mideo, N., 2021. Differential drivers of intraspecific and interspecific competition during malaria-helminth co-infection. Parasitology 148: 1030-1039. DOI: 10.1017/S003118202100072X.

Wale, N., Sim, D.G. and Read, A.F., 2017. A nutrient mediates intraspecific competition between rodent malaria parasites *in vivo*. Proceedings of the Royal Society B 284: 20171067. DOI: 10.1098/rspb.2017.1067.

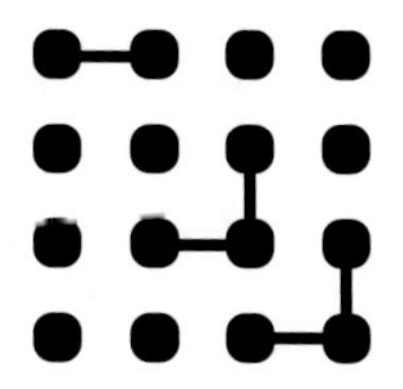

EPIDEMIC SPREAD OF PATHOGENS

Chapter 3 provided insight into the life cycles of various pathogens, of which a larger, or smaller, part of the life cycle is strongly associated with human life. The multiplication and dispersal of a pathogen among people may increase, resulting in epidemic spread of the pathogen. The term 'epidemic' is used as the spread is reflected in what we call 'disease epidemics'. Disease epidemics are determined by the spatial dynamics of pathogens interacting with the spatial dynamics of man and its environment. We will get understanding of epidemics in this chapter by relying on, general theory, modelling, empirical data, and the translation of knowledge of plant pathogens to human ones, like we also did in the preceding chapter. We will highlight the fundamental difference in epidemics that spread as travelling and dispersive waves, respectively.

6.1 Disease epidemics

Pathogens use the human body in various ways to complete the life cycle, as outlined in Chapter 3. Man may also be used as resource for a relatively quick build-up of a pathogen population. Such a build-up may go quite unnoticed. The build-up may also be accompanied by symptoms of disease among the people being used as resource. The build-up of pathogen populations becomes then, at least partially, visible in an increase of the number of people diagnosed with the disease. A disease epidemic, therefore, is an expression of the population dynamics of the pathogen that causes the disease. The expression may be partial, as pathogens may also survive and multiply outside the human body, which may affect the population dynamics considerably. It may also be partial because a pathogen may inhabit a human body without the accompanying disease symptoms. We have seen it for COVID-19 caused by the SARS-CoV-2 virus. People could test positive using a PCR-test, while not being ill at all. In addition, a negative result of such a test does not mean that the SARS-CoV-2 virus is absent from the body. It may just not be detectable. The exponential, and predominantly 'silent', spread of SARS-CoV2 among people stresses the shortcomings of using disease incidence, or prevalence, data solely to understand the population dynamics of a pathogen and the accompanying disease epidemics among man of it. Disease epidemics, however, get our attention commonly.

Fluctuations in abundance of infectious diseases caused by various pathogens have been observed (Martinez, 2018). We may observe it, for example, with respect to, measles and the Measles Morbillivirus, cholera and the bacterium *Vibrio cholerae*, trypanosomiasis and the protist *Trypanosoma brucei*, malaria and the chromist *Plasmodium falciparum*, schistosomiasis and the helminth *Schistosoma mansoni*, and histoplasmosis and the fungus *Histoplasma capsulatum*. The fluctuations may be called seasonality, so far these can be associated with specific seasons. Diseases caused by prions were not listed by the author. We look a bit closer to the seasonality by way of a study directed to five infectious diseases in Ethiopia (Menghistu *et al.*, 2018). The study was executed in the Northern region of Ethiopia, which is inhabited by about 1.1 million people. Medical records indicated about 1,3 million cases of disease in total during the period of 2012-2016. Nearly 54,000 cases could be attributed to the five diseases of interest, (1) rabies caused by the Rabies lyssavirus, (2) tuberculosis caused by the bacterium *Mycobacterium tuberculosis*, (3), visceral leishmaniasis caused by protists of the genus *Leishmania*, (4) schistosomiasis caused by the helminth *Schistosoma mansoni*, and (5) helminthiasis caused by other worms, helminths, than those of the genus *Schistosoma*. Helminthiasis was the dominant one among the five diseases encompassing 95% of the cases. Tuberculosis was ranked as second with a frequency of about 4% of the cases only. Each of the other three diseases encompassed less than 0.5% of the cases. All five diseases were diagnosed in each of the four seasons, except visceral leishmaniasis. Leishmaniasis was not diagnosed in autumn at all. A significant seasonality could be detected for rabies, schistosomiasis and helminthiasis. The abundance of rabies was highest in autumn, whereas it was summer for schistosomiasis and helminthiasis. Abundance of leishmaniasis also was highest in summer, but it did not deviate statistically significant from abundance in winter and spring, respectively. The number of cases was probably too low to detect a statistically significant effect. A significant seasonality could neither be detected for tuberculosis, despite the relatively high number of cases in comparison with rabies and schistosomiasis.

Fluctuations in abundance of infectious diseases, which can be attributed to seasonality, or not, are common. It is less common in human epidemiology to consider these as 'epidemics'. The term 'epidemic' has been associated especially with disasters of disease. This association pops up in the description of it in dictionaries, *e.g.*, an epidemic is a disease spreading quickly among many people in the same place for a time. It triggers questions of, what is 'quickly', what is 'many', what is 'same place'? We may avoid such questions adopting a neutral, scientific, definition that is common in botanical epidemiology (*cf.* Frantzen, 2007):

An epidemic is an increase of disease, limited in time and space, in a community.

We may, subsequently, quantify an epidemic by filling out the terms of, period of time of monitoring, the area of monitoring, and the community that is, of course, linked to the area of monitoring. In addition, we may set thresholds to the increase in order to distinguish endemics from epidemics. Endemics are actually epidemics that show fluctuations that we expect. Such an expectation is, of course, subjective. In addition, the term 'endemic' may suggest that the impact on people is relatively low, which may not be true. We may, for example, refer to malaria, which is commonly characterised as an endemic disease, whereas it has a relatively high impact on societies. We, therefore, do not use the term 'endemic' here and we refer to an 'epidemic of an indigenous pathogen' further. In addition, we abandon subjective definitions of epidemics and we just quantify these. We may, subsequently, determine the impact of the quantified ones on societies, as we will do in Chapter 7. Similarly, the term 'pandemic' triggers the feelings of a disaster, whereas it is simply a disease epidemic occurring on more than one continent.

We may use various approaches to the quantification of disease epidemics. Expressing epidemics by, standardised, numbers of cases per 100,000 inhabitants is a very basic one. It provides relatively little understanding of an epidemic. Actually, we are looking for a quantification providing meaningful insight in an epidemic as well. Such an insight may be generated by way of describing the progress of epidemics in time using various types of models. We may call it the traditional approach to understanding, and quantifying, disease epidemics. We will deal with it in the next section. We will, subsequently, go to a more advanced approach focusing on the progress of epidemics in both, time and space, in section 6.3.

REFLECTIONS

We may go back to Chapter 3 and look at the life cycle of, a prion, a virus, a bacterium, a protist/chromist, a fungus, and a helminth of choice. Which stage of the life cycle of each of these may cause an epidemic from a point of view of humans? May we consider that stage of epidemic spread among humans as pivot in the population dynamics of the pathogen? Or, is it just a facultative stage in the life cycle of the pathogen?

6.2 Progress in time

General description

We observe, in general, a typical sigmoid curve plotting the increase of the proportion of individuals diseased versus time for a specific area and community (Fig. 6.1). The onset is a rather stochastic process. First of all, the spatial dynamics of a pathogen needs to fit the spatial dynamics of man enabling contact between both. Pathogen and man also have to be in an infectious and sensitive state, respectively. This hinges on both, innate factors and environmental conditions. In addition, the strain of the pathogen should be virulent, or aggressive, and the genotype of a subject susceptible. All this is required to have a first case. A first case from which a pathogen may disperse to another subject repeating the process of causing disease. We go from a so-called monocyclic to a polycyclic disease process.

The onset of an epidemic is, in general, invisible. The initial number of cases may be too low to get noticed. In addition, the latent period may be shorter than the incubation period. We may remember from Chapter 4 that the latent period is the time between establishment of a pathogen and the start of its dispersal units leaving the human body. It includes growth and multiplication of the pathogen. The incubation period is the time between establishment of a pathogen and the appearance of clinical symptoms of a disease. A shorter latent than incubation period implies that a pathogen is dispersing, while an infected person is not ill, or not yet. COVID-19 and AIDS are striking examples of diseases, in which such a 'silent' spread of the pathogen occurs because of a shorter latent than incubation period.

A phase of accelerated spread of the pathogen, epidemic spread as we may call it, follows the silent onset of an epidemic, as expressed in a substantial increase of the fraction of people diseased in a relatively short time (Fig. 6.1). The slope reflecting this phase of an epidemic may

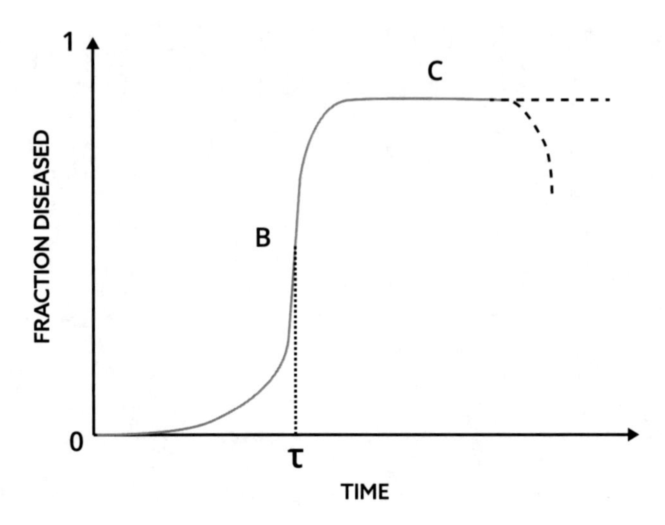

Figure 6.1. Hypothetical curve of an epidemic progressing in time within a specified area. Fraction of people diseased within an area versus time, which may be expressed in days, weeks, months, or even years, depending on the pathogen. The steepness of the curve, which is indicated by parameter *b*, may vary even as the maximum of the fraction of people diseased, which is indicated by parameter C. The mid-disease time, τ, is at the time that half of the maximum of people diseased, C, is reached. The epidemic may, subsequently, stabilise the more, or the less, at the level of C for a while, or it may decrease, as indicated by the dotted lines. See text for further explanation.

be steeper, or flatter. We may subdivide the phase of epidemic spread in an initial one of real exponential growth and a later one of restricted growth, as the number of susceptible people declines in a community and/or the number of recovered, immune, people, increases. The epidemic settles at around a maximum fraction of people diseased at the end of the phase of epidemic spread. The level of the maximum depends on, the abundance of resistance and tolerance in a community, environmental conditions, and medical interventions.

We may abstract the progress of an epidemic, as drawn in figure 6.1, by applying log-logistic regression to data of disease prevalence during a certain period of time using, for example, the equation,

(1) $Y_t = C/1 + e^{-b\ln(t/\tau)}$

in which Y_t is the fraction of people diseased at a time *t*, C the maximum of fraction of people diseased, *b* a slope parameter indicating the steepness of the curve, and τ the mid-time, *i.e.*, the time that the fraction of people diseased is 0.5C. The rate of epidemic progress, *v*, at the mid-time may be calculated as,

(2) $v = b/(4\tau)$.

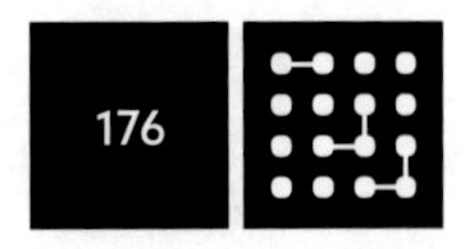

We may notice that such parameters provide more information than the traditional measures in human epidemiology, *i.e.*, the number of cases per 100,000 inhabitants of a community. In addition, a traditional measure may suggest absence of an epidemic at relatively low number of cases per 100,000, *e.g.*, 5, whereas it is full expanding already.

An epidemic may expand after the phase of onset like a fire that is caused by a burning cigarette thrown away in a dry forest. So, noticing and declaring an epidemic as such at numbers of 100 cases per 100,000, or more, implies actually that an epidemic is, in general, beyond a point of control. In addition, we may notice that 100 cases per 100,000 is a fraction of 0.001 of inhabitants of a community. A fraction that does not match the common sense of epidemics as disasters destroying communities. In addition, we may fairly state that the maximum of the fraction of people diseased is, in general, by far not as high as indicated in figure 6.1.

The phase of epidemic spread of an epidemic is followed by a phase that we may call the one of 'decline'. The fraction of people diseased of a community decreases after some time due to natural, or disease induced, mortality, recovery of diseased people while getting immune, remaining people are resistant to the pathogen, or multiplication and dispersal of the pathogen is limited in environmental conditions. A decline of an epidemic does not mean necessarily that the disease vanishes. It may persist in a community at a lower fraction of people diseased.

The log-logistic equation is one way of modelling epidemics until the phase of decline. We may see it as a statistical model. We apply it to data of a disease in an iterative curve fitting process generating estimates of the parameters. We may, subsequently, infer from a relatively low value of b and a high one of τ that an epidemic may expand relatively slowly offering opportunities of timely management. In contrast, a relatively high value of b and low value of τ urges us to monitor a disease very concisely to detect the onset of an epidemic as early as possible, as it may, subsequently, expand very fast. The statistical model turns into a predictive model. The beauty of the model results from the single equation included, whereas it also is the weakness providing relatively little understanding. We go now to so-called SIR-compartmental, models to get more insight.

SIR-compartmental models

We may assign people to the categories of, (S), susceptible for the disease of interest, (I), infectious in the sense that dispersal units in an infectious state leave the body, and, (R), removed from the epidemic by way of becoming immune for the disease, passing away due to any cause of mortality, or getting in complete isolation from the pathogen. We may restrict the removal from the epidemic to people getting immune. If so, the R-category stands for 'recovery' from the disease. The SIR-model turns into a SIS-model, if recovery is not accompanied by immunity and infectious people do not pass away. People get in the category of 'susceptible' again. We may also fit an additional category between those of susceptible and infectious, the one of exposed (E). People in this category of people have catch the pathogen, but they are not 'infectious' yet. The pathogen is latent within the body. The SIR-model turns into a SEIR-model. We focus her on the SIR-model, in which the 'R' stands for 'recovery', providing a basic understanding of the so-called compartmental models and the use of these.

We assume a community of N individuals, who are all susceptible for a specific pathogen (Bjørnstad *et al*., 2020). We have the virtual compartment 'susceptible', which is indicated by S, encompassing N individuals before the onset of the epidemic. An individual carrying the pathogen in an infectious state enters the community. It is the first individual in the virtual compartment 'infectious', which is indicated by I, and the number of individuals in the compartment 'susceptible' changes to S=N-1. We assume that each individual of the community contacts *k* other individuals during a specific unit of time, assuming homogenous mixing of individuals. Each individual in the infectious state will, therefore, contact *k* susceptible individuals during the specified unit of time. The probability of transmission upon contact, π, multiplied by *k* results in a rate of transmission of the pathogen, β. It is assumed that the rate of transmission is, on average, constant. The number of susceptible individuals starts to decrease upon pathogen transmission, while the number of infectious ones increases. Infectious individuals get, subsequently, removed from the epidemic by way of immunisation at a rate γ. They constitute the virtual compartment 'recovered' which is indicated by R. We assume that γ is, on average, constant. The inverse of γ is the average infectious period of an infectious person. We have denoted the infectious period with the symbol *i* in Chapter 4.

The basic assumption of the SIR-model is random, homogeneous, mixing of individuals within a community. The community also needs to be large enough to justify the use of averages of the transmission parameter β and the recovery rate γ. The rates of changes in each of the, virtual, compartments are, subsequently, expressed in three coupled differential equations,

$$(1) \quad \frac{dS}{dt} = -\frac{\beta SI}{N},$$

$$(2) \quad \frac{dI}{dt} = \frac{\beta SI}{N} - \gamma I,$$

$$(3) \quad \frac{dR}{dt} = \gamma I,$$

in which *d* stands for difference in the subsequent parameter and *t* indicates time. These equations are solved numerically, which means that the resulting solutions are approximations rather than, mathematically, exact ones.

An epidemic takes off as soon as the number of infectious individuals increases in time, *i.e.*, dI/dt > 0. It implies that the ratio between the total number of individuals of a community and the number of susceptible individuals is less than the ratio of the transmission parameter and the recovery rate, *i.e.*, $N/S_{(0)} < \beta/\gamma$. If $S_{(0)}$ is approximately equal to N, which we assume in general, the ratio β/γ expresses the basic reproductive number, R_0. It is the number of newly infected individuals related to an originally infected one, as defined in Chapter 4. Here, we need to understand the term 'infected' as 'infectious'. If so, the latent period of the pathogen would be zero, which is, in general, unrealistic, and we would need to turn to a SEIR-model. Anyway, we continue here with the SIR-model for ease of presentation.

The basic reproductive number R_0 is larger than one at the onset of an epidemic to fulfil dI/dt > 0. We may remember from Chapter 4 that R_0 is defined for a completely susceptible community except one, infectious, individual, indeed. The assumption of S approximates N may be regarded as valid at the onset of an epidemic. It becomes certainly questionable when an epidemic progresses in time. We, however, continue to follow the common sense with respect to the SIR-based models (Bjørnstad *et al.*, 2020). The initial increase in the number of infectious people is then exponential and the epidemic growth rate, *r*, of the group of infectious individuals is given by,

$$(4) \quad r = \frac{lnR_0}{T_g},$$

in which T_g is the generation time, the time needed for a pathogen in an infectious stage to leave an individual and becoming infectious again on a next individual. We get a tool to estimate R_0 by re-arranging equation (4),

$$(5) \quad R_0 = e^{rT_g},$$

as we may derive estimates of *r* and T_g by quantifying real epidemics using logistic regression. Alternatively, we may use estimates of β and γ derived from observational data to arrive at an estimate of R_0. Anyway, we cannot estimate R_0 directly, leaving it as a rather theoretical parameter. We mentioned it already in Chapter 4. A parameter that has appeal, nevertheless, expressing a wealth of epidemiological understanding in a single value. We, therefore, work out the SIR-model further with respect to R_0.

We may infer from equation (2) that the number of infectious individuals (I) will decrease at a certain moment, because the availability of susceptible ones decreases, while infectious ones continue to recover. The number of infectious individuals is at maximum at the time that the number of newly infectious individuals equals the number of newly recovered ones. The maximum may be estimated by,

$$(6) \quad I_{max} = N\left(1 - \frac{1 + \log R_0}{R_0}\right),$$

which will be at the time that the number of susceptible individuals is,

$$(7) \quad S = \frac{N}{R_0}.$$

The SIR-model also implies that a certain number of susceptible individuals does not turn into the stage of infectious on the long term. Long term means in mathematics that the time approaches infinity. So, the number of remaining susceptible individuals at infinity may be estimated by,

$$(8) \quad S_\infty = N(e^{-R_0(1-S_\infty)}).$$

We may simulate epidemics now by, for example, fixing the total number of individuals and the infectious period, while varying R_0 (Fig. 6.2). The representation of the results of the simulations requires actually that we turn from numbers to proportions of each of the three types of individuals regarding the total number of individuals. We see the typical logistic increase in the proportion of infectious individuals, say diseased people, in time, which is steeper at a value of R_0 of 3 compared with a value of 2. The proportion of infectious individuals at maximum is twice as high at a value of R_0 of 3 compared with a R_0 of 2, *i.e.*, 30% versus 15%, and the proportion of susceptible individuals not affected at all by the pathogen is about three times lower, *i.e.*, 6% versus 20%. The proportion of infectious people at maximum occurs earlier in time for a value of R_0 of 3 than a value of 2, *i.e.*, day 54 versus day 95.

R_0 is pivotal in the progress of epidemics, as simulated by SIR-compartmental models. We may also use such models to estimate the number of people that needs to be vaccinated to control an epidemic. It is the number of susceptible individuals, at which an epidemic starts to vanish. This density depends on R_0 according to equation (7). We, therefore, adapt equation (7) to arrive at,

$$(9) \quad S_{vac} = N\left(1 - \frac{1}{R_0}\right),$$

in which the parenthetical term indicates the proportion of susceptible people to be vaccinated. This proportion is 0.5 for a value of R_0 of 2, while it is 0.67 for a value of 3. Such calculations assume an efficacy of 100% of the vaccines, which is unrealistic, so far. In addition, the complexity

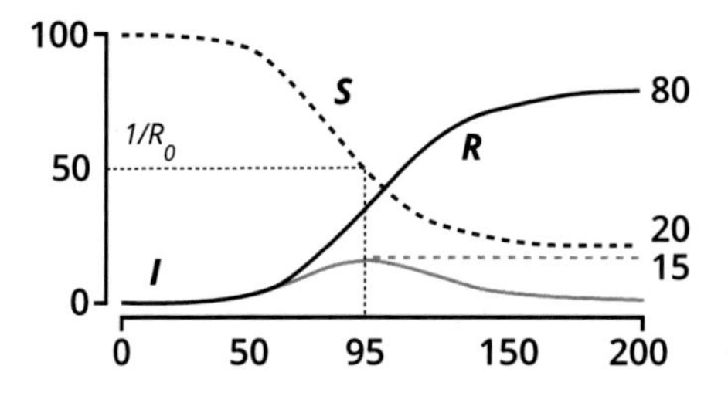

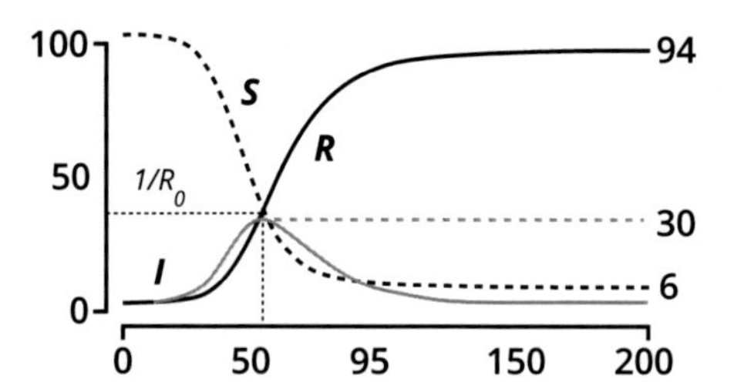

Figure 6.2. Simulations of an epidemic using an infectious period, *i*, of 14 days and a basic reproductive number (R_0) of 2 (left panel), or 3 (right panel). The ordinate indicates the percentage of community members belonging to each of the three types of individuals, susceptible ones (S, dotted black line), infectious ones (I, continuous grey line) and those recovered (R, continuous black line). The maximum of I is indicated by a dotted, grey line. The abscissa indicates the time in days.

of the processes, of which R_0 is an expression, hinders a proper estimation of it (Delamater *et al.*, 2019). The R_0 expresses contagiousness, which results of an interaction between a pathogen and man. The outcome of the interaction is determined by specific characteristics of both, pathogen and man. These show, in general, a high variation in space and time, as outlined In Chapter 5 already. The resulting estimate of R_0 may, therefore, vary considerably as well. The estimates of the R_0 reported for the Measles Morbillivirus / measles, for example, is in the range of 3.7 to 203.3 indeed. Calculating S_{vac} using equation (9) is superfluous for such variable estimates of R_0. We feel replacing R_0 by R_{eff} for such calculations, as suggested by Delamater *et al.* (2019), seems proper from a point of view of theory, as the use of R_0 is restricted to the very early stage of an epidemic. We face then the problem of estimating the R_{eff} properly, which also expresses contagiousness of a highly variable interaction between a pathogen and man, although in a later stage of an epidemic. It remains a parameter characterised by conceptual and statistical challenges (Pellis *et al*, 2022).

The SIR-compartmental models may be adapted to express heterogeneity in a community. The transmission rate β for measles, for example, may be set higher among children than adults. The parameter β may then be replaced by a matrix expressing the rates of the various subgroups in the model. The three basic compartments are then sub-divided to express such a heterogeneity. The assumption of random mixing within a specific (sub-)compartment, however, remains. So, individuals within a specific (sub-) compartment have a similar probability to acquire the target pathogen. Such an assumption is no more needed turning to individual, stochastic, models, like the network ones.

Network models

The study of networks is based historically in sociology and mathematics. Communities may be seen as complicated networks, in which the probability of contacting somebody else varies considerably, depending on all kinds of factors. Human networks become also extended the more and more, due to urbanisation and globalisation. Sociologists use the term 'actor' for an individual and the term 'relation' for the interaction between two individuals, which includes more than just contact. Mathematicians investigate connections, which they call 'edges', between objects, which they call 'vertices', in an abstract way developing so-called graph theory and, as an extension of it, a theory of percolation. We, subsequently, arrive at epidemiological network models by combining mathematics with social knowledge. In addition, we may adapt the terminology with respect to epidemics. We have then 'individuals' characterised by their role in dispersal of the target pathogen and 'contacts' expressing the likelihood that a pathogen disperses from one individual to another, becoming again infectious. The role of each individual in an epidemic depends on the mechanisms of resistance and tolerance, respectively, as described in chapter 5. We may also notice that dispersal of pathogens shows a huge variety, as outlined in Chapter 3. The term 'contact' in modelling, therefore, encompasses a large variety as well. It ranges from an intimate bodily contact between two individuals, as it is for sexually transmitted diseases, to indirect contact by way of the environment, like water contaminated by the bacterium *Vibrio cholerae* that causes cholera.

We adopt the terms 'individual' and 'contact' here and define 'neighbourhood' as the set of contacts of an individual, which has a certain 'degree', say the size of the neighbourhood of an individual (Keeling and Eames, 2005). A contact needs to be understood as an infectious one, *i.e.*, dispersal of the target pathogen from one individual to another, becoming infectious again on the second individual. The probability of such a contact between any of the pairs of individuals may be set on a value in the range of 0-1. Setting it to zero, or one, may be mandatory from a point of view of computer calculation power. If so, we may fill an adjacency matrix, denoted with A, for all pairs of individuals denoting each pair with $A_{ij} = 0$, or $A_{ij} = 1$, for transmission of the pathogen, or not, respectively. The position of an individual within a network may be defined in a spatial sense, or a sociological one. In addition, the contact between two individuals may be reciprocal ($A_{ij} = A_{ji}$), or not ($A_{ij} \neq A_{ji}$). The latter may, for example, occur when a pathogen passes during a blood transfusion. It can, then, go from the donor to the recipient only. We may notice that individuals are categorised using the same terminology as used for the compartmental models, *i.e.*, susceptible, infectious, exposed, and removed/recovered.

We may distinguish various types of networks. A random network assumes that contacts are established at random, like we do in the standard SIR-compartmental model. We may think of a pathogen dispersing at a dance party, in which dancers are mixing quite randomly, or its' dispersing in the environment of a community. We may notice that the epidemic spread of a pathogen is reduced according to the network model, as compared with a standard SIR-model, anyway. A reduction we may observe the more, or the less, for each type of network model, as compared with a SIR-compartmental model. We may indicate two reasons for such a reduction. Firstly, an infectious individual catches the pathogen from one of its potential contacts, reducing the number of contacts to which the pathogen may, subsequently, be passed. We may, for example, assume that an individual has, on average, four contacts of passing a pathogen. If so, the number of potential contacts is reduced to three by receiving the pathogen from one of these. Secondly, passing the pathogen, subsequently, to a contact reduces the number of non-infectious contacts, to which the pathogen may be passed, further, *i.e.*, two in our example. So, we may observe points of depletion of susceptible contacts that inhibits an epidemic in network modelling, in comparison with a SIR-compartmental model.

A network model based on a lattice is quite the opposite of the random type. Individuals are positioned in a regular, two-dimensional, grid and individuals are connected to their neighbours only. The epidemic spread of a pathogen is, therefore, reduced relatively strongly, as compared with the type of random mixing. In addition, the modelling results in a wave-like spread from the point of introduction of the pathogen in the network community. We will return to such spatial waves of epidemic pathogen spread in section 6.3. All in all, the lattice-type of network modelling may be characterised by a relatively high clustering of the pathogen and relatively short paths. A pathogen disperses step by step from the point of origin to the individuals at larger distance. In contrast, the random type is characterised by a relatively low clustering and relatively long paths. The pathogen may jump across a community.

Small-world networks may be generated by adding just a few random connections to a grid-type of network. Such small-world networks remain to be dominated by a high clustering due to short paths, but some longer paths are possible, accelerating epidemic spread. We may state it in other words, small-world networking incorporates some opportunity of long-distance dispersal of a pathogen, albeit a relatively small opportunity. We will explore the consequences of long-distance dispersal for epidemic spread of a pathogen further in section 6.3.

The sizes of the neighbourhoods, the degrees, of individuals show little heterogeneity in the types of networks described so-far. We may add individuals now to such networks who preferentially connect to those having relatively many contacts already, say the popular girls and boys of a community. The degree distribution obeys then a so-called 'power law' distribution (Fig. 6.3). We see some individuals contributing extremely much to the total number of contacts. These are the so-called 'super-spreaders' in epidemiological terms. They may pass relatively quickly the pathogen within a community, as do the few long path contacts in the small-world networks. We call networks that include super-spreaders the scale-free ones.

The dependency of epidemic spread of a pathogen on the community structure is evident from the use of various types of networks in modelling. Determining a real, epidemiologically relevant, community structure is quite a challenge. We may, roughly, distinguish three approaches to determine it, (i) infection tracing, (ii) contact tracing, and (iii) diary-based tracing. Infection tracing by an expert is directed to all individuals of a community, of whom is known that they carry the target pathogen. These individuals are linked together in a network by way of estimating who is source of whom. It reveals an actual picture of an epidemic. The strength of this approach. It also is the weakness of it by way of ignorance of potential contacts that could have been resulted in dispersal of the pathogen. It eliminates stochasticity, which may be quite relevant, especially at the onset of an epidemic, reducing the general validity. In contrast, contact tracing by an expert is directed to all contacts of a source individual, irrespectively she, or he, caught the pathogen, or not. If not, we classify these as 'subjects at risk'. The inclusion of persons at risk enables some forecasting regarding the progress of an epidemic. The third approach, the one of diary-based tracing, may result in an over-estimation of contacts at risk. Subjects are requested to keep diary of all their contacts. The diary may, therefore, include contacts that are not relevant from a point of view of pathogen spread.

We need to validate network models to go beyond theoretical considerations and to use these in forecasting epidemics. It is a challenge. A model needs to be parametrised using data of a real community. It also needs to be validated using a real community that resembles the one offering the parameter estimates. We run in redundancy using a community for both, parametrisation and validation. In contrast, we may question the similarity in structure using two, independent, communities. A common practice of model validation is a bit in between the use of a single community and two independent communities. It is running a model using a range of values of the parameters and, subsequently, selecting the best fit to data of an epidemic observed (*e.g.*, Croccolo and Roman, 2020). But, does the model then have predictive value?

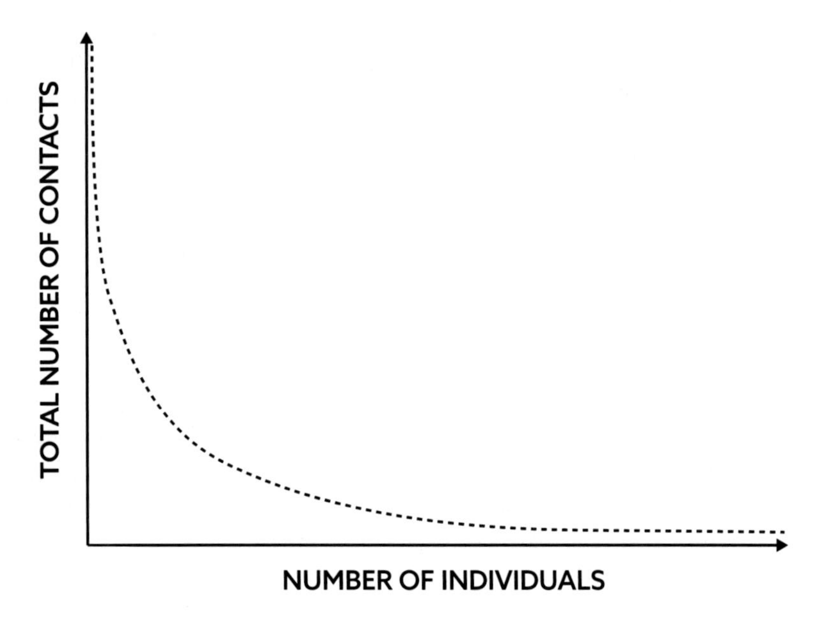

Figure 6.3. A power law distribution of individuals contributing to the total number of contacts in a network, in which relatively few individuals have relatively many contacts contributing strongly to the total number of contacts.

Besides a proper validation, a predictive use of network models may be limited in computational tractability (Rosenkrantz *et al.*, 2022). It does not preclude the use of these, but appropriate solutions need to be adopted and adapted to master the computational challenges. One such a solution is the introduction of a higher-level aggregation in a model. If so, it implies a return to compartmentalisation to a certain degree.

REFLECTIONS

Compartmental and network models using the SI(E)R categorisation have provided insight in the basics of epidemic spread of pathogens. The focus was, and is, on the parametrisation of the models rather than validation. We may reflect about the use of each of the type of models in forecasting epidemics and, subsequent, management by way of vaccination.

6.3 Spatial dynamics

Clustering of cases

The basic reproductive number R_0 may suggest a homogeneous distribution of a pathogen, and the subsequent disease among people, during an epidemic. A basic assumption of the SI(E)R-compartmental models. The R_0, however, is a number without any dimension. The distribution of the offspring of a reproducing pathogen may, therefore, take any distribution in time and space. It may then result in a clustering of cases of disease, or infected persons, as indicated by the output of network models incorporating a certain level of heterogeneity of contacts between individuals. In addition, clusters may be unstable due to removal of individuals from the epidemic. Such a dynamics hampers the detection of clustering in real life, but it is not impossible. We may, for example, use a space-time scan statistic, as outlined in the following.

Data of COVID-19 cases and their geographical location was retrieved from John Hopkins University's Center for Systems Science and Engineering GIS dashboard (Desjardins, *et al.* 2020). Data of the period between the 22nd of January and the 27th of March of 2020 was used in the analysis of clustering in the United States, so far as this could be allocated to one of the counties. The number of new cases was calculated by subtracting the number of cases of the previous day from the number of a current day. The prospective version of a Poisson space-time scan statistic, as implemented in SaTScan™, was used. It employs so-called 'cylinders' moving over the data. The base of a cylinder is the spatial window and the height of it the temporal one. The centroid of a county was the centre of the cylinder in the US-study. A cylinder expands until it arrives at the maxima set, which were scanning of 10% of the population at risk during 32 days in the present study. This period, therefore, included about 2-3 generations of the SARS-CoV-2 virus. These maxima were set to avoid very large clusters. In addition, the number of clusters was limited in the minima set, *i.e.*, 5 cases at least and two days of scanning. The expected number of cases within an area was calculated as the local population multiplied by the US-prevalence of COVID-19 at a specific day. The assumption was that local populations were stable during the period of the study. A maximum likelihood ratio was used to identify windows with an elevated risk of COVID-19,

$$(1) \quad \frac{L(Z)}{L_0} = \frac{\left(\frac{n_Z}{\mu(Z)}\right)^{n_Z}\left(\frac{N-n_Z}{N-\mu_{(Z)}}\right)^{N-n_Z}}{\left(\frac{N}{\mu(T)}\right)^{N}},$$

in which L(Z) is the likelihood function for cylinder Z, L_0 the likelihood function of the zero hypothesis that the number of cases follows a Poisson distribution, n_Z the number of cases in the cylinder, N the total number of cases of the United States during the period of the study and μ(T) the total number of cases expected in an area during the whole period of investigation. A specific cylinder has an elevated risk, if the ratio is larger than one. A relative risk was also calculated for each county as,

$$(2) \quad RR = \frac{c/e}{(C-c)/(C-e)},$$

in which c is the total number of cases in a county, e the expected number, and C is the total number of cases in the United States.

Twenty-six emerging COVID-19 clusters could be detected in the period from the 22nd of January to the 27th of March (Fig. 6.4). The largest cluster, C8, was detected in the North-Eastern part of the United States. It was assumed to be the epicentre of the epidemic. Smaller, intense, clusters were detected elsewhere. These foci of the epidemic may be targets for specific interventions rather than all counties with an elevated relative risk.

The spatial scale of the study of Desjardins, *et al.* (2020) was relatively large by encompassing the United States of America and by using data at the county level. In contrast, clusters of COVID-19 were also detected in the relatively small Swiss Canton of Vaud (Ladoy *et al.* 2021). The area of Vaud is about 3,200 km^2. The sizes of counties in the USA vary between 30 and 51,000 km^2. The area of Vaud may, therefore, be classified as relatively small indeed. Ladoy *et al*, (2021) used the same software, SaTScan™, as Desjardins *et al.* (2020), but they set different conditions. The maximum of the spatial window was set at 0.5% of the population at risk and that of the time window at 14 days. The minima were set at three cases at least and two days of scanning.

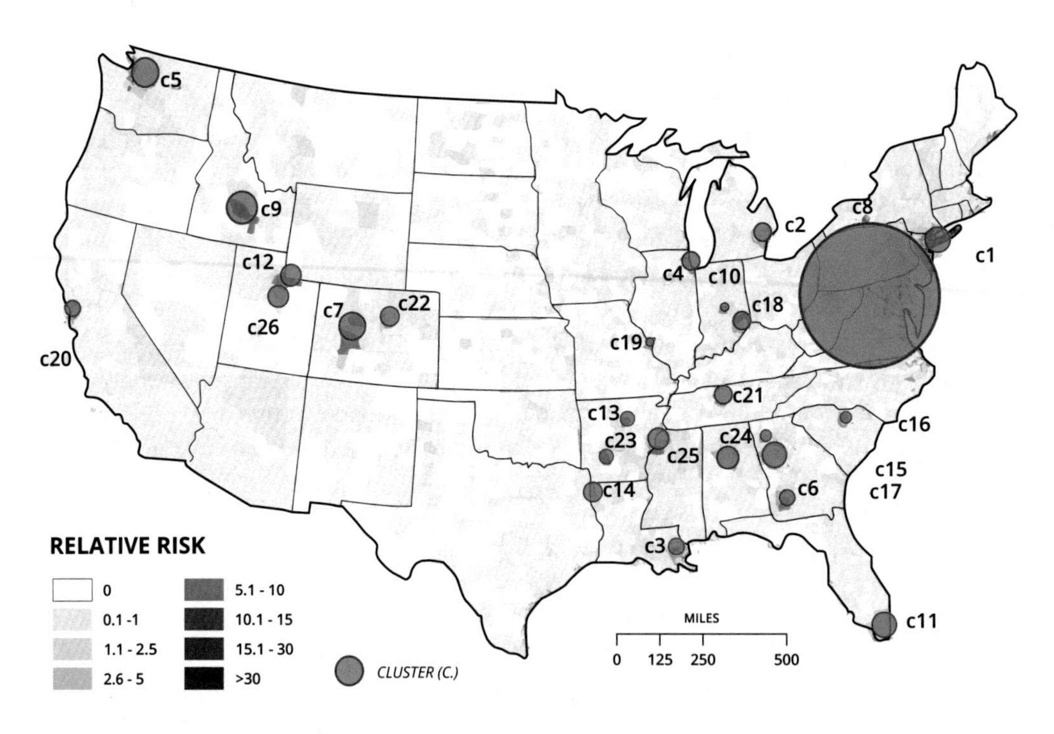

Figure 6.4. Dynamic clusters of COVID-19 cases detected in the United States at the county level in the period from 22 January until 27 March 2020. The size of a circle indicates the number of cases of a cluster. Numbers indicate the chronological order of appearance of the clusters. The relative risk per county is indicated by grey colours. See text for further explanation.

These specific conditions enabled the detection of, especially, small and homogenous clusters. The significance of the clusters detected, in a statical sense, was determined by using 999 Monte-Carlo permutations. The available data are re-ordered ad random in each permutation resulting in 999 orderings. These 999 ordering are compared with the one obtained by the space-time scan statistic. The level of significance could be down to a P of 0.001 using the 999 permutations. In addition, the diffusion dynamics of clusters was characterised using a Modified Space-Time Density-Based Spatial Clustering of Application with Noise (MST-DBSCAN). It actually monitors the fate of a cluster, taking into account the incubation period of the target disease. The centre of a cluster may move, or not, and the size of it may increase, stay stable, or decrease. The MST-DBSCAN does not provide statistical testing of the fates observed.

We focus here on 17 clusters of COVID-19 cases in Vaud that were selected for a subsequent study including genomics (Choi *et al.*, 2022). The first and second cluster appeared in the area of the City of Lausanne, whereas the third one appeared remotely western of the lake

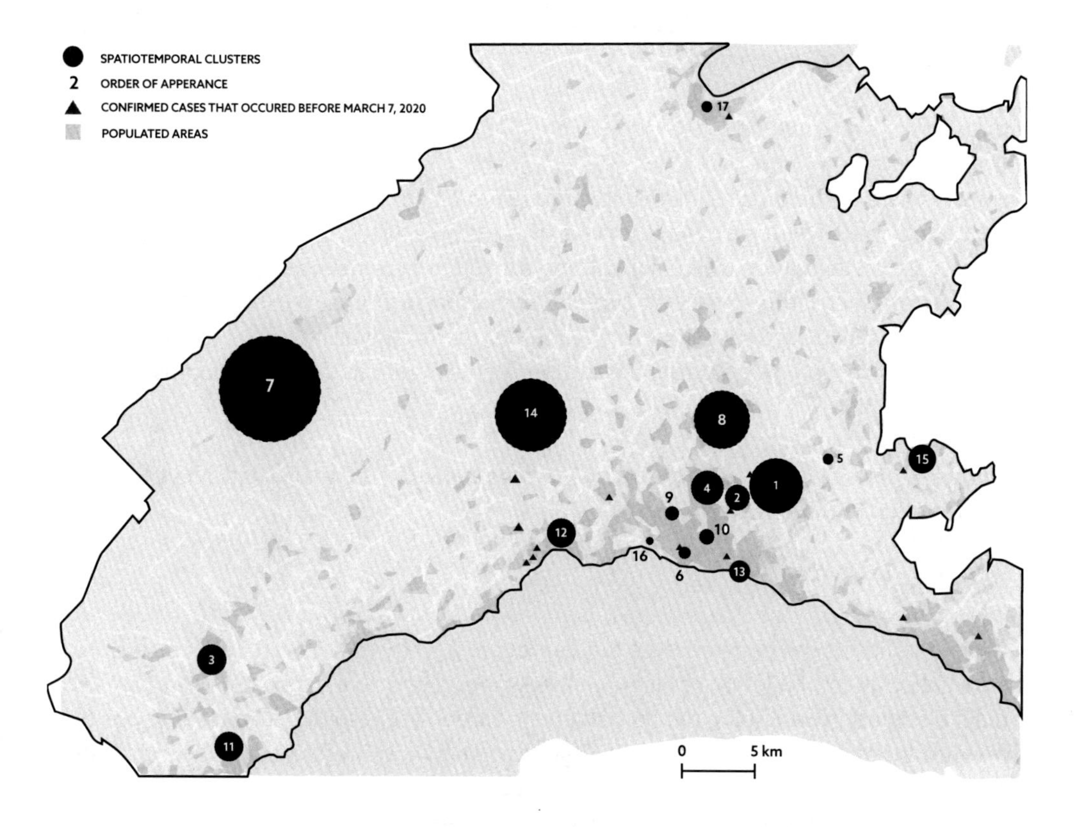

Figure 6.5. Dynamical clusters of COVID-19 cases detected in the canton of Vaud, Switzerland, in the period 2 March until 30 June 2020. The size of a circle indicates the number of cases of a cluster, whereas the number indicates the chronological order of appearance. See text for further explanation.

of Geneva (Fig. 6.5). We also see a relatively large cluster, number 7, emerging outside the Lausanne area relatively early. The clusters showed a fate of increase followed by a stabilisation, or a subsequent decrease. The clusters could not be linked straightforward to cases identified before the 7th of March, on which date the first cluster was detected. The results of genomics indicated genetic homogeneity of SARS-CoV-2 virus among cases in clusters of rural areas and less so in those of urban ones. The results also suggested two super-spreading events that started in the area of the City of Lausanne.

The results of both, the American and Swiss study suggested a link between the COVID-19 clusters by way of dispersal of the SARS-CoV-2 virus from a single focus over relatively long distances. We will, therefore, dive in the theory of spatial dynamics of organisms, and more specifically the one based on reaction-diffusion processes of dispersal, to explore this explanation. We will, subsequently, look at the validation of the theory, using data on dispersal of pathogens of plants, as we may arrange plants in any spatial design. In addition, we may use pathogens and plants with a rather unform genetic constitution, eliminating confounding factors as much as possible. Finally, we will return to the patterns of COVID-19 observed in the USA and Switzerland, respectively.

Waves of epidemic spread

We present here the theory of epidemic spread as a wave by assuming one-dimensional spread of a pathogen based on its generations rather than using a distribution of continuous time. It simplifies the mathematics, while highlighting the differences between two types of spatial spread of pathogens, the so-called travelling and dispersive waves, respectively. Travelling waves are also called 'solitons' in physics. We refer further to 'travelling waves', as it is common in botanical epidemiology. We use here the review of the theory, and its' validation, as provided in Frantzen (2007).

The number of lesions caused on plants by a subsequent generation of a pathogen, N_{t+1}, and the distribution of these is given by,

$$(1) \quad N_{t+1}=R_0 \int_{-\infty}^{+\infty} N_t(\xi)D(x-\xi)d\xi,$$

in which *t* indicates the generation of the pathogen, R_0 is the basic reproductive number, ξ the spatial position of a lesion resulting in a new lesion at a spatial position *x* and $D(x-\xi)$ a contact distribution describing the probability of a lesion at ξ resulting in a lesion at position *x*. We introduced contact distributions in Chapter 4 and we addressed the statistical hurdles to determine a proper contact distribution of a pathogen. We, therefore, explore here three different types of contact distributions and the consequences of these for the epidemic spread of a pathogen: (1) the double exponential, (2) the root, and (3) the modified power law.

The double-exponential contact distribution reads as,

$$(2)\quad D(x-\xi)=\frac{1}{\sigma\sqrt{2}}\,e^{\left(-\frac{\sqrt{2}}{\sigma}|x|\right)},$$

in which σ is the specific parameter to be estimated. We introduce this contact distribution in equation (1) and we arrive at,

$$(3)\quad N_t(x)=N_0R_0^t\,\frac{1}{\varnothing}\,\frac{e^{-\left(\frac{|x|}{\varnothing}\right)}}{2^{2t-1}\,(t-1)!}\sum_{j=0}^{t-1}\frac{2^j\,(2t-j-2)!}{j!(t-j-1)!}\left|\frac{x}{\varnothing}\right|^j,$$

in which N_0 is the number of lesions on the inoculum plant(s) and ø equals $\sigma/\sqrt{2}$. We determine the velocity of the epidemic spread of a specific number of *n* lesions over the generations by solving numerically $N_t(x) = n$. The front of the epidemic spread may be visualised by plotting the number of lesions, N, versus the distance from the inoculum source, x, for several generations.

The root contact distribution reads as,

$$(4)\quad D(x-\xi)=\tfrac{1}{4}g^2e^{(-g\sqrt{|x|})},$$

in which *g* is the specific parameter to be estimated. The number of lesions in generation *t* at a position *x* is then defined by,

$$(5)\quad N_t(x)\approx N_0R_0^t\tfrac{1}{4}g^2e^{(-g\sqrt{|x|})},$$

and we may repeat the procedures to depict the velocity of epidemic spread and visualising the front of the epidemic, as outlined with respect to the double exponential contact distribution.

The modified power law contact distribution reads as,

$$(6)\quad D(x-\xi)=\frac{h+1}{2l(1+\frac{|x|}{l})^h}$$

in which *h* is the specific parameter to be estimated and *l* is a length scale to have a similar estimate of *h* independently of the units of *x*. The number of lesions in generation *t* at a position *x* is then defined by,

$$(7)\quad N_t\,(x)\approx N_0R_0^t(h+1)2(1+|x|)^{-h}$$

and we may repeat the procedures to depict the velocity of epidemic spread and visualising the front of the epidemic, as outlined with respect to the double exponential contact distribution.

The three types of contact distributions were fitted to data of the rust fungus *Puccinia lagenophorae* infecting the annual plant *Senecio vulgaris*, as determined in a relatively small-scale experiment. The goodness-of-fit was similar among the three distributions, as expressed

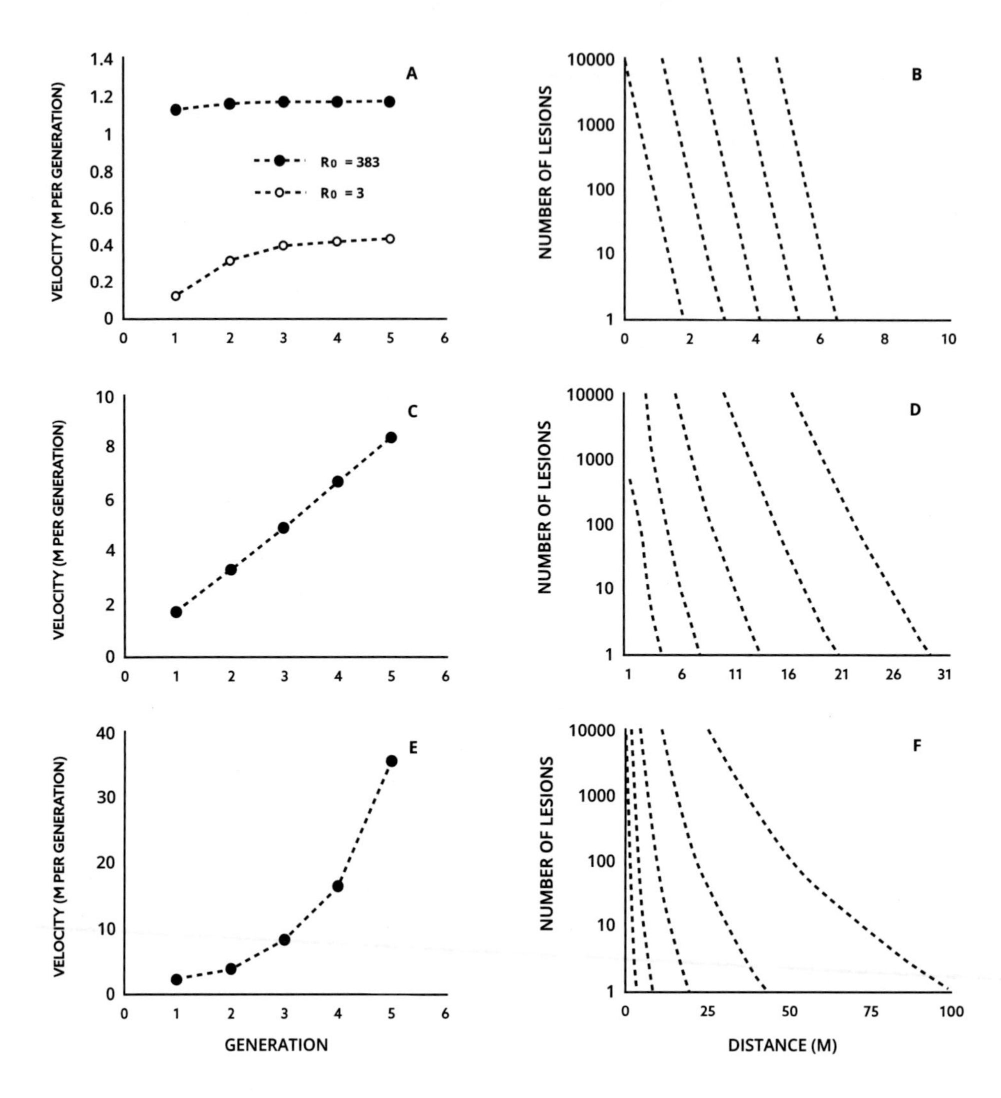

Figure 6.6. Velocity of epidemic spread of the rust fungus *Puccinia lagenophorae* infecting the plant *Senecio vulgaris* (A, C, E) over generations and the shape of the front of it (B, D, E), as modelled by assuming, the double exponential contact distribution (A, B), the root (C, D), and the modified power law contact distribution (E, F), respectively. See text for further explanation.

in values of R^2 between 0.9 and 0.94. The estimates were, 0.28m for σ and a standard deviation of 0.02m, 7.07m for *g* and a standard deviation of 0.34m, and 3.64 for *h* and a standard deviation of 0.17. The standard deviations thus indicated a quite well fit of each of the distributions, as well. The R_0 was determined in the same experiment, being 383. The estimates were used to explore the epidemic fate based on each type of contact distribution by using the equations (3), (5) and (7), respectively. Dispersal according to a double-exponential contact distribution resulted in epidemic spread of the pathogen at a constant velocity, after an initial phase

(Fig. 6.6 A). The duration of the initial phase is determined by R_0, *i.e.*, the higher the value of it, the shorter the phase. The velocity of spread calculated by using a fictive value of R_0 of 3 is indicated in the figure, as compared with the one of 383, to illustrate this. The front of the epidemic spread shows a stable pattern (Fig. 6.6 B). The epidemic spread is like a travelling wave. The waves we may see after throwing a pebble in a pond. In contrast, the velocity of epidemic spread increases for dispersal according to a root (Fig. 6.6 C) and modified power law (Fig. 6.6 E) contact distribution, respectively. The front of the epidemic spread flattens over the generations for epidemic spread, according to these contact distributions (Fig. 6.6 D, F). The epidemic spread is like a dispersive wave. We may notice especially the exponentially increasing velocity of the epidemic spread in the case of dispersal described by a modified power law contact distribution.

The theory predicting epidemic spread of pathogens as travelling and dispersive waves, respectively, was validated using various plant pathogens. Travelling waves could be, for example, determined for the spread of yellow rust, *Puccinia striiformis*, on common wheat, and downy mildew, *Peronospora farinosa*, on spinach. The velocity of spread observed in field plots corresponded fairly well with those calculated by assuming an exponentially bound contact distribution in the model used. In contrast to travelling waves, determination of dispersive waves is a challenge due to the relatively extensive and fast spread. One-dimensional spread of *P. lagenohorae* among *S. vulgaris* plants was determined in field plots at three sites. The spread was monitored up to 6 weeks after introduction of inoculum. The period of observation, therefore, encompassed 2-3 generations of the pathogen. The wave front showed a clear flattening in time at all three sites. The wave front resembled most the one that resulted from a model that included dispersal of the pathogen according to the modified power law distribution. The velocity of spread could not be calculated properly, as the pathogen passed the borders of the experimental plots too quickly. The scaling of the study in both, time and space, was too small. The US plant pathologist Christopher Mundt, and his team, designed a sequence of studies at a relatively large scale, using a different approach to investigate epidemics progressing like dispersive waves. He also provided clues to manage these. We will elaborate on the work of Mundt, as reported in 2019 (Severns *et al.*, 2019). The research was directed to the rust fungus *Puccinia striiformis* forma specialis *tritici*, dispersing on club wheat, *Triticum aestivum* sub-species *compactum*. Races and cultivars were used that could not interfere with cultivars grown commercially. Monocultures and mixtures of susceptible and resistant wheat were planted in rectangular plots. The plots were 7.6 m broad and these had a length between 73 and 171m. Plots were elongated according to the prevailing wind direction to maximise dispersal of the pathogen by wind. The design enabled monitoring of the spread of five to six generations of the pathogen. A few plants were inoculated in each of the plots to establish a focus of disease. A focus was defined as the area encompassing those plants with lesions, disease, that resulted from the spread of the pathogen from the inoculated plants, *i.e.*, the first generation of the pathogen. The size of the focal area, and the proportion of susceptible plants within it, varied in order to determine potential effects of the strength of a focus on epidemic spread. The spread of disease was monitored by weekly assessments of the disease prevalence along the central axis of each plot, both upwind and downwind. The velocity of spread was calculated by determining for a certain level of disease the distance from the focus over time. Increase of velocity, assuming spread of the pathogen as a dispersive wave, was described by,

$$(1) \quad v_x = \frac{x}{b},$$

in which v is the velocity of epidemic spread at distance x of the focus assuming that velocity is expressed in distance per unit of time t/t_0 and t_0 is the time of measuring the velocity of initial spread of the pathogen outside the focus and b the exponent of a power law distribution of dispersal. The equation implies that dispersal of the pathogen described by the same value of b, but a different rate of reproduction, will attain the same velocity of spread at a specific distance at a different time. A difference in susceptibility of the plant population may, for example, result in a different rate of reproduction of the pathogen. An exponential increase of the position of the front of the epidemic was predicted as well. If so, it should be described by,

$$(2) \quad x_t = x_0 \left(\frac{b}{b-1}\right)^{\frac{t}{t_0}},$$

The analysis of data was accompanied by simulations using an advanced version of EPIMUL, a compartmental model that is commonly used to simulate epidemics of plant diseases. A modified inverse power function was used in EPIMUL to relate the number of lesions, y, to the steepness of the curve describing the, effective, dispersal, of the pathogen, also indicated by b,

$$(3) \quad y = a(x)^{-b},$$

in which the constant a indicates the amount of inoculum introduced. We may notice the similarity with equation (1).

The typical flattening of the wave front of *P. striiformis* spread was determined consistently over time at various sites under various environmental conditions. The velocity of spread increased linearly over distance from the foci. It was independent of the proportion of susceptible plants in the plots, applying equation (1) to the data. The authors stated: "This increase in the velocity of disease spread over time and space strongly suggests that epidemics (of *P. striiformis*) move across the landscape as dispersive waves". And they explored the consequences of such spread at larger scale, including data of various pathogens. The West Nile virus, a human pathogen, and the Avian influenza virus were included as well. The maximum of distance of spread reported for these pathogens was in the range of 0.3 km up to 9,300 km. Fitting equation (3) to this data resulted in values of b of about 2, irrespectively the type of pathogen. We arrive then at an inverse square law type of dispersal, which is based on the physical process of diffusion, diluting infectious dispersal units over distance. It also implies a scale-invariance. The findings of epidemic spread at a relatively small-scale may be used at large scales as well, like that of spread at the continental level. We may then deduce, for example, that a dispersive wave of spread of *P. striiformis* from a single lesion during a single generation may result in one lesion within an area of 18 km^2 at 10,000 km of the lesion of origin later on. This example indicates both, the potential of long-distance dispersal and the unpredictability of spread of a pathogen as a dispersive wave. Just find one lesion somewhere within 18 km^2. In addition, the size of an epidemic, as expressed in disease prevalence, depends on the strength of a focus created by the spread of the first generation of a pathogen. So, we need to inhibit the build-up of strong foci to manage dispersive waves of pathogen spread on time. We will return to this topic in Chapter 8.

We may simulate the spread of an epidemic as a dispersive and travelling wave, respectively, in a two-dimensional plane (Fig. 6.7). We see then that new foci emerge during the spread as a dispersive wave (left), whereas we see the more, or the less, a coherent spread from the primary focus in the travelling wave (right), during the same period of time. We may also state this in terms of probability. The spread of a travelling wave is rather predictable, *i.e.*, it is an expansion of the focus, whereas it is not for a dispersive wave. In other words, epidemics spreading as dispersive waves are hardly to control, except in the very early stage of focus formation. We may repeat that the difference between both types of spread is expressed in the tails of the underlying contact distributions. If it is a 'fat-tailed' one, we are dealing with dispersive waves and, if exponentially bound, with travelling waves. It is just a, relatively slight, difference in the proportion of effective dispersal units dispersing at a relatively long distance. So, it is not just long-distance dispersal only. The infection efficiency of the dispersal units is the second, major, determinant of the type of spread of an epidemic. If the infection efficiency is relatively high, new foci may quickly emerge, as seen in the left panel of figure 6.7. If it is relatively low, no new foci emerge in short, as seen in the right panel.

We return now to the spatial patterns of COVID-19 observed in the United States of America (Fig. 6.4) and Switzerland (Fig. 6.5), respectively. We may conclude that the COVID-19 epidemics spread as dispersive waves rather than travelling waves, comparing the patterns observed over

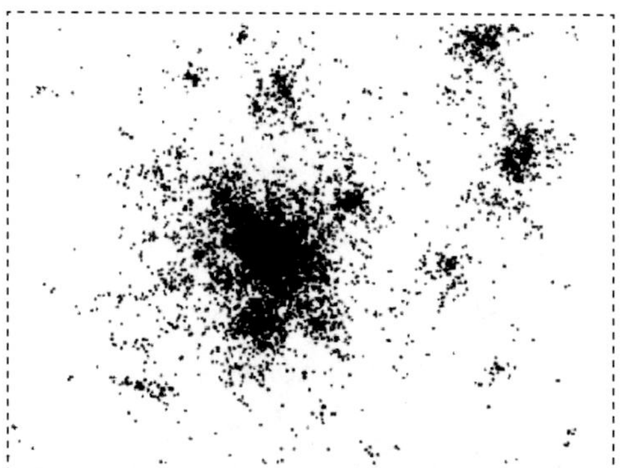
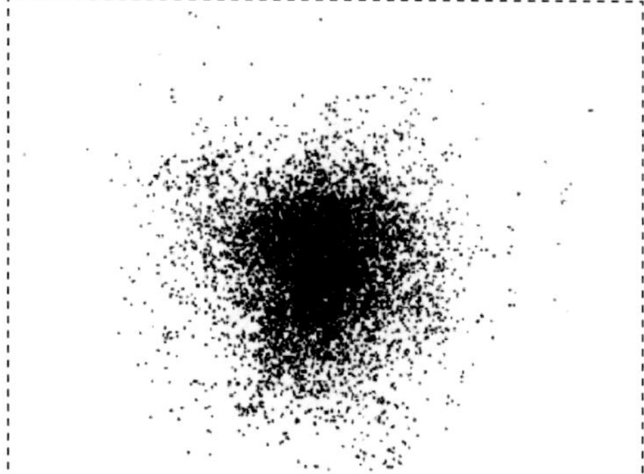

Figure 6.7. Simulation of an epidemic that spreads like a dispersive wave, left, and a travelling wave, right, from a focus at the centre. See text for further explanation.

there with those simulated here (Fig. 6.7). New foci, clusters, emerged relatively quickly, indeed, at quite a distance from the primary focus, which was assumed in the USA and Switzerland, respectively. But, may we understand this from a bio-physical point of view rather than a mathematical one? We look first at the dispersal of the SARS-CoV-2 virus by way of air flows, and more specifically, the exhalation of it by infected people.

Turbulent airflows in the respiratory tract may disrupt the mucus layer resulting in liquid droplets (Jarvis, 2020). The size of these ranges between less than 0.1 micro-meter up to 1 mm, or more. We call those droplets 'aerosols' that have a diameter up to 5-10 micro-meter and those above this threshold simply 'droplets'. We refer to these here as 'large droplets' to avoid any confusion. The threshold is not that sharp. It has been chosen as the large droplets obey to the force of gravity and these, therefore, settle relatively quickly, at least in still air. In contrast, aerosols may remain arial for a prolonged time. Aerosols originate especially from the lower respiratory tract and the larynx, whereas droplets originate predominantly from the oral and nasal cavities. We produce predominantly aerosols during breathing, and rather few large droplets (Fig. 6.8). In contrast, speech and cough generate abundantly both, aerosols and large droplets. Sizes of around two micro-metres and 100 micro-metres for aerosols and larger droplets, respectively, show the highest frequencies. We may, however, notice that the frequency distribution may vary largely among individuals.

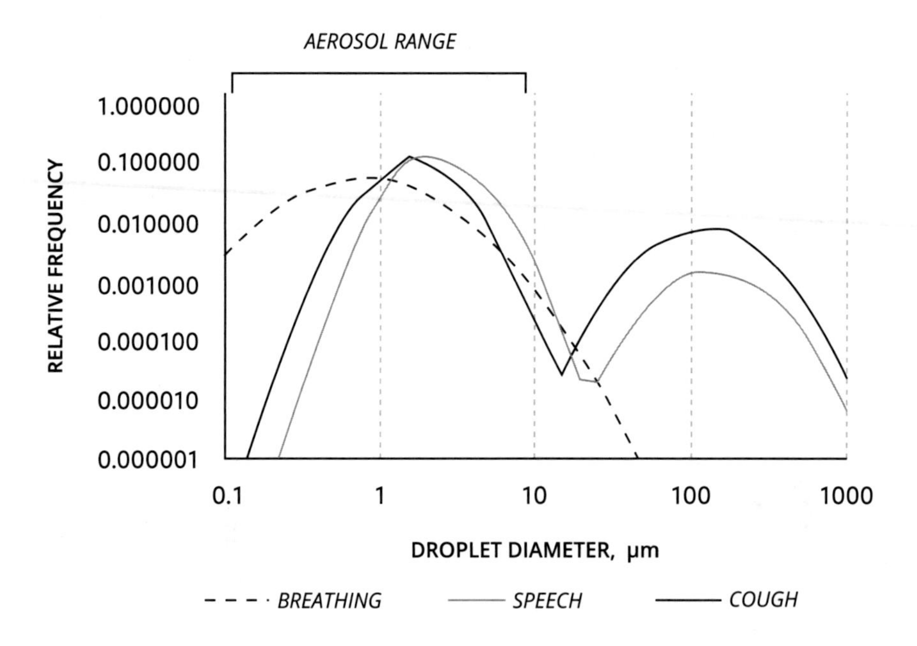

Figure 6.8. Frequency distribution of droplets generated in the respiratory tract of subjects without symptoms of respiratory diseases during, breathing (dotted black line), speech (grey line) and cough (black continuous line). Droplets smaller than 5-10 micro-meter are called aerosols. See text for further explanation.

The exhaled aerosols and large droplets become exposed to environmental factors affecting the size. Temperature and air humidity, for example, affect the velocity of dehydration. Large droplets may collide, producing smaller droplets and aerosols, or these may just fuse to larger droplets. Droplets may also adhere to all kinds of particles, like the ones of dust. Such interactions affect the fate of the droplets in the air. We, however, assume, in general, that aerosols may remain aerial for hours, whereas large droplets for seconds only.

The aerosols and large droplets enclose virus particles. The particles may have been inhaled by a subject and these may, subsequently, stick to the mucus, or the virus burst out of infected cells. Volumes and SARS-CoV-2 loads of both, aerosols and large droplets, show a large variation among cases, *i.e.*, a factor of 100,000, and more. The viability of the virus after exhalation depends on various environmental conditions affecting the aerosols and large droplets, or the virus particles directly. The half-life of the virus is, for example reduced towards 2-3 minutes by intense sunlight, which is likely due to a detrimental effect of ultra-violet radiation. The preceding indicates that large droplets constitute a risk of transmission between an infected and non-infected subject at a rather short-distance between these. The rule of thumb is a distance shorter than 1-2 metres. The risk is reduced by wearing a face mask, which catches the large droplets quite well. In contrast, it catches, in general, not properly aerosols. In addition, aerosols may disperse over a larger distance, especially out-of-doors. Dispersal of SARS-CoV-2 in aerosols over tens of kilometres, or more, is likely. Unfortunately, data of such long-distance dispersal is lacking, so far known.

Long-distance dispersal of SARS-CoV-2, or any other pathogen, may also result from long-distance travelling of man. Such a travelling is likely during an epidemic, if people are infectious without showing disease symptoms. Infectious SARS-CoV-2 particles may shed from people without symptoms of COVID-19, indeed (Meyerowitz *et al.*, 2021). Some people remain symptomless during the whole infectious period, others show symptoms zero to five days after onset of the infectious period. The incubation period of COVID-19 is thus clearly longer than the latent period of the virus. The peak of liberation of SARS-CoV-2 particles is, on average, one day before the onset of symptoms. In addition, the infectivity of virus particles at liberation wanes with time after this peak. All in all, the infectious period of SARS-CoV-2 is about 10 days and the number of infectious particles mounts steeply within the first two to three days of the infectious period and it decreases in the subsequent seven to eight days. Thereafter, non-infectious virus particles are liberated from infected people. All in all, COVID-19 patients are stronger sources of infectious virus particles before the onset of symptoms than after it.

We may conclude that a relatively high frequency of long-distance dispersal of SARS-CoV-2 is likely, either by air streams, or by travelling of COVID-19-symptomless people. The first pre-requisite of an epidemic that spreads as a dispersive wave. The dispersal should, subsequently, result in a relatively quick establishment of foci at the distant sites. The likelihood of establishing a focus depends on the specific site of settling. It is, for example, more likely for a cloud of virus landing at a home of elderly people than entering a fitness centre with young people. So, the frequency of resistance and tolerance at a site of settling determines whether foci emerge relatively quickly, or not. We may, however, state that SARS-CoV-2 has a rather high infectiousness overall. Epidemics of COVID-19, and the pandemic in total, thus spread as dispersive waves, combining a relatively high frequency of long-distance travelling of symptomless people and a relatively high infectivity of SARS-CoV-2 overall. Interestingly, a massive global spread of COVID-19 was forecasted in February 2020 already (Wu *et al.*, 2020). Data of confirmed COVID-19 cases of Wuhan and mainland China in the period December 2019 – January 2020 were used together with data of travelling from and towards Wuhan in mainland China. This data was used in a SEIR-meta-population model forecasting an exponential spread of COVID-19 in China. We may read this as spread like a dispersive wave, although the authors did not use this term. An outlook of them on world-wide spread indicated international travelling of, especially, symptom-less cases as a key risk factor for spread outside China. We could thus expect severe outbreaks in Europe at the start of 2020 already. We could, however, not predict that it would, for example, happen in the City of Milan (Italy). Unpredictability is the hallmark of epidemics spreading as dispersive waves. In contrast, epidemics spreading as travelling waves are relatively predictable.

We turn now to another virus, Ebola, of which epidemic spread may be characterised by travelling waves. Four species of the virus may cause Ebola among man. We presented the major one, the Zaire Ebolavirus, in Chapter 3 already. We see a clear focal spread of the disease in Africa in the past decades, indicating spread as travelling waves (Fig. 6.9). The typical spread is independent of species of the virus. Why do we see this epidemic spread as travelling waves?

The Ebola virus may jump from animals, like bats, to man (Sivanandy *et al.*, 2022). We call such a jump a spill-over event. The subsequent spread among people results from direct contact of mucous membranes of an uninfected subject with bodily fluids from an infected one. Blood, vomitus and faeces are considered as most infectious. The virus spreads systemically in the body and the accompanying pathogenesis may be characterised by four stages. The first stage manifests in quite general disease symptoms, like headache and fever. Ebola is then hardly to distinguish from other diseases without specific diagnostic tests. The symptoms of this stage are called 'dry' to distinguish these from the second stage. In that stage, extreme fatigue and gastroenteritis also appear. The gastroenteritis is characterised by the 'wet' symptoms of diarrhoea and vomiting. The third phase is characterised by remission of the disease, for some the start of a real recovery, for others a pseudo-remission resulting in the fourth phase, the lethal one due to organ failure. The distinction between the 'dry' and 'wet' symptoms is crucial in determining the latent and infectious period of the virus, respectively, and to relate these to the incubation period of the disease.

The latent period of an Ebola virus may be defined as the time between a detrimental exposure to the virus, say getting infected, and the onset of 'wet' symptoms (Velásques *et al.*, 2015). The onset of the 'wet' symptoms also is the start of the infectious period of the virus. This period lasts until a patient recovers, or passes away. We may even add a post-mortem period of infectiousness. The incubation period is defined as the time between a detrimental exposure to the virus and appearance of any symptom of disease, irrespectively a dry, or wet one. The average incubation period of Ebola disease reported was about 6.2 days, plus/minus 1.6 days. The latent period was 9.5 and 14 days in the two studies included in the systematic review, an average of 11.8 days. It is quite a few to calculate an average, but the transition from 'dry' to 'wet' symptoms, as reported in several studies, required on average 6.1 days plus/minus 2.4 days. We arrive then at 12 days, adding this period to the incubation period. So, a latent period of the Ebola virus of about 12 days may be a quite reasonable estimate and we may conclude that the latent period is, in general, longer than the incubation period of the disease. We may notice that the majority of studies included in the systematic review of Velásques *et al.* (2015) was directed to the Zaire Ebolavirus.

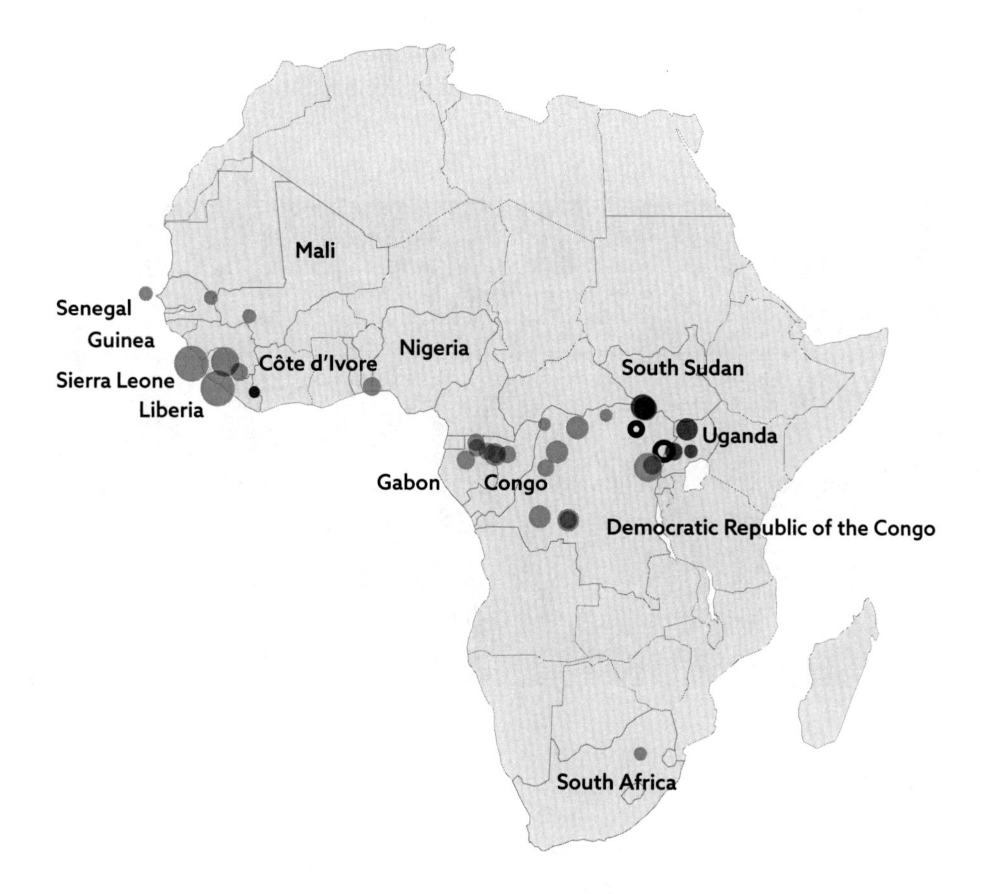

Figure 6.9. Cases of Ebola in Africa in the period of 1976 – 2019. The diameter of a circle indicates the number of cases, of which the smallest one represents 1-10 cases and the largest ones 10,000 cases, or more. Indicated are the various types of the virus, *i.e.*, Zaire (red-rose), Sudan (greyish, closed, circles), Forest (purple) and Bundibugyo (open circles). Online map of Centres for Disease Control and Prevention retrieved on the 9th of January 2023 at: https://www.cdc.gov/vhf/ebola/history/distribution-map.html.

People infected by an Ebola-virus may travel around during about one week without symptoms. They may go around even having some 'dry' symptoms, which is, on average, during another week. So, relatively frequent long-distance travel of the virus is certainly possible. The build-up of new foci is, however, limited in the lack of infectiousness during this period. Shedding of the virus from infected people starts at the time that the symptoms of the disease are quite obvious. A that time, isolation of diseased people is quite likely, restricting the spread of the virus considerably. In addition, the infectious period is relatively short. It was reported to be 5.3 days plus/minus 4.0 days for those passing away and excluding post-mortem dispersal (Velásques *et al.*, 2015). The infectious period was 9.4 days plus/minus 5.5 days for those recovering. So, we may conclude that the epidemic spread of the virus is limited in the severity of the disease that it causes. It, therefore, results in spread as travelling waves from a primary focus, or foci. The spread is quite predictable and it is, therefore, relatively easy from a point of view of management. We will return to this topic in the part 'Managing Epidemics' and, especially, Chapter 10.

REFLECTIONS

We have tools available nowadays to detect swiftly clustering of disease cases. Whether such a clustering results from epidemic spread of a pathogen as a travelling wave, or a dispersive wave, is less easily to determine often. In addition, we would like to forecast properly the type of spread rather than to determine it retrospectively. We may reflect about an approach to forecasting properly the type of epidemic spread.

6.4 Outlook

Epidemic spread may onset at any place, at any site. Mutations and recombination of a pathogen, mutations and recombination of humans, changes in environmental conditions, changes in human behaviour, or that of pathogens and vectors, these may all initiate a relatively fast spread of a pathogen among man, a disease epidemic as we call it commonly. Site and time of onset of an epidemic are rather unpredictable, although a lot of effort is spent on monitoring the well-known diseases, at least, as we will see in Chapter 8.

Epidemic spread implies a relatively quick and extensive dispersal of a pathogen among people. Statistical and mathematical tools are used to describe and understand the spread. The approach to characterising epidemics evolved from a rather descriptive one by calculating prevalence and incidence to one of using advanced mathematical models. We also see a shift from focusing on the progress of epidemics in time to one of focussing on spatial dynamics. In addition, we might detect a progress of modelling from, non-linear regression, to compartmental models, network models, meta-population models and, finally, to reaction-diffusion models.

Modelling is limited in the use of reliable estimates of key parameters. These are especially the basic reproductive number R_0 and the contact distribution D. Reliable estimates of R_0 and D may be obtained retrospectively, but such estimates can hardly be used prospectively. Models using these parameters are thus valuable with respect to the understanding of epidemic spread, but the power of forecasting epidemics of such models is at least doubtful. We feel the way ahead may be a combination of modelling and insight in the processes that determine the parameters R_0 and D, respectively. We illustrated such an approach here, retrospectively, distinguishing epidemic spread as dispersive and travelling wave, respectively. We may, however, turn it into a prospective approach by analysing various types of human-pathogen interactions regarding the essential epidemiological parameters, latent period p, infectious period i, and the incubation period of the resulting disease, and linking these to patterns of epidemic spread observed *casu quo* those generated by modelling. If so, we may have some guidance in the spatial patterns of epidemic spread to be expected, which is based on biological processes rather than mathematics only.

Analysis of the spatial dynamics of pathogens, as done here, is a novelty in human epidemiology, so far known. It revealed that the role of super-spreaders, in the sense of a case associated with relatively many secondary cases (*e.g.*, Wong and Collins, 2020), is minor in epidemic spread as a dispersive wave, in comparison with events of long-distance dispersal of a pathogen, resulting in a relatively quick emergence of new foci. In contrast, such super-spreaders may have a major role in epidemic spread as a travelling wave (*cf.* Lau *et al.*, 2017). It, all in all, underpins the relevance of assigning epidemic spread to dispersive and travelling waves, respectively. We, therefore, expect increasing interest in investigating epidemics on spatial dynamics, in which the mathematical analysis of spatial dynamics goes hand in hand with a biological view.

References

Bjørnstad, O.N., Shea, K., Krzywinski, M. and Altman, N. 2020. Modeling infectious epidemics. Nature Methods 17: 453-456. DOI: 10.1038/s41592-020-0822-z.

Choi, Y., Ladoy, A., De Ridder, D., Jacot, D., Vuilleumier, S., Bertelli, C., Guessous, I., Pillonel, T., Joost, S. and Greub, G. 2022. Detection of SARS-CoV-2 infection clusters: the useful combination of spatiotemporal clustering and genomic analyses. Frontiers in Public Health 10: 1016168. DOI: 10.3389/fpubh.2022.1016169.

Croccolo, F. and Roman, H.E. 2020. Spreading of infections on random graphs: a percolation-type model for COVID-19. Chaos, Solitons and Fractals 139: 110077. DOI: 10.1016/j.chaos.2020.110077.

Delamater, P.L., Street, E.J., Leslie, T.F., Yang, T. and Jacobson, K.H. 2019. Complexity of the basic reproduction number (R_0). Emerging Infectious Diseases 25: 1-4. DOI: 10.3201/eid2501.171901.

Desjardins, M.R., Hohl, A. and Delmelle, E.M. 2020. Rapid surveillance of COVID-19 in the United States using a prospective space-time scan statistic: detecting and evaluating emerging clusters. Applied Geography 118: 102202. DOI: 10.1016/j.apgeog.2020.102202.

Frantzen, J., 2007. Epidemiology and Plant Ecology, Principles and Applications. World Scientific Publishing, Singapore, 172 pp.

Jarvis, M.C. 2020. Aerosol transmission of SARS-CoV-2: physical principles and implications. Frontiers in Public Health 8: 590041. DOI: 10.3389/fpubh.2020.590041.

Keeling, M.J. and Eames, K.T.D. 2005. Networks and epidemic models. Journal of the Royal Society Interface: 2: 295-307. DOI: 10.1098/rsif.2005.0051.

Ladoy, A., Opota, O., Carron, P-N., Guessous, I., Vuillemier, S., Joost, S. and Greub, G. 2021. Size and duration of COVID-19 clusters go along with a high SARS-CoV-2 viral load: a spatio-temporal investigation in Vaud state, Switzerland. Science of the Total Environment 787: 147483. DOI: 10.1016/j.scitotenv.2021.147483.

Lau, M.S.J., Dalziel, B.D., Funk, S., McClelland, A., Tiffany, A., Riley, S., Metcalf, C.J.E. and Grenfell, B.T. 2017. Spatial and temporal dynamics of superspreading events in the 2014-2015 West Africa Ebola epidemic. Proceedings of the National Academy of Sciences USA 114: 2337-2342. DOI: 10.1073/pnas.1614595114.

Martinez, M.E. 2018. The calendar of epidemics: seasonal cycles of infectious diseases. PLoS Pathogens 14: e1007327. DOI: 10.1371/journal.ppat.1007327.

Menghistu, H.T., Hallu, K.T., Shumye, N.A. and Redda, Y.T. 2018. Mapping the epidemiological distribution and incidence of major zoonotic diseases in South Tigray, North Wollo and Ab'ala (Afar), Ethiopia. PLoS One 13: e0209974. DOI: 10.1371/journal.pone.0209974.

Meyerowitz, E.A., Richterman, A., Gandhi, R.T. and Sax, P.E. 2021. Transmission of SARS-CoV-2: a review of viral, host, and environmental factors. Annals of Internal Medicine 174: 69-79. DOI: 10.7326/M20-5008.

Pellis, L., Birrell, P.J., Blake, J., Overton, C.E., Scarabel, F., Stage, H.B., Brooks-Pollock, E., Danon, L., House, T.A., Keeling, M.T., Read, J.M., JUNIPER Consortium and De Angelis, D. 2022 Estimation of reproduction numbers in real time: conceptual and statistical challenges. Journal of the Royal Statistical Society: Series A (Statistics in Society) 2022: 1-19. DOI: 10.1111/rssa.12955.

Rosenkrantz, D.J., Vullikanti, A., Ravi, S.S., Stearns, R.E., Levin, S., Poor, H.V. and Marathe, M.V. 2022. Fundamental limitations on efficiently forecasting certain epidemic measures in network models. Proceedings of the National Academy of Sciences USA 119: e2109228119. DOI: 10.1073/pnas.2109228119.

Severns, P.M., Sackett, K.E., Farber, D.H. and Mundt, C.C. 2019. Consequences of long-distance dispersal for epidemic spread: patterns, scaling, and mitigation. Plant Disease 103: 177-191. DOI: 10.1094/pdis-03-18-0505-fe.

Sivanandy, P., Jun, P.H., Man, L.W., Wei, N.S., Mun, N.F.K., Yii, C.A.J. and Ying, C.C.X. 2022. A systematic review of Ebola virus disease outbreaks and an analysis of the efficacy and safety of newer drugs approved for the treatment of Ebola virus disease by the US Food and Drug Administration from 2016-2020. Journal of Infection and Public Health 15: 285-292. DOI: 10.1016/j.jiph.2022.01.005.

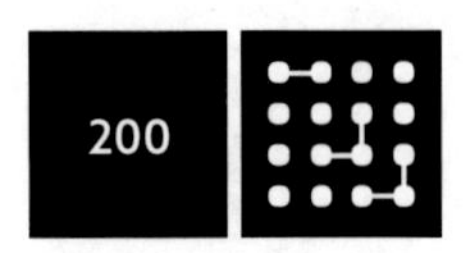

Velásquez, G.E., Aibana, O., Ling, E.J., Diakite I., Mooring, E.Q. and Murray, M.B. 2015. Time from infection to disease and infectiousness for Ebola virus disease, a systematic review. Clinical Infectious Diseases 61: 1135-1140. DOI: 10.1093/cid/civ531.

Wong, F. and Collins, J.J. 2020. Evidence that coronavirus super-spreading is fat-tailed. Proceedings of the National Academy of Sciences USA 117: 29416-29418. DOI: 10.1073/pnas.2018490117.

Wu, J.T., Leung, K. and Leung, G.M. 2020. Nowcasting and forecasting the potential domestic and international spread of the 2019-nCoV outbreak originating in Wuhan, China: a modelling study. Lancet 395: 689-697. DOI: 10.1016/S0140-6736(20)30260-9.

7

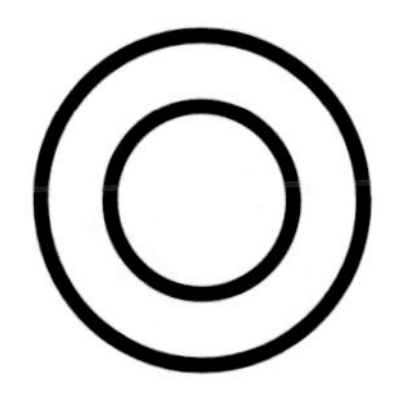

IMPACT OF DISEASE EPIDEMICS

We have distinguished, body, mind and environment, as three entities of human life in Chapter 2, whereas these are actually indistinguishable. Similarly, we will describe the impact of disease epidemics on four separate human domains in this chapter, whereas these cannot be regarded independently. We will pass across the bodily domain, the cognitive, the inter-personal, and the behavioural one, to highlight the impact of various types of pathogens on man. Determining the impact from a physical, bodily, point of view is common by using data of mortality and morbidity. Addressing other domains of impact is not common, as determining the impact in these domains is less tangible. We will provide some guiding here and we provide an example of each of the five categories of pathogens described in Chapter 3, the PrP prion of variant Creutzfeldt-Jakob, the SARS-CoV-2 virus of COVID-19, the bacterium *Vibrio cholerae* of cholera, the chromist *Plasmodium falciparum* of malaria, and various species of the genus *Schistosoma* of the helminth, worm, that causes schistosomiasis.

7.1 Four domains of impact

We adopt, and especially adapt, here a model proposed earlier to understand fear during the COVID-19 pandemic (Schimmenti *et al.*, 2020). They indicated four domains of fear, (i) the bodily, (ii) cognitive, (iii) inter-personal, and (iv) behavioural one.

We work out the bodily domain of impact by using data of morbidity and mortality that is attributed to a specific epidemic among a community. We need to take into account differences in resistance and tolerance within a community, even as variation in environmental conditions, including the social ones, as well. It results in profiles of impact with respect to communities. Such profiles may be used to compare various communities regarding the impact of epidemics caused by a specific pathogen.

The cognitive domain of impact is elaborated by taking into account that the impact depends on both, a proper estimation of our bodily risks and the willingness, mental capacity, to pick up the information included in bodily risk profiles. We will see again variation within and among communities with respect to the cognitive perception of epidemics. The perception hinges on, one's narrative of life, trustfulness of governmental bodies, (social) media, one's position within the community, and so on.

We will deal with the inter-personal domain of impact by translating the cognitively perceived risk profiles to inter-personal relationships. People may face a trade-off in their care of beloved-ones. Visits may be needed from a point of view of mental wellness and physical care of them, whereas a visit may pose a risk of transmission of a pathogen to the beloved one. We get in typical situations of flight-or-fight and, if no decision is taken, freezing. We get then actually in the fourth domain of impact, the behavioural one. How do we deal with an epidemic at both, the individual and societal level? And again, we will see variation in behaviour within and among communities with respect to epidemics caused by a specific pathogen. Behaviour that may be expressed in various concerted interventions. We will elaborate on such interventions in Chapter 10.

We will highlight five examples of disease epidemics that each may represent a category of pathogens described in Chapter 3. It will be, the PrP prion of variant Creutzfeldt-Jakob, the SARS-CoV-2 virus of COVID-19, the bacterium *Vibrio cholerae* of cholera, the protist *Plasmodium falciparum* of malaria, and various species of the genus *Schistosoma* of the helminth, worm, that causes schistosomiasis. We may note that just one example does not represent necessarily a whole category of pathogens. The examples may, however, visualise the impact of an epidemic across the four domains.

REFLECTIONS

We may reflect about an approach to determine the impact of epidemics. Which is the common approach to determine the impact of an epidemic? Do we have a common approach anyway? If not, are the four domains indicated useful to develop a proper approach to the determination of epidemic impact? If so, could we improve the common approach by a specification of the impact for each of the four domains?

7.2 Bodily domain

General

We use data of morbidity and mortality to express the impact of epidemics on the human body. We are aware of the effects of mental well-being on the bodily expressions of an infectious disease. We are, also, aware of the challenge to attribute clinical symptoms and mortality to a specific pathogen. We have to deal with a myriad of human pathogens and co-morbidity of non-communicable diseases. In addition, the social and natural environment may affect human susceptibility to a pathogen and the severity of the resulting disease. If so, we might talk even about a syndemic, rather than an epidemic (*cf.* Horton, 2020). A syndemic is a clustering of two, or more, diseases in a specific group of people, of which the specificity is determined by social status. The use of the term 'syndemic' points to the need of distinguishing people within a community who have an elevated risk of getting, severely, diseased. Similarly, we may distinguish communities with higher, or just lower, morbidity and mortality due to a specific disease epidemic. The subsequent five examples may illustrate the potential impact of epidemics on human bodies. First, the PrP prion as causal agent of the neurodegenerative disease of variant Creutzfeldt-Jakob.

Creutzfeldt-Jakob

The neuro-degenerative disease of variant Creutzfeldt-Jakob disease is caused by PrP prions. It is a rare and fatal disease. A definitive diagnosis requires autopsy of the brain of a patient. The disease got well-known in the nineties due to an epidemic that was linked to an epidemic of Bovine Spongiform Encephalopathy (BSE) among cattle in the United Kingdom. It turned out that the same strain of the PrP prion was responsible for both, BSE, which is called the mad cow disease, and a specific variant of Creutzfeldt-Jakob disease (Ritchie *et al.*, 2021).

The first cow, which succumbed to BSE, was recorded in 1984. The epidemic of BSE arrived at a maximum of new cases of about 37,000 cases in 1992. It was about 0.3% of the British herd. The epidemic was attributed to feeding cattle with meat and bone meal that was recycled from prion-contaminated animal carcasses. A ban on such a feed, and other interventions, started in 1988 to manage the BSE-epidemic. The number of BSE-cases, however, continued to increase to a maximum in 1992/1993, it, subsequently, declined, and it approached zero twenty years after detecting the first BSE-case.

The first, fatal, case of variant Creutzfeldt-Jakob disease was recorded in the United Kingdom in 1995. The number increased to a maximum of incidence of 28 cases in 2000. It, subsequently, declined and the last case was recorded in 2016. An epidemic of circa 20 years and it included 232 clinical cases in total. The consumption of beef, or other food derived from infected cattle, was a major route of dispersal of the PrP prion from cattle to humans. In addition, a minority of cases could be related to blood transfusion among people. Some cases of variant Creutzfeldt-Jakob outside the United Kingdom, during that period, may be associated with the British epidemic as well. Relatively many cases were reported among people aged 20-30 years old (Watson *et al.*, 2021). The incubation period of the disease was estimated at about 7 years by comparing the peak of the BSE-epidemic with the one of Creutzfeldt-Jakob. The mean period of survival was estimated at 14 months from the first presentation of neurodegenerative symptoms. The BSE / Creutzfeldt-Jakob epidemic also resulted in the slaughtering of 4 million cattle in the United Kingdom, at least, to contain the epidemic. Such a slaughtering is not an issue for SARS-CoV-2 and COVID-19, to which we turn now.

COVID-19

The specificity of variant Creutzfeldt-Jakob disease and the *post mortem* obduction of cases facilitate the determination of the primary cause of fatality. We are pretty sure about the attribution of the mortality primarily to the PrP prion. In contrast, we may state that SARS-CoV-2 as primary cause of mortality is unlikely, although it may be a very strong secondary factor in mortality. We are facing thus a huge challenge to estimate the bodily impact of SARS-CoV-2 properly with respect to mortality. Let us, therefore, look at a study that, on the one hand, warns for the pitfalls of estimating mortality and, on the other hand, tumbles in the pitfalls itself (Pifarré i Arolas *et al.*, 2021). Determining the case fatality rate is limited in the diagnosis of COVID-19. We know that people may be infected by SARS-CoV-2 without, or with relatively mild, symptoms of COVID-19. Are such people counted as cases, or not? Similarly, do we assign the label COVID-19 mortality to such, infected, people, if they pass away, or not? We are in short of world-wide standards in assigning mortality to specific causes while differentiating between primary and secondary causes. The question is then, whether determination of excess of mortality may be a proper alternative? It is limited in having a valid baseline and it is prone to confounding, as we may add the latter to the specific statements of Pifarré i Arolas *et al.* (2021) here. Interestingly, data of mortality was, subsequently, used by the researchers to calculate years of life lost due to COVID-19, despite the previous critical remarks about the validity of it. In addition, data of expected, average, life-time was used to arrive at estimation of life-years lost, although we may expect that subjects succumbing to COVID-19 would have been a shorter expected life-time than the average, *i.e.*, they should have been relatively fragile.

The number of hospital admissions related to COVID-19 may provide another route of determining the bodily impact of SARS-CoV-2. It focuses on COVID-19 cases that are most likely to result in death. In addition, the registration of hospital admissions is, in general, quite adequate. We may, however, miss fatal cases of people who do not have access to hospitals. We, therefore, get an, relatively valid, but limited estimation of bodily impact of COVID-19, using data of hospital admissions. It is a kind of estimation of the minimum of bodily impact. Let us look first at the data of hospitalisations due to COVID-19 in some countries (Fig. 7.1). Italy had a peak in number of hospital admissions immediately at the start of the epidemic in Europe. It was followed by a peak in autumn 2020, which was the highest for Italy in the period between March 2020 and February 2023. At the same time, autumn 2020 thus, a peak was recorded in Poland, which was followed by another one in spring 2021. The one in spring was the highest one among all countries recorded in the database of Our World in Data (https://ourworldindata.org). In Malaysia, the first, and highest, peak was observed late summer 2021. In Japan, hospitalisations showed a gradual, oscillatory, increase. Hospital admissions in Canada were at a relatively low level during the years, except a relatively moderate peak in January 2022.

We may put the data of hospital admissions in perspective by comparing it with those related to influenza. We turn firstly to British Columbia, Canada. Two types of retrospective cohorts of COVID-19 hospital admissions were created, using a public health surveillance platform (Setayeshgar *et al.*, 2023). One, an annual cohort that covered, partially, the first year of the epidemic, the period between March 2020 and February 2021. Less than 10% of the population in British Columbia was vaccinated against SARS-CoV-2 at that time. Interventions directed to public health were in operation. Two, a so-called peak cohort including people admitted in the period between 2 January and 12 March 2022. A period characterised by the highest rate of hospital admissions related to COVID-19, abundance of the omicron variant of SARS-CoV-2, a large proportion of people vaccinated once, or twice, and fewer restrictions of public life. Non-vaccinated subjects older than 18 years were excluded from the peak cohort. Nine categories of age were distinguished within a cohort, from birth to 70 years and older. Both cohorts, the annual and the peak one, were compared with annual and peak cohorts, respectively, of influenza. Cohorts of the influenza epidemics, and peaks of these, of the years, 2009/2010, 2015/2016, and 2016/2017 were used. So, each COVID-19 cohort was compared with three cohorts of influenza. The influenza epidemic of 2016/2017 may be regarded as the most severe one looking at the number of hospital admissions, especially in the age category of 70 years and older. The status of vaccination against the influenza virus was not known.

The admission rates in British Columbia showed, overall, an exponential increase with age for both, COVID-19 and influenza. The rate of the annual COVID-19 cohort was lower than those of the annual influenza cohorts until an age of 30 years. The rates were the more, or the less, similar between 30 and 50 years, and the rate of COVID-19 admission was higher between

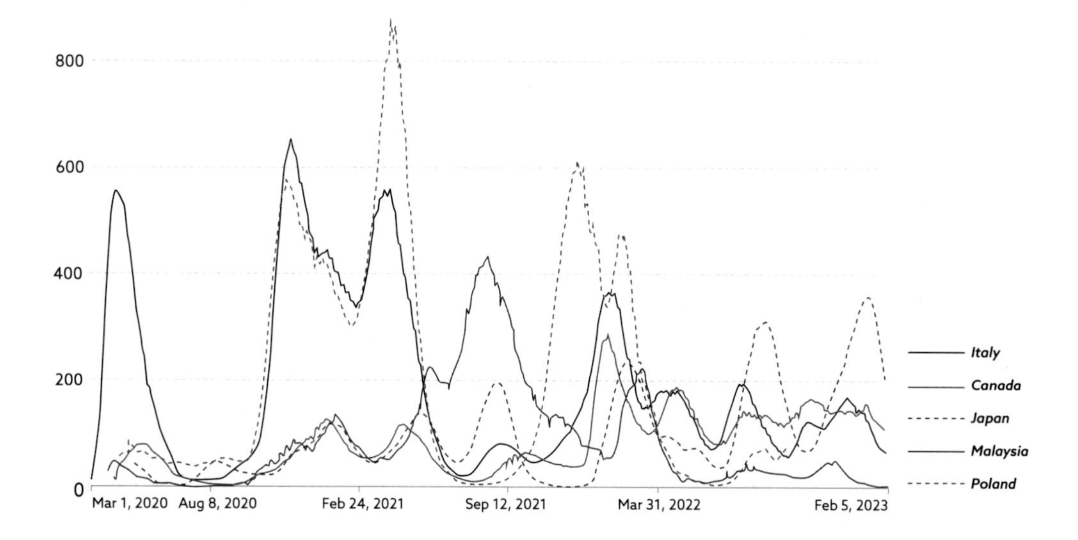

Figure 7.1. Weekly, new, hospital admissions per million inhabitants, which are related to COVID-19 (ordinate), in five countries in the period of the first of March 2020 until the fifth of February 2023 (abscissa). Data from Our World in Data (https://ourworldindata.org), which uses official sources.

50 and 70 years. The admission rate of the COVID-19 cohort in the age category of 70 years, and older, was similar to the one of the 2016/2017 cohort of influenza, and it was significantly higher than those of the two other influenza cohorts. The admission rate of the peak COVID-19 cohort was lower than those of the influenza cohorts for all age categories of 18 years and older, except the age category of 60-69 years. We may notice that alle subjects in the peak cohort were vaccinated and the omicron variant of SARS-CoV-2 was abundant during that period of time. A similarity between the COVID-19 cohorts and those of the 2016/2017 influenza cohorts was the relatively high frequency of four, or more, co-morbidities among the subjects of these cohorts.

The study of Setayeshgar *et al.* (2023) was directed to admission rates rather than morbidity and mortality in the hospital after admission. Other studies also included the clinical stay in comparing COVID-19 and influenza. In France, a retrospective study was conducted using nation-wide data (Piroth *et al.* 2021). This data serves primarily the funding of public and private hospitals. A cohort of hospital admissions of COVID-19 was created for the period of the first of March until the 30th of April 2020. This cohort was compared with a cohort of admissions of influenza in the period of the first of December 2018 until the 28th of February 2019. Nine categories of age were distinguished within each cohort. No vaccination against SARS-CoV-2 was in operation and the status of vaccination against influenza was not reported. The admission rates were higher for COVID-19 than influenza in each category of age, except the category of 17 years and younger. The number of influenza admissions was a sevenfold of those of COVID-19 in this category of age. The percentage of admissions was also relatively high at this age with *c.* 20% of all admissions due to influenza. The highest proportion was attributed to the age category of 81-90 years. It was about 25%. This proportion was similar to the one of contributing to COVID-19 admissions by this category of age. The rate of in-hospitality death was similar, and less than 5%, in the cohorts of COVID-19 and influenza up to patients belonging to the age category of 50-60 years. The rate, subsequently, increased to about 14% and 36% in the influenza and COVID-19 cohort, respectively, for patients aged 90 years and over.

A retrospective study comparing COVID-19 and influenza cohorts was conducted in Japan, as well (Taniguchi *et al.*, 2022). Data of 350 acute care hospitals was included in the study. These hospitals used a commercial business support system providing the data. No vaccination against SARS-CoV-2 was in operation and the status of vaccination against influenza was not reported. The COVID-19 cohort encompassed all subjects admitted to an hospital in 2020. This cohort was compared with influenza cohorts of 2017, 2018, 2019, and 2020, respectively. Data of the influenza cohorts were pooled for analysis of the in-hospital mortality data. Adjusted multi-factorial regression models were used to generate odds ratio of mortality, in which the influenza cohort served as a reference category. The number of admissions due to influenza reflected the typical winter seasonality in Japan (Fig. 7.2). In contrast, the frequency of COVID-19 admissions was highest in summer. The highest number of monthly admissions, overall, was recorded for influenza in January 2019. The number of hospital admissions due to COVID-19 was higher than those due to influenza for all categories of age, except for those of nineteen years and younger (Fig. 7.3). The number was just a bit higher for COVID-19 than the influenza cohorts in the category of age of 90 years and older, except the 2017 influenza cohort. The number of admissions of this cohort was quite low anyway at this age.

The in-hospital mortality rate among patients was less than 1%, and similar for COVID-19 and influenza, up to an age of 29 years. The rate was higher for influenza than COVID-19 up to an age of 60 years. The rates of both cohorts were equal for the age category of 60-69 years being 4%. The rate of COVID-19 was higher than the one of influenza for patients older than 70 years. The rate was highest for COVID-19 in the age category of 80-89 years, being about 13%, and it was highest for influenza in the age category of 90 years and older being about 7%. COVID-19 patients had, overall, a significantly higher probability to pass away in the hospital than influenza patients, as indicated by an odds ratio of about 1.8. The higher probability of mortality was primarily associated with the category of age of 70 years and older. We will leave COVID-19 here and turn to another pandemic, that of cholera and the causal agent *Vibrio cholerae*. It started in Asia, like COVID-19, and it is ongoing since the sixties.

Cholera

Cholera is caused by specific strains of the bacterium *Vibrio cholerae*, as we have seen in Chapter 3. We may call it an ectoparasite, as it remains in the gut of people. The pathogenesis is based on the excretion of a toxin damaging the epithelium in the gut. The life cycle of the bacterium is clearly bound to water being the environmental reservoir of it. Man is a typical side-host.

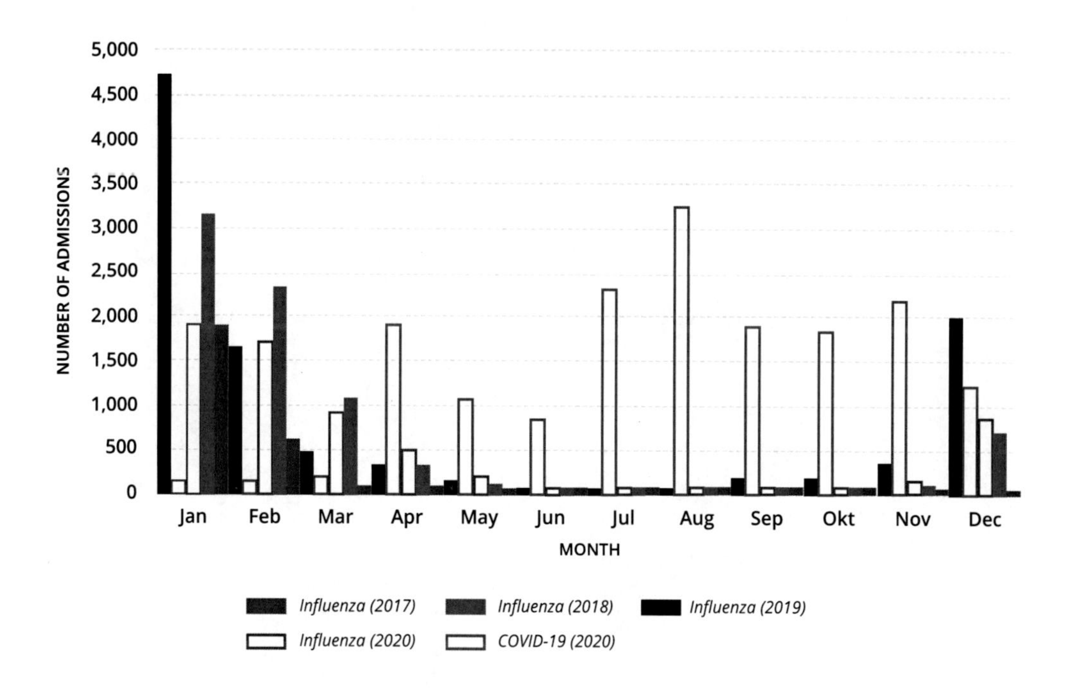

Figure 7.2. Number of hospital admissions per month due to COVID-19 in Japan in 2020 and influenza in 2017, 2018, 2019, and 2020, respectively.

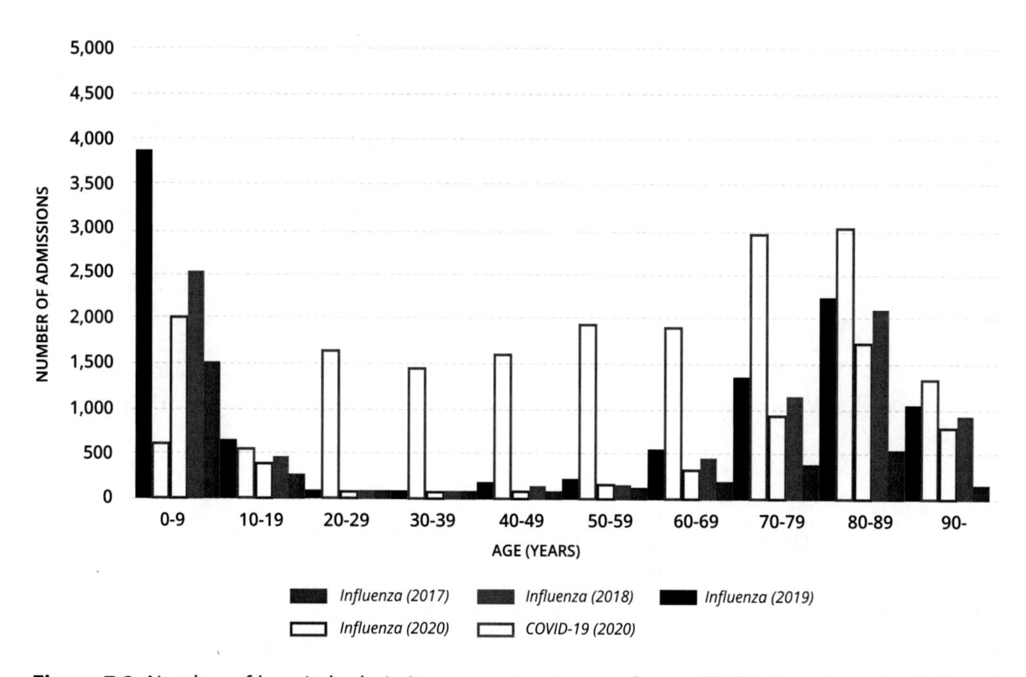

Figure 7.3. Number of hospital admissions per age category due to COVID-19 in Japan in 2020 and influenza in 2017, 2018, 2019, and 2020, respectively.

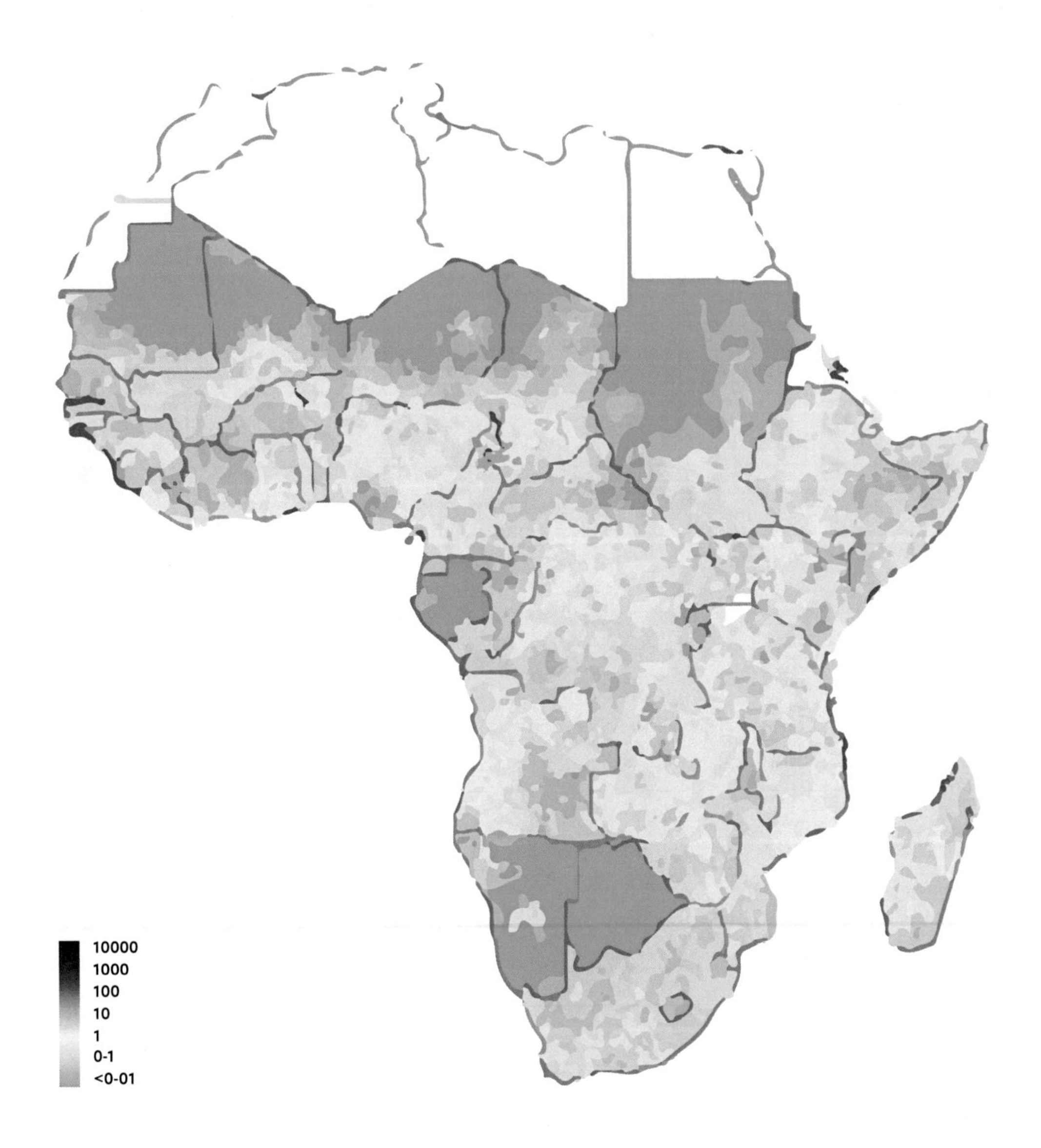

Figure 7.4. Annual incidence of cholera expressed as number of cases per 100,000 inhabitants using a Bayesian modelling framework and a grid of 20 by 20 km cells. The scale of colour indicates the number of cases per 100,000. See text for further explanation.

The World Health Organisation reported about 64,000 fatal cases of cholera world-wide between the years of 2000 and 2015, of which the majority (83%) was reported for sub-Saharan Africa (Lessler *et al.*, 2018). Incidence of cholera was investigated further for sub-Saharan Africa in the period 2010-2016 to identify high-risk areas, which would need priority in controlling the bacterium by way of, sanitation, promotion of hygiene, and vaccination. Data of, suspected, cholera cases, which was collected at various temporal and spatial scales, was unified at one annual and spatial scale. The data was integrated with key environmental and socio-economic risk factors using a Bayesian modelling framework. Djibouti and Eritrea could not be included as data was missed. Areas of annual incidence rates of cholera of more than 100 cases per 100,000 were classified as high-risk ones, between 10 and 100 as moderate-risk ones, between 1 and 10 as mild-risk ones, and those with less than 0.1 cases per 100,000 as negligible- risk areas. No classification of risk was used for areas with incidence rates between 0.1 and 1 case per 100,000. The various categories of risk are indicated in figure 7.4. The high-risk areas had about 22 million inhabitants, *i.e.*, the total number of persons at a relatively high risk of cholera. The number of inhabitants of areas of moderate and mild risk was 65 million and 126 million, respectively. High numbers of people at risk may also be noticed for malaria and the causal *Plasmodium* species, which we will address now.

Malaria

Species of the genus *Plasmodium* cause malaria, especially, *Plasmodium falciparum* and *P. vivax*. These pathogenic chromists are indigenous in the southern hemisphere (Fig. 7.5). It could be eliminated from various countries since the year 2000, which was certified by the World Health Organization (WHO, 2023). In contrast, the estimated total number of malaria cases remained fairly constant between the year 2000 and 2021 being around 245 million cases worldwide. The proportion of cases due to *P. vivax* decreased from about 8% in the year 2000 to 2% in 2021. The estimated number of fatal cases decreased in the same period from about 900,000 to about 600,000 per year. The disease became less fatal, overall, but we see exceptions taking a closer look on data per WHO-define region.

Malaria shows the highest abundance in the WHO African region. We see an increase of the disease from an estimated 211 million cases in 2000 to 234 million in 2021. We may, however, notice that the rate of incidence decreased in the same period until the year 2020 and it seemed to stabilise subsequently. We have to deal with a very high prevalence of the disease in Africa. The prevalence varied relatively strong within the region, from as high as circa 70 million cases in Nigeria to zero in Cabo Verde in 2021. The mortality attributed to malaria in the WHO African region decreased from about 840,000 in 2000 to a bit less than 600,000 cases in 2021.

The estimated number of malaria cases decreased in the WHO South-East Asia region from about 23 million in the year 2000 to about 5 million in 2000-2021. The rate of incidence decreased as well in this period, but it seemed to stabilise, or a bit to increase even, in 2020-2021. A similar trend of a relatively strong decrease, which was followed by a stabilisation, was observed in the rate of mortality. The mortality, overall, decreased from 36,000 to 9,000 in the period 2000-2019.

The trend of morbidity and mortality, both in prevalence and incidence, in the WHO Eastern Mediterranean region was one of a decrease between the year 2000 and 2010, which was followed by a subsequent increase until 2022. The estimated number of cases and fatal ones

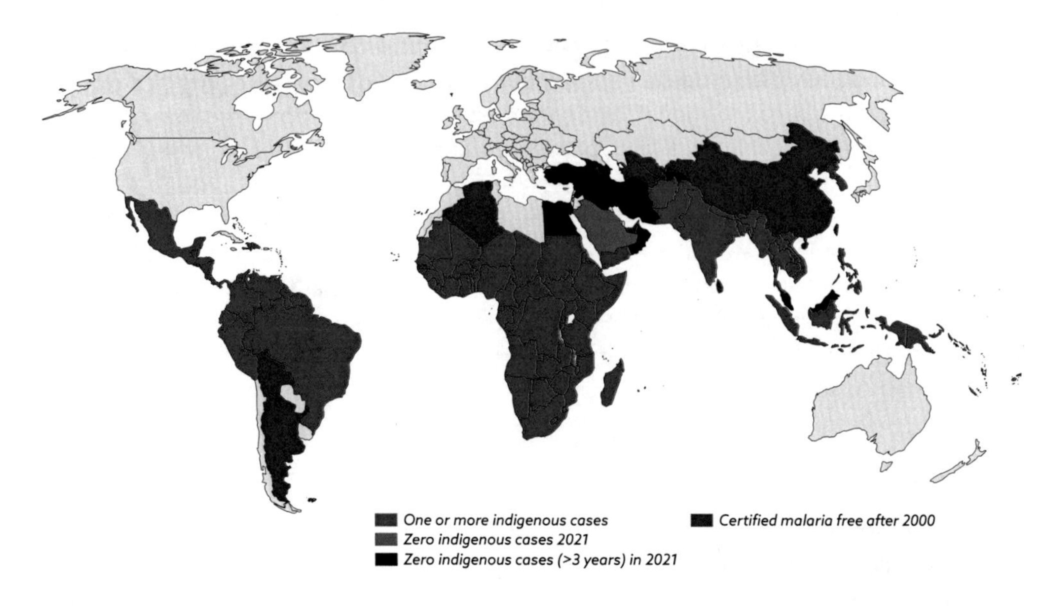

Figure 7.5. World-wide distribution of indigenous malaria, status of 2021, according to WHO (2022). See text for further explanation.

was lowest by 3,6 million and 6,900, respectively, in 2009. In 2021, these numbers were 6,2 million and 13,400, respectively, which was the more, or the less, similar to the numbers of morbidity and mortality in the year 2000.

In the WHO Western Pacific region, we see a continuous decrease in morbidity and mortality in terms of both, prevalence and incidence. In 2021, the estimated numbers of cases and fatal ones was 1.5 million and 2,600, respectively. A similar trend was observed in the WHO region of the Americas. The numbers, however were lower, as compared with the Western Pacific region, being about 600,000 cases of malaria and 334 fatal ones in 2021. In the WHO European region, the last case of indigenous malaria was observed in Tajikistan in 2014. No fatal cases were reported for this region in the period 2000-2021.

Malaria had a larger geographical distribution in the past. It was, for example, common in Europe, which indicates that the environmental conditions were suitable for both, the pathogen and its vector, which are mosquitos of various species of the genus *Anopheles* (Boualam *et al.*, 2021). We may notice that the word 'malaria' is derived from the Italian words 'mal aria'. It means 'bad air'. Abundance of malaria was historically related to swamps and stale air. It is, of course, not the air that is bad from a point of view of health, but the mosquito transmitting the pathogen. And yes, mosquitos are, in general, abundant in such areas. The cultivation of those areas has, subsequently, contributed to the elimination of the mosquitos, and therefore, malaria in Europe. The abundant use of insecticides, like DDT, completed the elimination. Europe is completely free of indigenous malaria since the eighties.

The term 'endemic' is associated with malaria often. We abandoned the use of this term in Chapter 6 already. We illustrate the abandonment here a bit further with respect to malaria. Endemic suggests that malaria occurs at relatively low, and stable, numbers. We cannot support the use of this term looking at the numbers mentioned above. In addition, the numbers may vary considerably in time. The annual rate of testing positive on malaria, for example, varied between about 10% and 70% in the district of Abobo, Ethiopia, in the period between 2008 and 2019 (Haileselassie, 2022). We feel addition of the term 'indigenous' to a pathogen is more appropriate to indicate that the pathogen is established in a community, in an area. So, we refer to malaria epidemics caused by indigenous *Plasmodium* species. Indigenous also are the various *Schistosoma* species that cause schistosomiasis. We continue with these.

Schistosomiasis

The life-cycle of the *Schistosoma*-species encompasses several stages, as we have seen in Chapter 3. Man serves as a major host and water snails as intermediate host. Occasionally, and depending on *Schistosoma*-species, other mammalian hosts may be involved. People catch the pathogen in water that is colonised by infected snails. The worm is released from the snails in the form of cercariae. The pathogen returns to the water by way of human faeces and urine containing cysts of it. The completion of the life-cycle of the helminths that includes man, thus, hinges strongly on inappropriate sanitation. Schistosomiasis is present in Africa, South-America and the Caribbean, China and Southeast Asia. The global burden of schistosomiasis was estimated at 1.4 million Disability-Adjusted Life-Years (DALY) in 2017 (GBD 2017 DALYs and HALE Collaborators, 2018). DALYs are calculated by way of adding the estimated number of years of life lost and years lived with disability, respectively, due to a specific disease. We may put the DALY of 1.4 million years of schistosomiasis in perspective. It was, for example, similar to the one of the, non-communicable, disease of myocarditis in 2017. It was a thirtyfold less than malaria and a thirtyfold more than Ebola, just to compare schistosomiasis with some infectious diseases as well.

REFLECTIONS

We have highlighted the impact of epidemics of quite distinct pathogens on man. May we rank these now with respect to the impact on humans in the bodily domain? Did, for example, the COVID-pandemic have a stronger impact on people, globally, than the ongoing cholera one? Or, the continuous malaria epidemics? We may reflect about the answer, taking into account the pitfalls in determining the impact of epidemics in the bodily domain.

7.3 Cognitive domain

General

The knowledge generated by experts in the bodily domain is, in general, transmitted to the general public, or specific groups at risk of the (infectious) disease. We enter then the cognitive domain of impact. The transmitted knowledge incorporates inevitably some uncertainty. We may distinguish two types of it, the epistemic and aleatory one (van der Bles *et al.*, 2019). The epistemic uncertainty is due to lack of sufficient knowledge of past and current phenomena. It is knowledge that we do not have yet, but that we could gain, at least in theory. It is actually the uncertainty inherent to the common scientific process. In contrast, the aleatory uncertainty is related to future events that are inherently unpredictable. We have to deal with both types of uncertainty communicating health risks related to epidemics.

The process of communicating knowledge and its associated uncertainty may be presented by the phrase 'who communicates what, in which form, to whom, and which effect is intended' (Table 7.1). Two types of people are involved in the part 'who' of communication. First of all, the assessors of the uncertainty, the experts. Epidemics are trans-disciplinary phenomena involving scientists of various disciplines, which hampers a proper assessment of the uncertainty associated with the knowledge. Assessing, for example, the uncertainty in case-fatality rates of an infectious disease is quite different from assessing uncertainty in epidemic spread of the causal pathogen. And thus, you have to deal with various uncertainties in the communication of epidemiological knowledge. A communication that is, in general, executed by other persons than the experts themselves. These may be people trained in science communication, or not. Whether, or not, trained in science communication, the uncertainty communicated may deviate from the uncertainty as determined by experts, because of a different interpretation and re-wording of it by the communicators.

Epidemics are the typical objects of the part 'what' in the communication of uncertainty of knowledge (Table 7.1). The causal pathogens specify the epidemics. We have seen above already that various experts are involved in assessing uncertainty. We, consequently, have various sources of uncertainty. It is an inherent complication in communicating epidemiological knowledge. In addition, we may distinguish, in general, some specific sub-sources of uncertainty, (i) variability in samples and repeated measurements, (ii) computational and systematic inadequacies in measurements, (iii) insufficient knowledge of underlying processes, and (iv) disagreement among peers. Uncertainty may be expressed in numbers and percentages, which we may call the 'direct' level of uncertainty, or in a narrative dealing with the quality of the knowledge. We may call it the 'indirect' level of uncertainty. We express both types of levels of uncertainty in magnitudes.

Shaping the form of communicating uncertainty starts by choosing the type of expression. It may range from providing an explicit full probability analysis to an explicit denial of uncertainty. The format may be one of, visuals, words, or numbers. The format depends, amongst others,

on the medium of communication like, social media, broadcast, newspapers, meetings, flyers and consults of experts. These are all possible means to communicate knowledge and its associated uncertainty.

Communication to 'whom' addresses actually a triadic interaction between, the characteristics of the audience addressed, the relation of the audience to the content of the message, and the relation to the subject who communicates. Communication, for example, of a science communicator to an audience of higher-educated people is less sensitive to misunderstanding and omissions than a communication to a less educated audience. Or, another example, epidemiologists will pick up easily essential information about an epidemic in the newspapers and detect errors and omissions in it, whereas non-epidemiologists do not.

We finally arrive at the key question of communicating uncertainty. Which effect does it have on the audience? We do, in general, focus on the cognitive receipt of the information. A receipt that may vary widely among people due to the huge variety of life narratives, which include, for example, education and social status. Narratives that also include past experiences and, therefore, emotions. Trust in the senders of the message is also incorporated in the experiences of the past life. Trust that may increase, or decrease, due to the knowledge, and its related uncertainty, communicated. And last, but certainly not at least, which effect does the communication have on the behaviour of the audience? A question that has been noticed before with respect to risk assessment in general.

Strong public concerns may be triggered by a minor risk of bodily impact, which has not been foreseen in a technical risk assessment, as people in modern societies seem to view themselves as quite vulnerable (Kasperson *et al.*, 1988). A feeling that is not addressed in a technical approach to risks. Such an approach is actually based on a multiplication of the probability of a detrimental event by, physical, consequences of such an event. We also encounter in such

Table 7.1. Parts of the process of communication and the topics that are addressed by each. Entries based on van der Bles *et al.* (2019).

PART	ADRESSING			
	1	2	3	4
Who	accessor uncertainty	communicator uncertainty		
What	object	source	level	magnitude
Form	expression	format	medium	
Whom	characteristics	relation to what	relation to who	
Effect	cognition	emotion	trust	behaviour

an approach the contradiction that a low probability/high consequences event has a similar risk than a high probability/low consequences event, although the perception of both events by people may be completely different. The Social Amplification of Risk Framework (SARF) was introduced to integrate a technical risk assessment, like that of an epidemic, and the response structure of individuals and communities. A response structure that may amplify, or attenuate, the public perception of a risk. SARF is a multi-factorial framework with a narrative structure and it appeals one's intuition.

We may characterise SARF in terms of objectivity and subjectivity. The input is by way of the, quite objective, risk analysis, the output is the subjective perception by people. In between, we see an iterative process of amplification, or attenuation, of a risk assessment at the individual and societal level. The process is governed by four key attributes of information, (i) volume, (ii) disputation, (iii) dramatization, and (iv) symbolic connation. A relatively high coverage by media, or a daily repetition of it, increases the volume of a message. The higher the volume, the higher the receipt of a message. In contrast, the receipt may diminish the more a message is disputed. A quite neutral message of "thousands of people might be at risk" may be dramatized in using the phrase "thousands of people threatened". A symbolic connation of COVID-19 to the common flu is something quite different than one to the Spanish flu. In addition, people may have their own, personalised, connation of a disease.

SARF may be used to highlight the role of trust in perceiving risks (Bearth and Siegrist, 2022). The focus of SARF is on trust in both, sources of information and the regulation of risks. Trust in a source determines, first of all, whether we pay attention to some information at all. We have plenty of sources that are responsible for our daily flows of information. Some people rely completely on information provided by social media, others on scientific journals, newspapers, broadcasting, or just on-site meetings with other people. We may also see a selection of sources owned by public bodies, or just private ones. So, we are adapted to one, or more, sources of information and we, therefore, trust these. In addition, the owners of the sources adapt to their audience, framing the information accordingly. We get in a conflict of trust, if we use various sources and we receive different information of these regarding, for example, an epidemic. Which one do we trust? We see then a differentiation according to one's narrative. Younger people may, for example, rely more on social media than older people. In one country, people may have more trust in the public bodies than in another country. Some people do trust the information received by social contacts only, like family, or physicians. We can, therefore, not talk about the general trust of a community. A huge variation in trust within, and among, communities is common, in the past and nowadays. It also holds for the trust in the regulation of risks. We may then encounter even a risk paradox. People may be too trustful, assuming a risk will be eliminated by others, say the government, and they give up their own protective measures. We may also notice that trust is not static. It may increase, or decrease, due to new experiences. So, we have to deal with various sources of information, and their specific audience(s), to get a piece of epidemiological information, and its inherent uncertainty, cognitively received by a whole community, if possible, at all.

We may adopt and adapt SARF with respect to infectious diseases and epidemics (Fig. 7.6). We see then that the process underlying the cognitive domain may broaden the impact of an epidemic, if so-called ripple effects occur (*cf.* Kasperson *et al.*, 1988). People not at (substantial) risk of an infectious disease are affected due to individual and societal responses, as triggered by an amplified risk perception. The reverse, attenuation, might also happen, although we estimate the probability of it as rather low once a health risk has been picked up by media.

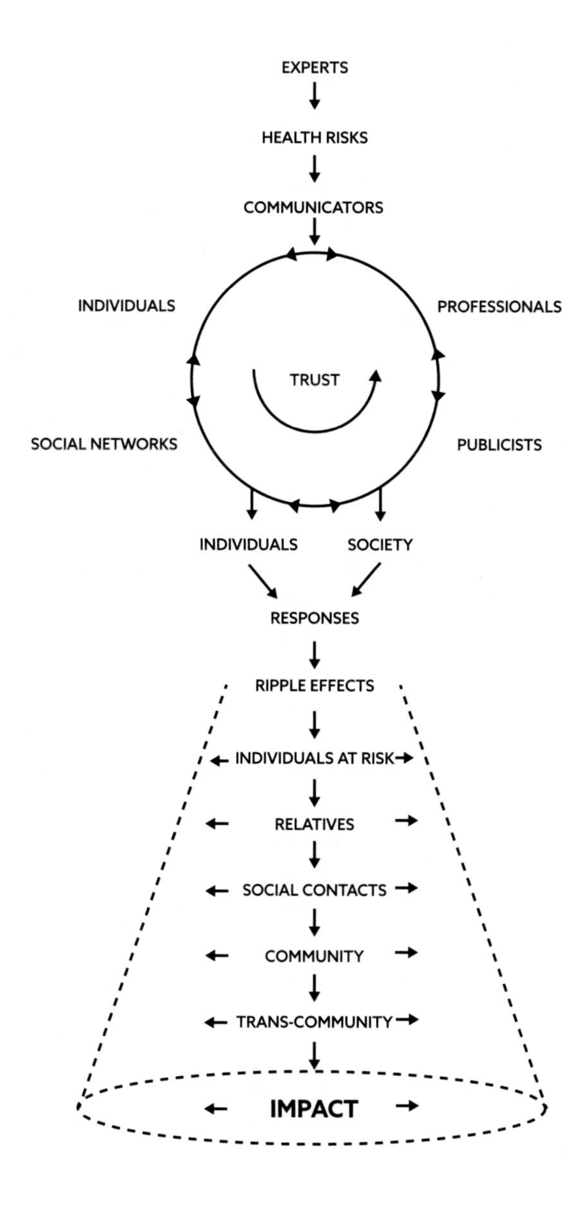

Figure 7.6. Input and output of the circle of cognitive adoption, and adaption, of health risks, determining the overall impact of an epidemic.

We return to the individual and societal responses, and the resulting impact of an epidemic, in the sections 7.4 and 7.5, dealing with the inter-personal and behavioural domain, respectively. Here, we will work out the cognitive circle with respect to the five examples selected. First of all, the PrP prion as causal agent of variant Creutzfeldt-Jakob disease.

Creutzfeldt-Jakob

The variant Creutzfeldt-Jakob disease may be characterised by a lack of well-funded risk analysis throughout its epidemic spread during decades. One reason may be that the disease was relatively rare and the cases were distributed over a period of about 20 years. We had 128 fatal cases in total in the United Kingdom. A risk of about 1 out of 500,000, taking into account the size of the British population in the year 2000. It is less than the probability that we have to be victim of a fatal traffic accident. Another reason may be the length of the incubation period, which was estimated, on average, at about 7 years. We may notice that an association between BSE and variant Creutzfeldt-Jakob was denounced by public bodies several times before the first fatal case was diagnosed in 1995. In contrast, the first warning of BSE as a risk of Creutzfeldt-Jakob disease was published in the British newspaper the Guardian of the second of August 1985 (Bauer *et al.*, 2006). The Guardian remained in the lead of warnings in the subsequent three years. In 1988, the Southwood Working Party was established to investigate BSE. It concluded the risk of BSE for humans was minimal, but if the estimates were wrong, the implications would be very serious. We, subsequently, see some attention in the newspapers at a rate of, on average, about one article per week until 1995-1996. Then, at the time of the first fatal human case, the papers exploded. It was not in the British newspapers only. The affair became breaking news in foreign countries, as well. The start of the variant Creutzfeldt-Jakob epidemic got full attention, whereas the BSE-epidemic could not.

Analysis of the framing of the BSE-crisis, in conjunction with the variant Creutzfeldt-Jakob disease, in the newspapers indicated that existential topics, like food ethics and trust, got relatively little attention (Fig. 7.7). Scientific expertise also received relatively little attention and the attention decreased even in the post-1995 period. National interest, costs-benefits of the crisis, and food safety-public health were major topics in the newspapers. Costs-benefits of the crisis got more attention in the post- than pre-1995 period, whereas it was the reverse for food safety-public health. We may notice that newspapers were a major source of information at the end of the 20th century and at the start of the 21st. About six million copies of the four largest British newspapers were, for example, sold in 1998. We did not enter the area of social media yet.

We will leave Creutzfeldt-Jakob here with the question whether the public opinion forced finally the governments of the United Kingdom, and other countries, to drastic interventions? The public opinion did it certainly with respect to SARS-CoV-2 and COVID-19.

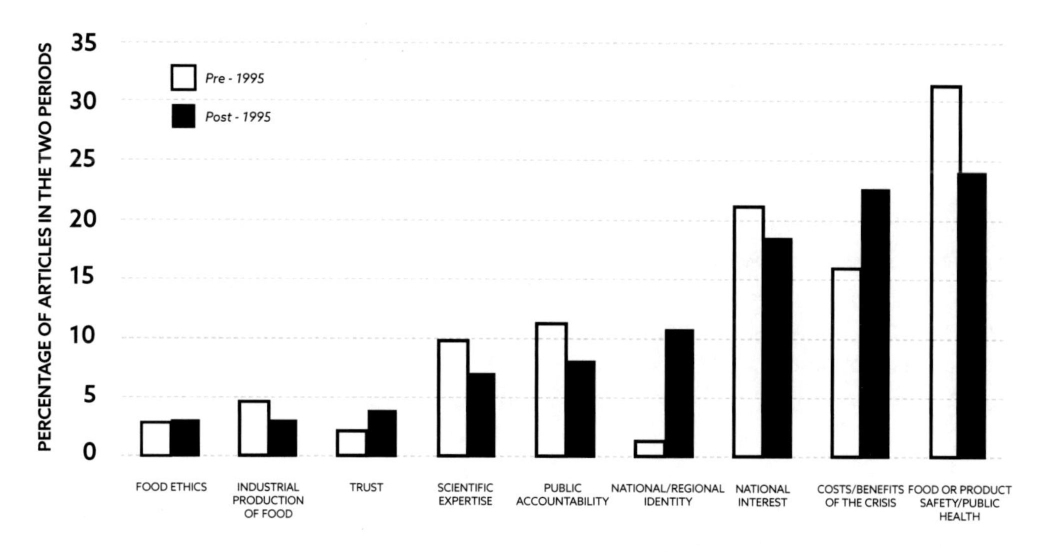

Figure 7.7. Framing of the BSE/Creutzfeldt-Jakob crisis in the United Kingdom as expressed by the proportion of articles in the newspapers directed to a specific interest. Framing is indicated for the period before announcing the first fatal case of variant Creutzfeldt-Jakob in 1995, and thereafter.

COVID-19

Perception of risk related to COVID-19 was investigated from the start of the pandemic on, as reviewed by Cipolletta *et al.* (2022). The systematic review included 77 studies, which were directed to investigating COVID-19 risk perception in the period of March 2020 until February 2021. It was the more, or the less, the first year of the epidemics worldwide. The studies had a cross-sectional set-up, except one. The quality of the studies was evaluated using the Joanna Briggs Institute critical appraisal checklist. Overall, studies met four out of the eight items of the checklist. The studies did not meet the item of 'objectivity', as these were based on self-reporting of subjects. Confounding was neither excluded. Validity of results reported was further comprised by using exclusively online-collection of data. It may have resulted in exclusion and under-representation of specific categories of people, like elderly and less-educated people. The terms 'high risk perception' and 'low risk perception' are used in the review. We may understand it as a risk is perceived as high, or low, but it may also mean that the perception of the 'real' risk is perceived highly, in the sense of 'good', or lowly, in the sense of 'bad'. We, however, feel that an absolute, real, risk of COVID-19 cannot not be provided unequivocally. We, therefore, interpret the results of the systematic review here as relative within a specific category of people. Some may perceive the risk as relatively high, whereas others do it as relatively low. The 'real' risk may, actually, be, high, low, or in between these subjective terms. We will continue with the results of the systematic review of Cipolletta *et al.* (2022) keeping this in mind.

Elderly people perceived risk, in general, higher than younger people. It, therefore, reflects the actual difference in bodily impact on older and younger people, respectively, as indicated in section 7.2. Some studies indicated perception of a higher risk by women than by men. Other studies, just reported the reverse. The level of education of subjects showed a similar ambiguity among studies. In contrast, employees of health institutions inevitably had a perception of higher risk than non-employees. It was especially reported for nurses. An optimistic life style, positivity, was associated with a perception of lower risk. It was also associated with happiness, serenity and well-being. In contrast, a perception of high risk was, positively, correlated with psychological complaints. Experience with COVID-19 resulted in perceiving a relatively high risk. Risk perception did also differ between countries. The differences may be associated, at least partially, with real differences in bodily impact recorded for countries, as outlined in section 7.2.

The cognitive perception of COVID-19 risk may affect our emotions. It may cause emotional distress, as determined in various studies. The link between risk perception and emotional distress was further explored in a study, to which 1700 under- and graduated students were invited online at two universities in Beijing (Feng *et al.*, 2022). About 79% responded positively to the invitation, *i.e.*, filling out a questionnaire and providing consent. Thirty-four students were excluded, as they did not fulfil the inclusion criterion of consciously filling out the inviting questionnaire. The study population was dominated by women (73%), Han ethnicity (85%) and undergraduate students (95%). Enrolled subjects filled out various questionnaires to determine their, (1) exposure to sources of COVID-19 information, (2) perception of COVID-19 risk, (3) emotional distress, and more specifically symptoms of anxiety and depression, and (4) psychological resilience. We have seen in Chapter 2 that resilience is defined as an appropriate act of counter-balancing the impact of a stress factor, a stressor, on the functioning of our body and mind. Here, the focus is on the psychological counter-balancing. Responses to the questionnaires and demographic variables, like age and gender, were analysed using a 'model'. The model was not described, but a kind of regression, or logistic regression, is likely. The study was conducted in the period of the second of February until the third of March 2020. The epidemic had just passed a peak in China and a nation-wide order of self-isolation was in operation.

Emotional distress was correlated positively with perception of the risk and the exposure to information. Thus, the higher the risk was perceived by subjects, and the more intensive the exposure to COVID-19 information was, the more severe the emotional distress. In contrast, psychological resilience showed a negative correlation with emotional distress. Emotional distress was less, the more students showed resilience. This mediation effect was especially pronounced when the COVID-19 risk was perceived as high. We may see the weaknesses of the study. The input of the model was generated by self-reporting only and the final model was not validated using an independent group of subjects. The external validation is limited in the specific study population, which may be characterised by young, female, undergraduate students in Beijing. Anyway, the results of the study indicate some points of attention in communicating the risk of COVID-19. One, over-exposure to information amplifies the effects

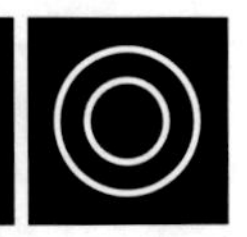

of risk perception on emotional distress and more specifically anxiety and depression. Two, psychological resilience may just attenuate these effects. Over-exposure to information does certainly not apply to another pandemic, the one of cholera caused by *Vibrio cholerae*. We look at it now.

Cholera

The bacterium *Vibrio cholerae* may be regarded as the 'founding pathogen' of human epidemiology, as outlined in Chapter 1. The Turkish novelist and Nobel prize winner Orhan Pamuk provided in his book 'Nights of Plague' an anecdote about the, actual, last sultan of the Ottoman Empire, Abdülhamit II (1842-1918). He was very impressed as he heard from the story of John Snow, who managed the cholera epidemic in London in 1848 just by conscious monitoring. One epidemiologist behind a desk may be more valuable than thousands of physicians in the hospitals. The sultan proclaimed an epidemiologist as one "like Sherlock Holmes", the fictional detective created by the British novelist Arthur Conan Doyle. Whether the anecdote is true, or not, epidemics of cholera are well-known since ancient times.

The origin of *Vibrio cholerae* is presumed to be somewhere at the Bay of Bengal in Asia (Lessler *et al.*, 2018). We saw seven times an extensive, global, spread of the bacterium in the last two centuries, which we may call a pandemic. The last one started back in 1961 and it is ongoing still, especially in sub-Saharan Africa. The ongoing pandemic is characterised by relatively low and stable incidence rates alternating with relatively high ones in both, a temporal and a spatial sense.

The variability of the risk of cholera in space and time may be perceived at the very local scale. The risk perception of citizens of the city of Beira in Mozambique was investigated from October 2002 until December 2005, using a set of research tools (Williams *et al.*, 2010). The study started with a questionnaire administered to 991 households along the 22 Bairros (neighbourhoods) of Beira. Twenty-two households were, subsequently, selected, in which cases of cholera had been experienced. The households needed to cover the various areas of the city. Cholera-centred interviews, which were in the language of preference, resulted in a cholera narrative of each of the households. The narratives were analysed qualitatively. In addition, eighteen focus groups of 8-25 subjects each were established.

Dynamics of cholera was associated by the subjects with rainfall. The risk of cholera was perceived as highest during periods of heavy rain, which occur in Beira especially during the hot and humid season between December and March. Cholera was also associated with places of dirt. Dirt was formulated in a rather broad sense meaning, rubbish, litter, faeces or mud. Flies were seen as going from 'dirt' to food causing cholera. Contaminated food was, in general, regarded as a cause of cholera. A risk of cholera was attributed to water wells in the vicinity of latrines as well. Hygiene was touched on in a general sense of cleanliness and smell. The results suggested that the risk-perception was, overall, based on the notion of contamination and it was linked to visuals in the environment of daily life. Environment of daily life is contaminated in a quite different way with respect to malaria and the various *Plasmodium* species, to which we turn now.

Malaria

Elimination of malaria is on the agenda of the World Health Organisation since decades. The latest target is becoming free of malaria in 2030. Elimination of the pathogen may proceed along varies paths. Elimination of the mosquitos of the various *Anopheles* species transmitting it is one way. It requires appropriate water management to minimise the habitat of mosquitos, in which these reproduce. We may also protect people sufficiently against bites of the mosquitos by way of the use of insecticide-coated nets. A preventive path also is vaccination of people against the pathogen, or the use of, preventive, anti-malaria drugs. In practice, the various paths are followed simultaneously. Control of malaria hinges on both, the adoption of measures by people at risk and the political pressure they may exert on governments to employ interventions at the societal level. It is preceded by people's cognitive perception of the severity and causes of malaria in daily life. We look at two quite different approaches to determining the cognitive perception of malaria in daily life.

The perception of malaria was part of a study directed to adoption of anti-malaria measures in Nigeria (Duodu *et al.*, 2022). Data of the cross-sectional 2018 Nigeria Demographic and Health Survey was used. Women in a reproductive age, which was defined as an age between 15 and 49 years, were selected for the survey. Of these 42,000 women, in total, 99% filled out the questionnaire. About 3% of the respondents were excluded from the study presented here, as they could not answer the question "whether malaria could result in death" with a clear 'yes', or 'no'. Respondents answering with 'yes', or 'no', were eligible for the study only. One third of the women denied that malaria could result in death, which indicates an underestimation of the severity of the disease. A similar proportion was determined in a village of the Khatyad Rural Municipality in Nepal (Awasthi, 2022). People in this village are at a relatively high risk of malaria. Twenty-five inhabitants were interviewed in a so-called 'quality study'. A variety of subjects was included, *i.e.*, farmers (13), priest/traditional healers (2), female community health volunteers (2), school teachers (2), students (2), self-employed/service (2) and local leaders (2). Interviews were based on a semi-structured interview guide. Nepali was used as language. The interviews of 20-30 minutes were executed online, due to existing travel restrictions related to COVID-19. Transcripts of the interviews were translated into English and, subsequently, analysed anonymously. We will return to other results of this study, and the one in Nigeria, in section 7.5., which is directed to the behavioural domain. Now, we pay attention to the various *Schistosoma* species that cause schistosomiasis.

Schistosomiasis

In the preceding, we dealt especially with the cognitive perception of disease epidemics, as highlighted for four pathogens. Here, we focus on the impact of these on the cognition itself. Worms that cause schistosomiasis also impair cognition of, especially, children, as summarised by a systematic review (Ezeamama *et al.*, 2018). It included 30, eligible, studies that were reported in the period of 1948 until 2016. A total of 39,000 children aged 5 to 19 years was included in the studies. Most of these, twenty-six, were directed to children in an African country, three to children in an Asian country, and one to children in St. Lucia in Mid-America. A minority of the children, about 2400, were subject of a randomised controlled trial, or a cohort study. The majority participated cross-sectional, or case-control, studies. The outcomes of the studies were assigned to one, or both, of the categories, (i) psychometrically assessed cognitive function, and (ii) education. The former was sub-divided into four specific attributes, (1) memory, (2) learning, (3) attention and reaction time, and (4) intelligence. The latter was sub-divided in, (1) school attendance and (2) school achievement. The studies were assessed on quality and, more specifically, potential bias. Three studies were judged to be at very high risk of bias and 16 to be at high risk. Two types of comparisons were included. The one was between children with confirmed worm infection and those without. The other was between children treated with praziquantel and those not treated. Praziquantel is a chemo-therapeutic drug, as we have seen in Chapter 3.

Overall, worms had a negative effect on the memory of children, even as the capacity of learning. The latter was supported by the negative impact of worms determined on the achievements at school. No, statistically significant, negative effects could be detected on the attention and reaction time of children, neither on their intelligence. In contrast, school attendance was associated negatively with worm infection.

REFLECTIONS

Cognitive perception of epidemic risks determines, overall, the impact of an epidemic and the responses of individuals and communities triggered by it. Our cognitive perception depends on a wealth of factors, of which the real risk is solely a minor one. We may reflect about an efficacious communication of epidemiological risks serving a proportionate, human, management of an epidemic.

7.4 Inter-personal domain

General

We have seen in Chapter 2 that a person may be subjected to two impulses that seem to conflict with each other. The one is the self-maintenance of the body, as any organism is assumed to do from point of view of biology. We also have seen the drive of full-filling the desires of others. A drive that may be that strong that we may offer our own bodily life even. We see then, during an epidemic, a reflex of protecting our own bodily life, whereas we also like to aid other people, and especially our beloved-ones. This aid, however, also is a complicated one, as we may support dispersal of a pathogen while serving others. We may constitute a risk ourselves. We get in a conflict of emotions, which is hardly to solve cognitively. Inter-personal relationships are largely determined by something irrational as love. Love that cannot be captured objectively using metrics. The impact in the inter-personal domain, therefore, needs to be addressed using a rather qualitative approach to it. We will do it in the following using the selected examples. We start with Creutzfeldt-Jakob disease.

Creutzfeldt-Jakob

The specific PrP prion causing both, BSE and variant Creutzfeldt-Jakob disease, does not pass among people, so far known, except by way of blood transfusion (Ritchie *et al.*, 2021). The inter-personal domain is, therefore, not characterised by direct contacts between individual patients and persons at risk. We do not see a conflict of emotions between helping a beloved one and posing a risk to her, or him, as outlined above. In contrast, we face emotions like the trust in, anonymous, blood donors and sellers of potentially contaminated meat, or other products of cows. We get in the conflicting emotions of desiring, or needing even, products and keeping ourselves safe with respect to infection by the prion. We have a product-oriented conflict of interests in the inter-personal domain. In contrast, we have a person-oriented one regarding SARS-CoV-2 and COVID-19, to which we turn now.

COVID-19

We saw in section 7.3 already that COVID-19 impacts on emotions just by cognitive perception of the risk. Overexposure of COVID-19 information may result in emotional distress, which may be mediated by psychological resilience. Emotional distress was also investigated in a broader context in the United States at the onset of the epidemic (Heffner *et al.*, 2021). It included the inter-personal domain as well. Subjects were recruited using the commercial online participant platform Prolific (https://www.prolific.co). A total of 948 subjects were enrolled in the study. The average age was 49 years and a standard deviation of 16 years. Gender was in balance and various ethnicities were included. The proportion of ethnicities ranged from a high one of white people (73%) to a small one of Middle Eastern people (0.3%). Income ranged from less than $ 20,000 for the largest proportion of participants (19%), up to more than $ 200,000 of the minority (3%). An emotional distress index was created by way of combining the scores of a subject on emotional well-being, in general, and COVID-19, specifically. Subjects, subsequently, had to fill out a set of questionnaires directed to variables associated commonly with emotional distress. Variables belonged to the areas of, mental health, personality, and emotion regulation. The Interpersonal Regulation Questionnaire was one of the questionnaires used to determine the regulation of emotion by way of aid of others, an aspect of the inter-personal domain. In addition, the knowledge of subjects of COVID-19 was determined even as the intensity of media consumption. Various demographic variables of subjects were also collected, like age and political preference. Estimates of 30 variables were used in a subsequent stepwise procedure to determine the best multi-variable, linear, regression model predicting the scores of subjects on the emotional distress index. Data of a quarter of the subjects were not included in the development of the model. This data was used to validate the model.

The model could predict 46% of the scores on the emotional distress index, using the data of the test sub-population. The model included 16 out of the 30 variables, of which 14 contributed significantly. The strongest predicting variable was anxiety, *i.e.*, the more anxious the more emotional distress. The probability of emotional distress was higher among women than men. Gender was second in predictive power among the variables included. Inter-personal emotion regulation contributed significantly as well, *i.e.*, the less able to regulate, the greater the emotional distress. Intolerance to uncertainty also was a significant, predictive, variable of emotional distress. The more subjects were intolerant to uncertainty the more their emotional distress. We may notice that intolerance of uncertainty may also express a lack of psychological resilience of a subject. If so, COVID-19 may be the trigger of long-term mental disorders. Intolerance of uncertainty was assigned to the area of 'personality', but we may also see it as a variable of inter-personal emotion regulation. Relying on help of others implies actually to accept some uncertainty with respect to the response of the other.

Studies, like the one presented above, are limited in number of people that can be included. In addition, self-reporting may be biased in the sense that subjects provide investigator-desirable answers on questions. Computational linguistics might provide a scalable, relatively objective, alternative by way of detecting emotions in text of social media (Kleinberg *et al.* 2020). A study was executed among residents of the United Kingdom in April 2020. A time of lock-down. Subjects were recruited using the commercial online participant platform Prolific (https://www.prolific.co). A subject needed to be user of Twitter, besides being a resident in the United Kingdom. The average age of the 2500 subjects included was 34 years and a standard deviation of 22 years. The majority of subjects was female (65%). Subjects were first of all assessed on their emotional well-being with regard to COVID-19, as expressed in one, or more, of the following eight emotions: anger, anxiety, desire, disgust, fear, happiness, relaxation, and sadness. Subjects were invited to attribute a score to their feelings about each of the emotions using a nine-points scale, on which 1 indicated 'not at all' and nine indicated 'very much'. In addition, they had to choose one of the emotions that represented their feeling best at that moment. Subjects were then invited to write a text of a minimum of 500 characters and a Tweet text with a maximum of 240 characters, respectively, about their feelings about the Corona-situation. Subjects were also asked how well they could express their feelings in the long and short text, respectively.

In addition, the frequency of use of Twitter was recorded. The self-reported emotions were related to the emotions expressed in the long and short text, respectively. The emotions were extracted from the text using, (i) a common psycholinguistic lexicon and (ii) the meaning of words, as expressed by the 'Term Frequency Inverse Document Frequency' (TFIDF), and part-of-speeches (POS). The TFIDF expresses the meaning of a word in a text, which is proportional to the frequency of it, as related to the frequency of the word overall in the corpus (dataset). A POS is a group of related words expressing the same meaning in a text.

An association between self-reported emotions and those extracted from text could be established at best, among the eight emotions tested, for the emotion of anxiety. Establishment of the association was better using long text of subjects than short text. The explained variance in self-reported anxiety, however, was relatively low with about 11% and 16% for the lexicon and TFIDF-based extraction of the long text, respectively. It indicates that either emotions are poorly expressed in text, or the set-up of the study suffered from methodological shortcomings. One of the shortcomings is obvious, the use of averages and standard deviations of the various categorical variables in the analysis. In addition, we face the general question, whether self-reporting of emotions by way of metric scales is valid. We pointed to the pitfalls in using metrics in psychology in Chapter 2 already, referring to Desmet (2018). Anyway, we scale down and look at cholera and *Vibrio cholerae* in the following.

Cholera

We return here to the study of cholera in Beira, Mozambique, which we described in section 7.3 already (Williams *et al.*, 2010). The bacterium disperses by way of the faecal-oral route among people. People in Beira were aware of it and they complained about those defecating in the open. And more general, people complained about other inhabitants producing 'dirt' in public areas. So, cholera triggers some tension, friction, with respect to inter-personal relations. And what about malaria and *Plasmodium* species?

Malaria

Malaria does not directly impact on inter-personal relations. The pathogen, and its vector, resemble an omnipresent environmental condition like temperature. People seem to live with the disease without blaming each other for the abundance of it. They neither seem to use inter-personal relations to promote adoption of governmental interventions to control malaria, except health decisions that are taken within the context of a family, as we will see in section 7.5. We, however, go firstly to the *Schistosoma* species that cause schistosomiasis.

Schistosomiasis

Schistosomiasis does not directly impact on inter-personal relations. Getting ill cannot be attributed directly to a close relative, or neighbour. It is similar for malaria, as we have seen above. Inter-personal relations are, however, pivot in controlling schistosomiasis. Distribution of

the chemo-therapeutic drug praziquantel is primarily based at schools. Children are especially at risk of schistosomiasis and their cognitive capacities are threatened, as we have described in section 7.3. We may see it actually as the use of the educational relationship between teacher and pupil for purposes of public health. Pupils benefit, of course, primarily of this medical intervention. It, however, is a medical intervention that is actually enforced by using the hierarchical school structure. Anyway, it is a strategy that is rather efficacious, as we will see in the next section.

REFLECTIONS

We understand certainly the impact of an epidemic in inter-personal relations for a pathogen that directly disperses between two subjects, like the SARS-CoV-2 virus. We have, however, seen that indirect dispersal may, also, affect inter-personal relationships, although in a bit hidden fashion. We may reflect about methods to determine the impact of epidemics on inter-personal relations, considering the five examples presented here.

7.5 Behavioural domain

General

We have seen that the cognitive cycle of adoption, and adaptation, of health risks results finally in responses of individuals and communities (Fig. 7.6 in section 7.3). Responses that are primed by emotions that characterise the inter-personal domain (previous section). Responses that may range from ignoring risks to complete isolation of oneself. The target of the responses may be the people at (substantial) risk of the specific disease. It may also include the relatives of these, their social networks, the whole community, or it may sur-pass the community even. The ripple effects in figure 7.6. We will see inevitably deviations between interventions arranged at the societal and the individual level, respectively, as the design and execution of interventions depend on individual risk perception and (inter-)personal emotions.

The behavioural domain is one of incorporating the impact of both, the pathogen itself and the responses to it. We cannot distinguish these two types of impact, as both may trigger a cascade of bodily and mental processes. Processes that are intertwingled, as we have seen in Chapter 2. We may, for example, see an elderly person who becomes afraid of a pathogen. The pathogen triggers fear, besides posing a bodily threat. The mental well-being of the person is, therefore, affected by the pathogen. In addition, she, or he, may also decide to go in isolation due to the fear. An isolation that may reduce the mental well-being further. The decrease in mental well-being may have bodily repercussions, as well, and these may ultimately be fatal in conjunction with co-morbidity. This points to the feedback between the behaviour triggered by an epidemic and its impact on bodily life. On the one hand, the bodily impact may be reduced by way of preventive measures and, on the other hand, it may be increased by way of side effects of changes in behaviour. So, we need to quantify side effects regarding their bodily components, if possible, and add these to the direct, bodily, impact of an epidemic to arrive at a proper estimation of the impact in that domain.

Epidemics may also trigger changes in behaviour that remain after an epidemic. We may list, for example, an increase of the inter-personal distance, adoption of healthy life-styles, high frequency of online-meetings, permanency of emergency laws, deprived privacy of (health) data, and so on, and so on. Epidemics may, therefore, affect the structure and dynamics of a society profoundly. It all depends on the, politics, culture, economics, and environment, of a society of concern. Anyway, let us see now what we may infer from the five selected examples with respect to the behavioural domain, taking into account the bodily domain, as well. We start with variant Creutzfeldt-Jakob disease and the PrP prion.

Creutzfeldt-Jakob

The BSE-epidemic in the United Kingdom triggered governmental interventions, which intensified at the onset of the variant Creutzfeldt-Jakob epidemic. A ban on the use of ruminant protein in the feed of cattle, a ban of specific bovine offal in food, a ban on British beef export by the European Union, cows older than 30 months banned from food chain, general control of carcasses on BSE, and last, but certainly not least, massive slaughtering of cattle.

Consumers of beef also seemed to respond to the BSE/Creutzfeldt-Jakob epidemic and the related governmental interventions (Van Wezemael *et al.*, 2010). The consumption decreased from 19 kg to 15 kg per capita in the United Kingdom between 1991 and 2003. The consumption, subsequently, increased to 21 kg in 2007. In France, the consumption showed a decrease of about 3kg from 1991 to 1999 and the consumption seemed to stabilise in the period 1999-2007. We may notice that France recorded a relatively high number of fatal cases of variant Creutzfeldt-Jakob disease, *i.e.*, 28 in total. The highest number recorded worldwide, apart from the United Kingdom. The steepest decline in meat consumption was observed in Germany. It declined by 9 kg between 1991 and 2003. It seemed to be stabilized in 2007. The consumption in Spain did not show a clear change and, if so, it was a slight increase even. Data of the 2007 census was, however, missing for Spain. Perception of beef quality determines the purchase attitude of a consumer. She, or he, uses various attributes to perceive the quality. Search attributes are typically used in pre-purchase inspection of the beef. Price, colour, and labels of the beef product are examples of search attributes. Experience attributes, like flavour and taste, are based on the actual consumption of the beef. These are clearly post-purchase attributes.

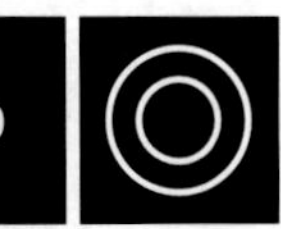

Safety is a credence attribute. It can neither be detected by visual inspection of beef products neither by consumption of these. The safety of the product may, of course, be indicated on a label referring to risk assessments and safety checks by public and private parties. Consumers may, however, have a different perception of food safety than experts, as we have concluded generally about risk perception in section 7.3 already.

The Creutzfeldt-Jakob epidemic resulted in 232 fatal human cases. The behavioural response of governmental bodies resulted in massive death of cattle, as well, which we need to add to the human victims to determine properly the impact in the bodily domain. We can attribute the additional mortality of cows only partially to the Creutzfeldt-Jakob epidemic, as the interventions were also directed to the protection of cows against mad cow disease. The question is then, which proportion? We do not have an answer to this question.

The societal and individual behaviour triggered by BSE changed as the epidemic waned. We saw the meat consumption returning to pre-epidemic levels. Awareness of food safety, however, remained, like the control of cattle cadavers on BSE and bans on blood donation by citizens of the United Kingdom, who might be exposed to BSE in the past. We may see the BSE-crisis also as a stimulus of thinking about human health in a broader context, the concept of 'One Health'. A concept that we will explore further in Chapter 8. Now, we look at COVID-19 and SARS-CoV-2.

COVID-19

The responses to COVID-19 were unprecedented in human history with respect to the strength of these. A lockdown was, for example, characterised as the world's biggest psychological experiment. Ripple effects (*cf.* Fig, 7.6 in section 7.3) surpassed the persons at risk widely, getting at the level of trans-community even. Complete communities got in lock down affecting social life in all its facets. The responses were especially strong in countries having a relatively high income per capita, like those in Western Europe and North-America. Trading was affected worldwide, including countries that did not face severe epidemics of COVID-19, or not yet. So, we cannot ignore the impact of the interventions directed to COVID-19, and the side effects of these, in treating the societal impact. We may state even that the societal impact of the interventions, like social distancing, exceeded by far that of the virus in a narrow sense. We will, therefore, return to the societal impact of COVID-19 in Chapter 10, in which we elaborate on societal responses. Here, we turn to cholera and *Vibrio cholerae*.

Cholera

We return here again to the study of cholera in Beira, Mozambique, which we described in section 7.3 already (Williams *et al.*, 2010). The majority of people involved in the study was aware of the need of personal hygiene to prevent cholera. They were also in favour of improving sanitation, but they missed the support of the government of Beira. Cholera seemed not to have priority in the governmental policy, at least at that time. In contrast, the World

Health Organisation aims at eliminating cholera by 2030. The goal needs to be achieved by way of, providing access of persons at risk to clean water, having appropriate sanitation, and personal hygiene. We have seen in section 7.4 that people are, in general, aware of all this, but they lack the governmental support to organise appropriate sanitation and clean water. The distribution of oral vaccines, which is also part of the roadmap to elimination of cholera, might also be limited in governmental support. The results of a study of Lessler *et al.* (2018), which is presented in section 7.2 already, indicated, for example, that the number of persons at high risk exceeded the amount of vaccine available at that time. We also face a lack of vaccines in the control of malaria and the causal *Plasmodium* species.

Malaria

Programmes to control malaria are running at the international, national, and regional lever for decades already. Adoption of the interventions by sufficient people seems to be Achilles' heel in achieving control of the disease. If so, we need a change in the behaviour of people at risk regarding malaria. A change that may be triggered by a cognitive perception of the severity of malaria, as expressed by, for example, the mortality caused by it. Indeed, the probability of using insecticide treated nets was higher among Nigerian women aware of the potentially fatal character of malaria than those who were not aware of it (Duodu *et al.*, 2022). We have looked into that study in section 7.3 already. The association between cognitive perception of the severity of the disease and a reduction of exposition to mosquitos was determined for women in rural areas. Such an association could not be detected for women in an urban setting. In an urban setting, the tendency was even that the likelihood of adopting pesticide-treated nets was lower among the women perceiving malaria as potentially fatal, as compared with those who did not.

We saw in section 7.3 that a third of the participants of a study in Nepal did not consider malaria as a potentially fatal study (Awasthi *et al.*, 2022). All of the 25 participants did, however, use insecticide-treated nets to prevent bites of mosquitos. Most of them did this during the 3-6 months of the year of a relatively high abundance of mosquitos. Three participants used the nets the whole year. The nets were provided by the government about two years before interviewing the respondents. Three people reported that the nets were torn, probably due to intensive washing. Indoor residual spraying, which was done every year, was less well accepted by most of the participants due to the negative side effects. Nine people, for example, reported that their bee-farming was affected by the insecticidal sprays. The use of health care by malaria patients was limited in the costs, except that of the public health practice in the vicinity. The services of it are free of charge and no, costly, travelling is required. Treatment of severe malaria, however, required travelling to, and use of, relatively costly, tertiary, health care facilities elsewhere. Health-related decisions were, in general, taken by those who were considered as the 'leaders' of family. And finally, we turn to schistosomiasis and the causal *Schistosoma* species.

Schistosomiasis

Schistosomiasis impairs the development of children, as we have seen in section 7.3. It, therefore, impacts on societies by way of reducing the human capital on the longer term. The disease may also impair economics of a country on a short term. Such an impairment may be less easily detected, as demonstrated in Burkina Faso (Rinaldo *et al.*, 2021). The economy of this low-income country depends strongly on agriculture. The so-called subsistence variety of agriculture dominates. It means that farmers use all of the crops, or livestock, raised to maintain their families, leaving little surplus for sale, or trade, if any. Projects have been executed since the eighties to improve water management of agriculture. These, however, created additional water sites favourable for snails, increasing the dispersal of worms to man. So, the water projects were on the one hand beneficial for agricultural yield, but on the other hand detrimental for peoples' health. The aim of the study was to quantify the impact of schistosomiasis on the economics of agriculture, taking into account the socio-economic status of farmers and the water infrastructure. Close-nit data was used with respect to, agricultural input and output, demographics and household characteristics, prevalence and severity of schistosomiasis, climate, and vegetation. Data had, in general, a country-wide distribution. If not, extrapolations were used to complete data. Severity of schistosomiasis was expressed by the number of worm eggs determined in the stool and urine. Modelling was used to integrate and link various data, while controlling for confounding factors.

Schistosomiasis caused, on average, a yield loss of 7%. The percentage of loss increased to 32% for households suffering most from the disease. The loss increased further to 45% for the poorest households. The vicinity of a large dam increased the yield of agriculture by 24%. Such a dam also resulted in an increase of schistosomiasis, which caused a subsequent decrease of the yield by 14%. The decrease attributed to schistosomiasis was observed up to a distance of 30km from a dam. The effects of a dam, or other water constructions, were not specified further with respect to socio-economic status.

The World Health Organisation has defined goals to control schistosomiasis and finally to eliminate it. The first goal would be to control the disease by 2020-2030, which means that severe schistosomiasis occurs at less than 5% of sentinel sites. It needs to be achieved predominantly by way of chemotherapy and, eventually, complementary interventions. The second goal would be elimination of the disease as a public health problem by 2025-2031, which means that severe schistosomiasis occurs on less than 1% of sentinel sites. It would require adjusted chemotherapy and the use of complementary interventions is needed. The third and last goal is elimination of man as host of the worms by 2025-2030. Intensification of chemotherapy is needed and complimentary interventions are essential. Massive chemotherapy is pivot in achievement of the goals of the World Health Organisation. The efficacy of it was estimated using data of previous campaigns employing massive chemotherapy (Deol *et al.*, 2019). Data of nine countries were used, Mali, Niger, Yemen, Burkina Faso, Rwanda, Uganda, Burundi, Tanzania and Malawi. These countries employed schistosomiasis control programmes based on annual, or bi-annual, rounds of massive chemotherapy application. In addition, the countries monitored the results of the interventions. Baseline was defined as the year of the first round of treatment. It ranged between 2003 and 2012 among the nine countries. Effects of two to four rounds on prevalence of schistosomiasis could be determined, depending on country.

Five out of the nine countries had eliminated schistosomiasis as public health problem at baseline already. Severe schistosomiasis thus occurred on less than 1% of the sentinel sites. The campaigns did not result in a complete elimination of schistosomiasis in these countries, the final goal, but the prevalence of moderate schistosomiasis was reduced. In another country, Mali, schistosomiasis got under control, *i.e.*, severe disease was recorded in less than 5% of the sentinel sites after two campaigns, and elimination of it as a public health problem was approached. Prevalence of severe schistosomiasis fluctuated around the limit of 5% of the sentinel sites in three other countries during the study period. The disease was not under control yet. The results presented indicate the need of tailor-made strategies to eliminate schistosomiasis worldwide. These also suggested that more time than foreseen will be needed to achieve the goals.

REFLECTIONS

The cognitive cycle of adoption, and adaptation, of health risks related to epidemics results finally in responses of individuals and communities. We have seen it for all of the five examples presented here. We may reflect about the appropriateness of the responses at the individual and community level. And what may we learn from a comparison of the five examples with respect to the societal impact of epidemics?

7.6 Outlook

We presented a figure in section 7.3 that seems to be pivot in determining the impact of an epidemic overall. It is presented here again (Fig. 7.8). Bio-medical experts provide an assessment of the potential, bodily, health risks at the onset of an epidemic. The assessment is inevitably based on estimates that include a certain level of uncertainty. The uncertainty is the larger, the earlier the assessment is provided during an epidemic. The risk assessment is, subsequently, communicated to a broad audience and a rather chaotic, societal, process starts up. The perceived risk may be amplified, attenuated, or remain unchanged, in this process, overall, at least. Heterogeneity in risk perception seems inevitably. On the one hand, the actual health risk differs among people. The risk of hospital admission due to COVID-19, for example, was a manifold for elderly people compared with children. But, the risk also differed among elderly people depending on their health status. On the other hand, we have a personalised perception of risk. The one may, for example, accept a health risk of 1 out of a million, whereas another does not. We may conclude that we cannot predict the outcome of this rather chaotic process, but we may get some grip on it. Grip that we need to avoid severe conflicts within a community. Grip that needs to be provided by humanities, rather than by bio-medical sciences, as we are dealing with a purely societal process. We feel COVID-19 epidemics provided a proper stimulus to reinforce the inclusion of humanities, social sciences, in the management of epidemics.

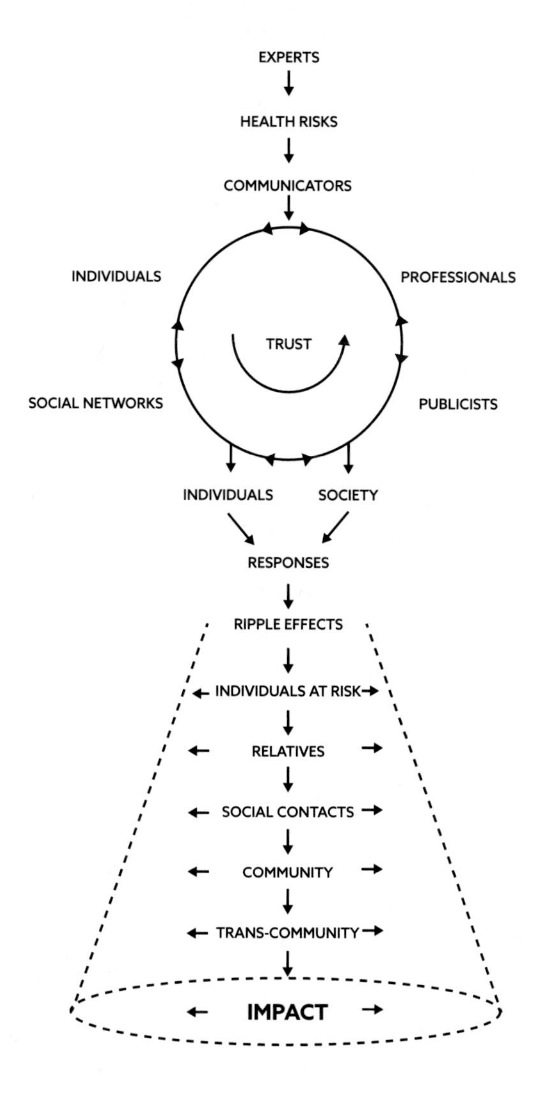

Figure 7.8. Input and output of the circle of cognitive adoption, and adaption, of health risks determining the overall impact of an epidemic, presented in section 7.3 as figure 7.6 already.

Responses of individuals and societies result from the circle of cognitive adoption and adaption of health risks, in theory. In practice, potential responses discolour that process. Some people may downplay the risks to avoid interventions, other may just play these up to benefit from interventions. The market-driven healthcare in Western countries is, especially, sensitive to such up-plays. Anyway, responses pop-up at the individual and population level at some time. Responses that are primarily directed to individuals at risk. But, who are those? Is it somebody who has a risk of one out of a million, or one at a risk of one out of 1000? And, should we include the severity of disease? If so, how do we define severity? If we are able to answer these questions properly, we arrive at a clear delineation of the category of individuals at risk (Fig. 7.8). If not, we may see emerge the first ripple effects, *i.e.*, people not at (substantial) risk are affected due to individual and societal responses triggered by an amplified risk perception.

We may accept ripple effects, if necessary to protect individuals at risk. Close-relatives, social contacts and the community, respectively, may be included in interventions to protect those at risk. Such interventions may finally result in impact on other communities that are not responding to, or not at risk of, an epidemic of the disease. We may imagine that adverse side effects increase from the level of relatives to the one of trans-community. The COVID-19 pandemic may illustrate this quite well. We arrive at a major question. How to manage an epidemic proportionately by maximising the protection of individuals at risk, while minimising detrimental side effects? We will look for an answer in the next part, managing epidemics.

References

Awasthi, K.R., Jancey, J., Clements, A.C.A. and Leavy, J.E., 2022. A qualitative study of knowledge, attitudes and perceptions towards malaria prevention among people living in rural upper river valleys of Nepal. PloS One 17: e0265561. DOI: 10.1371/journal.pone.0265561.

Bauer, M.W., Howard, S., Hagenhoff, V., Gasperoni, G. and Rusanen, M., 2006. The BSE and CJD crisis in the press. In: Dora, C. (ed.) Health, Hazard and Public Debate: Lessons for Risk Communication from the BSE/CJD Saga. WHO, Geneva, Switzerland, pp. 126-164.

Bearth, A. and Siegrist, M., 2022. The Social Amplification of Risk Framework: a normative perspective on trust? Risk Analysis 42: 1381-1392. DOI: 10.1111/risa.13757.

Bles van der, A.M., van der Linden, S., Freeman, A.L.J., Mitchell, J., Galvao, A.B. and Spiegelhalter, D.J., 2019. Communicating uncertainty about facts, numbers and science. Royal Society Open Science 6: 181870. DOI: 10.1098/rsos.181870.

Boualam, M.A., Pradines, B., Drancourt, M. and Barbieri, R., 2021. Malaria in Europe: a historical perspective. Frontiers in Medicine (Lausanne) 8: 691095. DOI: 10.3389/fmed.2021.691095.

Cipolletta, S., Rios Andreghetti, G. and Mioni, G., 2022. Risk perception towards COVID-19: a systematic review and qualitative synthesis. International Journal of Environmental Research and Public Health 19: 4649. DOI: 10.3390/ijerph19084649.

Deol, A.K., Fleming, F.M., Calvo-Urbano, B., Walker, M., Bucumi, V., Gnandou, I., Tukahebwa, E.M., Jemu, S., Mwingira, U.J., Alkohlani, A., Traoré, M., Ruberanziza, E., Touré, S., Basáñez, M-G., French, M.D. and Webster, J.P., 2019. Schistosomiasis – Assessing progress toward the 2020 and 2025 global goals. The New England Journal of Medicine 381: 2519-2528. DOI: 10.1056/NEJMoa1812165.

Desmet, M., 2018. The pursuit of objectivity in psychology. Borgerhoff & Lamberigts, Ghent, Belgium, 109 pp.

Duodu, P.A., Dzomeku, V.M., Emerole, C.O., Agbadi, P., Arthur-Holmes, F. and Nutor, J.J., 2022. Rural-urban dimensions of the perception of malaria severity and practice of malaria preventive measures: insight from the 2018 Nigeria demographic and health survey. Journal of Biosocial Science 54: 858-875. DOI: 10.1017/S0021932021000420.

Ezeamama, A.E., Bustinduy, A.L., Nkwata, A.K., Martinez, L., Pabalan, N., Boivin, M.J. and King, C.H., 2018. Cognitive deficits and educational loss in children with schistosome infection – A systematic review and meta-analysis. PLoS Neglected Tropical Diseases 12: e0005524. DOI: 10.1371/journal.pntd.0005524.

Feng, Y., Gu, W., Dong, F., Dong, D. and Qiao, Z., 2022. Overexposure to COVID-19 information amplifies emotional distress: a latent moderated mediation model. Translational Psychiatry 12: 287. DOI: 10.1038/s41398-022-02048-z.

GBD 2017 DALYs and HALE Collaborators, 2018. Global, regional, and national disability-adjusted life-years (DALYs) for 359 diseases and injuries and healthy life expectancy (HALE) for 195 countries and territories, 1990-2017: a systematic analysis for the Global Burden of Disease Study 2017. Lancet 392: 1859-1922. DOI: 10.1016/S0140-6736(18)32335-3.

Haileselassie, W., Parker, D.M., Taye, B., David, R.E., Zemene, E., Lee, M-C., Zhong, D., Zhou, G., Alemu, T., Tadele, G., Kazura, J. W., Koepfli, C., Deressa, W., Yewhalaw, D. and Yan, G., 2022. Burden of malaria, impact of interventions and climate variability in Western Ethiopia: an area with large irrigation based farming. BMC Public Health 22: 196. DOI: 10.1186/s12889-022-12571-9.

Heffner, J., Vives, M-L. and FeldmanHall, O., 2021. Anxiety, gender, and social media consumption predict COVID-19 emotional distress. Humanities & Social Sciences Communications 8: 140. DOI: 10.1057/s41599-021-00816-8.

Horton, R., 2020. Offline: COVID-19 is not a pandemic. Lancet 396: 874. DOI: 10.1016/S0140-6736(20)32000-6.

Kasperson, R.E., Renn, O., Slovic, P., Brown, H.S., Emel, J., Goble, R., Kasperson, J.X. and Ratick, S., 1988. The Social Amplification of Risk: a conceptual framework. Risk Analysis 8: 177-187. DOI: 10.1111/j.1539-6924.1988.tb01168.x.

Kleinberg, B., van der Vegt, I., & Mozes, M., 2020. Measuring emotions in the COVID-19 real world worry dataset. In: Verspoor, K., Bretonnel Cohen, K., Dredze, M., Ferrara, E., May, J., Munro, R., Paris, C. and Wallace, B. (Eds.), Proceedings of the 1st workshop on NLP for COVID-19 at ACL 2020 Association for Computational Linguistics, https://www.aclweb.org/anthology/2020.nlpcovid19-acl.11.

Lessler, J., Moore, S.M., Luquera, F.J., McKay, H.S., Grais, R., Henkens, M., Mengel, M., Dunoyer, J., M'bangombe, M., Lee, E.C., Harouna Djingarey, M., Sudre, B., Bompangue, D., Fraser, R.S.M., Abubakar, A., Perea, W., Legros, D. and Azman, A.S., 2018. Mapping the burden of cholera in sub-Saharan Africa and implications for control: an analysis of data across geographical scales. Lancet 391: 1908-1915. DOI: 10.1016/S0140-6736(17)33050-7.

Pifarré i Arolas, H, Acosta, E., López-Casasnovas, G., Lo, A., Nicodemo, C., Riffe, T. and Myrskylä, M., 2021. Years of life lost to COVID-19 in 81 countries. Scientific Reports 11: 3504. DOI: 10.1038/s41598-021-83040-3.

Piroth, L., Cottenet, J., Mariet, A-S., Bonniaud, P., Blot, M., Tubert-Bitter, P. and Quantin, C., 2021. Comparison of the characteristics, morbidity, and mortality of COVID-19 and seasonal influenza: a nationwide, population-based retrospective cohort study. Lancet Respiratory Medicine 9: 251-259. DOI: 10.1016/S2213-2600(20)30527-0.

Rinaldo, D., Perez-Saez, J., Vounatsou, P., Utzinger, J. and Arcand, J-L., 2021. The economic impact of schistosomiasis. Infectious Diseases of Poverty 10: 134. DOI: 10.1186/s40249-021-00919-z.

Ritchie, D.L., Peden, A.H. and Barria, M.A., 2021. Variant CJD: reflections a quarter of a century on. Pathogens 10: 1413. DOI: 10.3390/pathogens10111413.

Schimmenti, A, Billieux, J. and Starcevic, V., 2020. The four horsemen of fear: an integrated model of understanding fear experiences during the COVID-19 pandemic. Clinical Neuropsychiatry 17: 41-45. DOI: 10.36131/CN20200202.

Setayeshgar, S., Wilton, J., Sbihi, H., Zandy, M., Janjua, N., Choi, A. and Smolina, K., 2023. Comparison of influenza and COVID-19 hospitalisations in British Columbia, Canada: a population-based study. British Medical Journal Open Respiratory Research 10: e001567. DOI: 10.1136/bmjresp-2022-001567.

Taniguchi, Y., Kuno, T., Komiyama, J., Adomi, M., Suzuki, T., Abe, T., Ishimaru, M., Miyawaki, A., Saito, M., Ohbe, H., Miyamoto, Y., Imai, S., Kamio, T., Tamiya, N. and Iwagami, M., 2022. Comparison of patient characteristics and in-hospital mortality between patients with COVID-19 in 2020 and those with influenza in 2017-2020: a multicentre, retrospective cohort study in Japan. Lancet Regional Health – Western Pacific 20: 100365. DOI: 10.1016/j.lanwpc.2021.100365.

Van Wezemael, L., Verbeke, W., Kügler, J. O., de Barcellos, M.D. and Grunert, K.G., 2010. European consumers and beef safety: perceptions, expectations and uncertainty reduction strategies. Food Control 21: 835-844. DOI: 10.1016/j.foodcont.2009.11.010.

Watson, N., Brandel, J-P., Green, A., Hermann, P., Ladogana, A., Lindsay, T., Mackenzie, J., Pocchiari, M., Smith, C., Zerr, I. and Pal, S., 2021. The importance of ongoing international surveillance for Creutzfeldt-Jakob disease. Nature Reviews Neurology 17: 362-379. DOI: 10.1038/s41582-021-00488-7.

Williams, L., Collins, A.E., Bauaze, A. and Edgeworth, R., 2010. The role of risk perception in reducing cholera vulnerability. Risk Management 12: 163-184. DOI: 10.1057/rm.2010.1.

World Health Organization, 2023. World malaria report 2022. World Health Organization Geneva, Switzerland, 293 pp.

MANAGING EPIDEMICS

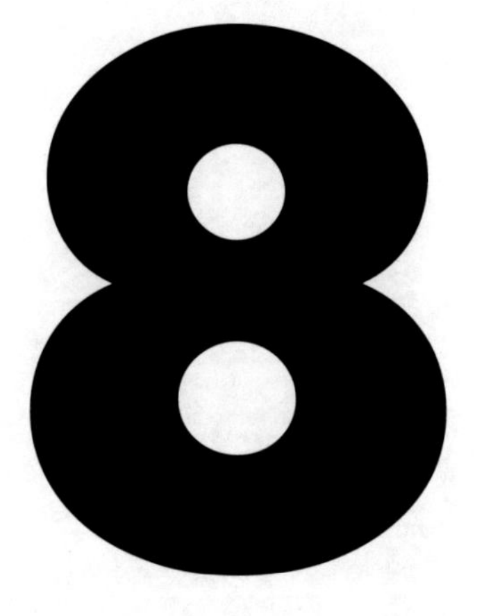

HEALTH RISK ASSESSMENT

This chapter introduces us to the so-called 'appropriate, human, management of epidemics' and, more specifically, the first step towards it, the assessment of epidemic risk and its dissemination. We have concluded the previous chapter with the circle of cognitive adoption, and adaption, of health risks that determines finally the overall impact of an epidemic. The function of the circle may be abstracted in the term 'societal health risk perception', on which we will elaborate in Chapter 9. The circle is fed by health risk assessments that are provided by experts and societal responses are the output of this process. We will pick up the responses in Chapter 10. In this chapter, we describe the basics of risk assessment of an epidemic by way of, (i) the identification of risks, (ii) the assignment of these to specific groups of people, and (iii) the dissemination of risk profiles to target groups. We will touch on the topics of 'One Health' and surveillance of pathogens. We will introduce the term 'epidemic seed'. It indicates an interaction between a human individual and a pathogen that might initiate an epidemic.

8.1 Identification of risks

Epidemic seed

The origin of an epidemic is basically at an encounter of a pathogen with man at the individual level. The pathogen needs to be virulent, or aggressive, and man needs to be susceptible and in a susceptible state. In addition, the environmental conditions need to be favourable for establishment of the host-pathogen interaction. We have elaborated it in Chapter 5. The pathogen may be, indigenous, exotic, or a new one with respect to man. We may notice that new strains of indigenous pathogens may emerge, as well. The distinction between the three types of pathogens is, therefore, a bit artificial regarding the establishment of a host-pathogen interaction. In addition, a change of (micro-)climate may also result in establishing a new interaction. We, therefore, introduce here the term 'epidemic seed' for a colonisation of a human individual by a pathogen that might result in an onset of an epidemic. It is like seed of a plant indeed.

Epidemic seed results from all the ongoing mutations, genetic recombination, and re-assortments within pathogens and ourselves. The genetic changes of the pathogens may occur in, environmental reservoirs, man, and other hosts. Epidemic seed may also be governed strongly by the dynamics of environmental factors, like temperature and rain. Factors that may affect both, susceptibility of the host and dispersal and pathogenicity of pathogens. Pathogenicity that is, of course, related to the susceptibility of people. Epidemic seed may be abundant within a community throughout the year, like that of *Plasmodium falciparum* causing malaria, or it may quite erratic, like the first COVID-19 case in Wuhan, China. Detection of common epidemic seed may not contribute much to management of epidemics, like the ones of malaria. In contrast, detection of the erratic ones may be essential in a timely management of resulting epidemics. An epidemic, or pandemic, that progresses like a dispersive wave is hardly to be contained once it sets on, as we have seen in Chapter 6. The COVID-19 pandemic is a typical example of missing epidemic seed. And, as we knew later on, it was high-risk seed. We then arrive at a major question. How do we distinguish between high- and low-risk seed? We may recognise here the common trade-off between sensitivity and specificity of diagnostic tests. We may miss epidemics, like the ones of COVID-19, having a too low a sensitivity in risk assessment, whereas we will be in a permanent emergency state by a too low a specificity of it.

Identification of high-risk seed may focus on novel infectious diseases, pathogens, emerging from non-human animals (Aarestrup *et al.*, 2021). The so-called zoonotic diseases. Diseases that have triggered a broader, holistic, view on infectious diseases. A view that is called One Health. The World Health Organization defines One Health as "an integrated, unifying approach that aims to sustainably balance and optimize the health of people, animals and ecosystems. It recognises that the health of humans, domestic and wild animals, plants, and the wider environment (including ecosystems) are closely linked and interdependent" (https://www.who.int/health-topics/one-health#tab=tab_1, retrieved March 2023). Genomic pathogen surveillance is at the centre of One Health, while focusing on zoonotic diseases (Aarestrup *et al.*, 2021). Sampling may be directed to human cases, animals serving as vectors and hosts of pathogens, and environmental reservoirs. Samples may be first cultured and then subjected to sequencing, or these may be sequenced directly. The former approach is the traditional microbiological one. The latter may be called the one of meta-genomics. Sampling results in a massive flow of sequence data that needs to be stewarded and to become available for sharing world-wide.

The International Nucleotide Sequence Database Collaboration (INSDC, https://www.insdc.org/) is the core infrastructure for sharing nucleotide sequence data, and the metadata of it, in the public domain (Arita *et al.*, 2021). The INSDC is set up and governed by a collaboration between, the DNA Data bank of Japan, the European Nucleotide Archive, and the Genbank in the United States of America. It serves as the registry of all nucleotide sequences data in the public domain. Data remains owned by the one who submits it originally. Updates are accepted from the submitter only. Access to INSDC is free of any restrictions, except the use of human genome data. This may be subject of restrictions resulting from the informed consent of the genome donors. We may see INSDC as a forerunner with respect to the principles of FAIR-data. Data stewarded by INSDC increases exponentially and it surpassed a storage of 9 petabytes of data in 2020 already. Reduction of (meta-)data, use of commercial cloud environments, and super-computing, are needed to keep the repository accessible in terms of, searching for data, transfer of it, and analysis. These constitute technological and financial challenges. Challenges that can be addressed by public and private parties in high-income countries. In contrast, parties in low- and middle-incomes countries may not have the resources to collect and use the data (*cf.* Aarestrup *et al.*, 2021). Inclusiveness is an issue in running repositories of data, besides aspects of privacy and technology.

Genomic pathogen surveillance is more than sequencing and stewardship of resulting data. We need to attribute a risk of epidemics to sequences. That turns out to be a real challenge, at least for viruses (Wille *et al.*, 2021). Three taxa of animals were selected for the study, (i) birds, (fish, and (iii) shrews. The number of new viruses detected among these animals was related to both, the ability of the viruses to cause disease among these animals and the compatibility of these with man. The number of viruses reported among birds increased by a factor 2.8 between 2008 and 2020, the increase was a factor 6.8 among fish, and 3.6 among shrews. Half of the viruses among birds were ratified as species by the International Committee on Taxonomy of Viruses (ICTV) by 2020, 18% of the viruses among fish, and 49% of those of the shrew. The ratification by the ICTV was thus behind the rate of publishing novel viruses.

In birds, 65% of the viruses caused disease of the host, as reported by 2008. In contrast, it was 30% for all viruses reported by 2020 and 55% for those ratified by ICTV. About 25% of the viruses recorded by 2008 could establish in man, whereas it was 10% of those reported by 2020. The proportion was higher for those ratified by ICTV, *i.e.*, 20%. In fish, 90% of the viruses caused disease of the host, as reported by 2008. In contrast, it was hardly 20% for all viruses reported by 2020, but about 90% for those ratified by ICTV at that time. Viruses of fish are not compatible with humans, so far known. In shrews, disease was not reported for any of the viruses, neither in

2008 nor in 2020. About 40% of the viruses recorded by 2008 could establish in man and 35% of those reported by 2020. The proportion was substantially higher for those ratified by ICTV, *i.e.*, 60%. We may conclude that an increase of number of viruses reported does not necessarily result in an increase of disease among the hosts that may serve as sources of spill-over to man, or an increase of epidemic seed among people.

The results of Wille *et al.* (2021) seem to indicate an inherent weakness of genomic pathogen surveillance in identification of risk of an epidemic. It is directed solely to the genomics of a pathogen, whereas an initial case, an epidemic seed, results principally from a pathogen-host interaction at the level of phenotype. We may notice that viruses are relatively simple pathogens in comparison with, bacteria, protists/chromists, fungi and helminths, and, if a direct relationship between genomics and pathogenicity does not apply for viruses, it will certainly not apply for other categories of pathogens. We may, therefore, doubt whether genomic pathogen surveillance is the tool to determine high-risk epidemic seed. We may go a step further even and we may question the efficacy of world-wide monitoring on individual seed. We, therefore, shift our attention from detection and prediction of individual cases, seed, to clusters of cases. In terms of epidemics, a pathogen needs to disperse from the human individual colonised firstly, which may be an erratic, stochastic, event, to others to become a source of an epidemic. In terms of plants, a seed needs to germinate and the emerging seedling needs to establish before it may develop into a plant.

Epidemic source

An epidemic seed turns into a source, if the pathogen disperses to other people, these are susceptible, and environmental conditions remain favourable for establishment on new individuals of the host. In addition, the pathogen may need to outcompete other pathogens, or strains of the same species. The pathogen, or the strain, needs to be fitter than others, at least in the specific environment. We have seen it in Chapter 5. An antibiotics-resistant strain of *Vibrio cholera*, for example, may have a competitive advantage, compared with non-resistant ones, in an environment, in which use of anti-biotics is abundant. In contrast, it may be at a competitive disadvantage in an environment without anti-biotics. Fitness of a pathogen is the result of genetics of itself and its host, or hosts, and the biotic and a-biotic environment. And yes, we may outline various evolutionary strategies of a pathogen to establish as an epidemic source, but we can, in general, not predict these. We have outlined this in Chapter 5. Inclusion of evolutionary dynamics of a pathogen in risk assessment is accompanied by nearly complete uncertainty, except for evolutionary dynamics under one-dimensional, uniform, conditions.

The focus on clusters of cases, the potential epidemic sources, offers opportunities to arrive at a proper risk assessment. First of all, the focus on clusters of cases increases the sensitivity of detection of epidemic sources considerably in comparison with individual cases, as a pathogen shows the potential of spread among people. We may add, data of demographics, the local landscape of human immunity, and environment, to determine whether the clusters identified are exceptional from a point of view of epidemics, or not. We need to take into account global changes to determine whether an epidemic source is exceptional, and thus risky, or not. Global changes affect local dynamics of people and the environment (Baker *et al.*, 2022). Urbanisation, for example, may result in higher densities of people that increase the risk of epidemics of respiratory disease. People may also become more sensitive for respiratory diseases due to air pollution. In contrast, urbanisation may result in improved sanitation, reducing risks on outbreaks of cholera, or helminths. In addition, access to health care may be facilitated. Access to health care may, for example, increase the rate of vaccination. We will return to appropriate, human, management of epidemics in a context of urbanisation in the next chapter.

Climate change will also affect the distribution of pathogens and vectors of pathogens. A comeback of malaria to Europe is, for example, expected due to establishment of mosquitos that may transmit *Plasmodium falciparum*. We may also see that the seasonality of epidemics reduces in temperate areas due to the global warming. We may anticipate the impact of climate change on the risk of epidemics by extrapolating current trends, as it has been done with respect to, for example, cholera in Iran (Asadgol *et al.*, 2019) and malaria in China (Wang *et al.*, 2022).

The focus on clusters of cases may also provide us initial knowledge about the epidemiological characteristics of a host-pathogen interaction. We have seen in Chapter 6 that the length of the latent period, in comparison with the incubation period, may be pivotal in epidemic spread as a travelling and dispersive wave, respectively. If the latent period is shorter than the incubation period, we may observe dispersive waves, whereas a longer latent period may result in travelling waves. In other words, do we observe pre-symptomatic, or even symptomless, dispersal of a pathogen among cases, or a post-symptomatic one? The SARS-CoV-2 virus clearly belongs to the former category (Baker *et al.*, 2022). The Ebola virus clearly in the latter. Influenza virus is just at the border of pre- and post-symptomatic dispersal. The SARS-CoV-1 virus is clearly in the category of post-symptomatic spread. The assignment of SARS-CoV-1 to the category of post-symptomatic spread may explain that virologists assigned SARS-CoV-2 initially to post-symptomatic, as well. This erroneously assignment was discovered later on by way of fitting models to data of epidemic spread. The models did not fit, assuming post-symptomatic spread, whereas these did by assuming pre-symptomatic dispersal of, on average, 2-3 days. Determination of such a key epidemiological parameter as the latent period is a real challenge in an epidemic source of a relatively few cases. If possible, we have a quite well predictor of the character of the subsequent epidemic spread at hand. In addition, monitoring cases in an epidemic source also provides clues about the severity of the resulting disease and, eventually, groups of people that are especially at risk. We will deal with the assignment of risks to specific groups further in the next section.

Surveillance of epidemic sources world-wide may be facilitated by way of digital-health technologies. Data capture and sharing got full attention during the COVID-19 pandemic to inform policy makers swiftly and evidence-based (Dron *et al.*, 2022). Digital-health technologies range from primary data collection in electronic health records to web-based Artificial Intelligence. Rough data may be used, like the one of records of patients in health care, aggregated data in publications and online platforms, or data of clinical trial registries. The first category of data is of interest only to detect and analyse epidemic sources on time. Digital platforms to share such data world-wide and real time do not exist yet, so far known. The set-up of such platforms requires, first of all, a substantial investment in data collection with respect to staffing and technology. We also need to address development of proper, standardised data formats. In addition, reproducibility of data needs to be guaranteed. An issue we indicated in Chapter 1 already. The last, but certainly not the least a topic, is the safeguard of privacy of patients. Establishment of digital surveillance of epidemic sources worldwide constitutes a long-term action of the world community, if possible, at all. Stability in politics and peace world-wide will be necessary.

Epidemic spread

Epidemic spread of a pathogen from sources may be characterised as a travelling wave, or a dispersive wave, if we can distinguish a clear initial source, a focus. We may also see a more generalised pattern, which is characterised by abundant epidemic sources without a clear focus. A malaria epidemic is a typical example of such a general one. A general epidemic may result from an abundant vector, like mosquitos, or an abundant environmental reservoir of a pathogen, like water is for *Schistosoma* species that cause schistosomiasis. We may notice that an epidemic progressing as a travelling, and certainly a dispersive, wave may get a general character after some time, due to the establishment of abundant secondary foci.

Epidemic spread of a pathogen may be at a rather local scale, as spread may be limited in specific environmental conditions, including the social ones. An outbreak of cholera in a high-income country, for example, is very unlikely due to proper, and robust, sanitation. Epidemic spread may, however, scale up from local to global, if it is not limited in environmental conditions. We have seen it with SARS-CoV-2. Travelling of people carrying pathogens over large distances facilitates such a global scale-up (Baker *et al.* 2022). Long-distance pathogen dispersal may not be facilitated by way of travel of people only. Pathogens may also be transported by goods and animals. We may notice that travelling and transporting at large scale became relatively massive during the last century. In addition, we see a substantial increase of the human population density and the climate change, as mentioned before. The risk of global epidemic spread of pathogens increased certainly the last century, and it will likely increase further.

REFLECTIONS

Identification of epidemic risk by way of genomic pathogen surveillance would be ideal from a point of view of management of epidemics. Genomics of a pathogen, however, is one factor only in establishing a human case. We, therefore, need to focus on cases, either at the individual level, which we called epidemic seed, or emerging clusters of cases, which we called epidemic sources. The urgency of an early identification of risk increases from an epidemic spread as a generalised one, to a travelling wave, a dispersive wave one, to a global one. We may reflect about the opportunities of proper risk identification at the stage of epidemic sources, taking into account the trade-off between sensitivity and specificity.

8.2 Assignment of risks

Everybody is at some risk of getting ill by each of the pathogens that are present in a specific environment, area, although the risk may be negligible. It is like participating daily traffic, or more philosophically, just by living. We need to assign levels of risk to specific groups to protects just those at a relatively high risk, whereas those at a relatively low risk should not worry unnecessarily. We have seen in Chapter 2 that, reasonable, concerns may escalate into, fear, anxiety and ultimately panic and all its associated, unpredictable, consequences. An Austrian physician provides a striking, and very sad, example of such an escalation. She promoted strongly, and publicly, the use of vaccines to protect against COVID-19. She was, subsequently, threatened anonymously for her statements. No wonder that she got worried about her life and we may understand even her development of fear. We may remember from Chapter 2 that fear is in some sense irrational, but it has a clear cause still, like it is for acrophobia. We may cope with fear still. The threats to the Austrian physician were anonymous, and so, everybody could be the potential source of violence, at least in the view of the physician. She became anxious and she felt threatened from all sides. She committed suicide finally, as she couldn't cope with the anxiety anymore.

The sad case of the Austrian physician warns us about the escalation of concerns into anxiety. A danger that is well-known among psychologists and psychiatrists. We may also see the danger of escalation with respect to infectious diseases, as pathogens are hardly visible and

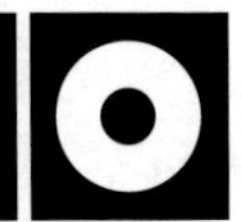

epidemics may have a rather unpredictable character. We, therefore, need to prevent escalation of worries and fears among both, people with a relatively low risk of disease and those with a relatively high one. It requires appropriate, human, management of epidemics, as we may call it. Human, as it needs to address the concerns, feelings, of a whole community. Appropriate, as the management is protective with respect to people at relatively high risk, while re-assuring those at relatively low risk that no specific interventions are necessary for them. In terms of diagnostics, management that combines high sensitivity and high specificity. Management that can be based on a proper assignment of risks only. How to achieve this before onset of epidemic spread of a pathogen? Let us look at the source of the COVID-19 epidemic in Wuhan, China, in December 2019.

Fifty-nine cases, who were suspected of the pneumonia of unknown cause, were admitted to a designated hospital in Wuhan by the 29th of December 2019 (Huang *et al*., 2020). The 2019-nCoV virus, which was renamed in SARS-CoV-2 later on, was detected among 41 of these 59 suspects. The larger proportion of the COVID-19 cases was male (73%). A relatively high proportion, 66%, had visited the seafood market of Huan, which was assumed to be the site of spill-over of the virus from non-human animals. Nearly half of the patients belonged to the age category of 25-49 years, 34% to the category of 50-54 years, and 17% was aged 55 years and older. Co-morbidity was determined among 32% of the cases. A minority of 7% was smoking at that time. These were the characteristics provided about the COVID-19 cases at that time. A profile arises of a high risk on COVID-19 among relatively young males, who visited the Wuhan seafood market, who were, overall, relatively health from a point of view of co-morbidity, and they were non-smoking. Such a risk profile does not match the one we established later on, except the relatively high abundance of COVID-19 among males. No wonder. The size of such a first cluster of cases is relatively small, which causes inevitably some bias in the risk profile. Besides such an inevitable bias due to timing, the initial risk profile also suffered from a lack of profound insight in the risks of the virus. We may simply see that the number of relatively young, healthy, males visiting the seafood market in Wuhan in December 2019 was substantially higher than the *c*. 30 cases admitted finally to the hospital. We have to address the question, why just this small group of men, out of such a large group, was admitted to the hospital. In addition, the question why a smaller group of women was admitted, as well? We have to go beyond the obvious interpretation of the characteristics of the cases to find the answers. We may use Latent Class Analysis to identify the real group, or groups, at a relatively high risk that are not observable straight away.

Latent Class Analysis is a statistical procedure to detect latent, unobserved, heterogeneity among a group of subjects (Weller *et al*., 2020). Observable variables, like age and gender, are used to detect hypothetical groups within a community that cannot be observed directly, like those being at a relatively high risk of COVID-19 within the community of Wuhan. The latent class, or group, is thus defined indirectly by a set of variables that may be determined, as these are observable. Such observable variables are, therefore, called the indicator variables. The indicator variables are categorical ones in the typical Latent Class Analysis. If these are continuous, we are talking about Latent Profile Analysis, which refers to a quite similar procedure. We continue here with the typical Latent Class Analysis. It hinges on the selection of the indicator variables to define the hypothetical, unobservable, class, or classes. In addition, the quality of the analysis increases the more subjects are included to quantify indicator variables, but a specific number cannot be indicated. The appropriate sample size depends on the number of classes and whether these are well-separated, or not. It also depends on the validity of the quantification of the indicator variables. It is quite easy to have a reliable estimate for a variable like gender, but it is less so for determining a mental variable like anxiety. The scores of all of the subjects on the indicator variables are, subsequently, used in an iterative process of modelling to arrive at an optimum of number of classes, sub-groups, and the probability that a subject belongs to a specific class. So, the indicator variables are used for both, constructing the classes and assigning subjects to these classes. We do not have a gold standard to check the validity of the assignment, the Latent Class Analysis overall. Paradoxically, Latent Class Analysis is used for validation of diagnostic tests, if these lack a gold standard, like it is for the diagnosis of extrapulmonary tuberculosis (*cf.* MacLean *et al.*, 2022). Anyway, the lack of a gold standard implies that the validity of a Latent Class Analysis depends completely on a proper choice of indicator variables and a valid statistical procedure.

Latent Class Analysis offers opportunities to include various types of indicator variables. So, we may also include variables expressing behaviour of people, besides those expressing physical characteristics. Visiting the seafood market of Wuhan turned out to be a variable of behaviour that determined, amongst others, the risk on COVID-19. Health-related behaviour depends, in general, on an interplay of physical, mental and environmental factors, as we have seen in Chapter 2. Such an interplay was investigated by way of a Latent Class Analysis directed to the sexually transmitted disease chlamydia (Van Wees *et al.*, 2019). It is a common disease among young heterosexuals, which is caused by the bacterium *Chlamydia trachomatis*. A total of 813 subjects was included in the study. The subjects filled out questionaries before they attended a clinic for testing on chlamydia. The responses were used to have scores on the binarized variables, age, sex, educational level, migration background, the number of sexual partners in the past four weeks, health goals, anticipated shame, and impulsiveness. These variables were used to identify predictors of a chlamydia diagnosis. Models were run including 1-6 latent classes. The 2-class model turned out to be the best fitting one, as expressed in the lowest value of the Bayesian Information Score, the lowest value of the Akaike Information Criterion, and a very high value of 0.99 of Entropy. The maximum of Entropy is 1. The two classes differed, statistically, significantly in proportion of subjects that tested positive for chlamydia, *i.e.*, 9% versus 13%. The probability of subjects belonging to the one, or the other, class was determined by three variables, health goals, anticipated shame, and impulsiveness, taking into account that the number of sexual partners in the past four weeks was similar, and relatively high, in both classes. A relatively high impulsiveness, high anticipated shame, and low, or medium, health goals increased the probability to be assigned to the high-risk class. So, the study added relevant knowledge of mental aspects to the known physical determinant of risk, *i.e.*, the number of sexual partners, which should be see in conjunction with less use of condoms. We may conclude that the study was executed properly from a statistical point of view and choice of the indicator variables. It had internal validity. In contrast, the external validity was poor, as only 7% of the subjects who were invited, responded positively with respect to participation of the study. So, we may doubt the study was representative of the catchment population of young, Dutch, heterosexual adults. Anyway, the study is a very well example of getting more insight in epidemic risks by using a Latent Class Analysis.

Latent Class Analysis has not been applied to epidemic sources yet to determine an initial risk profile, so far known. Let us, therefore, return to the epidemic source of COVID-19 in Wuhan in 2019, as described by Huang *et al.* (2020), to explore whether it could have been used for a Latent Class Analysis. The number of subjects of 41, of whom data was provided, was quite low and all of them had the diagnosis COVID-19. Addition of matched controls would have relieved this rather low sample size, while strengthening the identification of specific risk classes. We may think about a number of matched controls as twice as many as cases, or even more. In addition, the inclusion of indicator variables directed to health-behaviour, or risk-behaviour, would have empowered a Latent Class Analysis. A binarization of the variables would have been inevitably, due to the small sample size, even after inclusion of matched controls. If all this had been done, we would have had a more reliable risk profile of the incipient COVID-19 epidemic. A profile that would have needed certainly an up-date from time to time, as the epidemic, the pandemic, progressed.

REFLECTIONS

The value of assignment of epidemic risks to specific groups of people seems, in general, not to be recognised in the management of epidemics. We may wonder about it. We face, for sure, methodological challenges to generate a reliable risk profile at the onset of an epidemic, as we are limited definitely in data. We may reflect about the conditions for a proper Latent Class Analysis, or Latent Profile Analysis, at a stage of epidemic source and epidemic spread, respectively. Can we make the turn to appropriate, human, management from a point of view of a proper assignment of risks to people at the onset of an epidemic?

8.3 Risk profile dissemination

Set-up of a proper risk profile on time is one challenge for appropriate, human, management of epidemics. The dissemination of it to the various target groups is quite another. We pronounce here the term 'dissemination'. Dissemination is just more than communication. It is the communication of a message, a risk profile, to a target group in such a way that it impacts as desired. So, it is not sending out a message only, but it takes into account the receipt by a specific recipient. Sending a message between two experts of the same profession is quite different from one sent between, for example, a virologist and philosopher, or between a professional and a layman. Dissemination of a risk profile is a real challenge, as the cognitive perception of risk is quite personalised due to everybody's own narrative of life, as we have seen in Chapter 7. We have also seen that the societal process of cognitive adoption, and adaption, of health risks hinges on trust. Trust in governmental bodies, in professionals, in media, in neighbours, or whatever parties are used to be informed about risks of epidemics.

Trust may be seen as the willingness to believe that one can rely on the goodness of another. It includes some reciprocity. I trust you, and in turn, you should trust me. I listen to you, and in turn, you should listen to me. I give you credits, and in turn, you should give me credits. The reciprocity may be direct, *i.e.*, between two persons, two parties. It may also be indirect if one's trustful behaviour is recognised as 'good' by others than the person, party, that benefits directly (Schmid *et al.*, 2023). In addition, such a third party needs to consider adoption of the 'goodness' as a contribution to its own social standing, of which the underlying premise is to receive 'good' of others as well. We enter the field of reputation-based social dynamics, a fundamental feature of human experience.

The 'donation game' is primarily used to investigate indirect reciprocity. It is a relatively simple social interaction. A donor pays a small cost to confer a benefit from the recipient. Payment expresses the desire to collaborate. In contrast, no payment indicates defection. Other members of a community observe the behaviour of the donor and they update its reputation accordingly. The up-date depends on the existing social norm. Such a norm consists of two components, a rule of assessment of the behaviour and the resulting action. An assessment is called 'first-order', if the specific action between donor and recipient is considered only. The assessment may also take into account the existing reputation of the recipient and we talk about a 'second-order' assessment. If we also include the reputation of the donor, we arrive at 'third-order' assessments. We will focus here on the third-ones, as these are considered to have the potential of stabilising collaboration, social cohesion, in a community.

Eight, binary, social norms have been identified that may maintain full collaboration in a community, while resisting invasion of 'free riders'. The 'leading eight', as these are called, have in common that adoption of a norm by a whole community in a similar way is pivotal in stabilising collaboration. Subjects may, however, interpret a norm differently and, if so, different interpretations may proliferate within a community resulting in sub-communities. The 'leading eight' are, thus, sensitive for 'private' and 'noisy' information. The emerging sub-communities may call each other 'bad', although they employ actually the same social norm. We may recognise here what may have happened during the COVID-19 epidemics. The 'leading eight' also agree in dealing with 'good' and 'bad' donors and 'good' recipients. These, however, disagree with respect to the treatment of 'bad' recipients. The sensitivity of the 'leading eight' for 'private' and 'noisy' information may be reduced by adopting a quantitative assessment of 'good' and 'bad', instead of a strictly binary one, in conjunction with a threshold value. It means that a reputation may increase, or decrease, stepwise by each action of collaboration, or defection, passing at some moment the border, threshold, of 'good' and 'bad'. We may, for example, have a range of reputation from minus five, which is very bad, to plus five, which is very good. A donor may have a score of plus two being a good girl. It may become four after two actions of collaboration and she remains a good girl. In contrast, she may execute two actions of defection reducing her score of reputation to zero. She becomes a bad boy. Introduction of such a quantitative assessment with a threshold turned out to stabilise community collaboration for four out of the eight leading social norms.

The efficacy of quantitative assessment of a person's 'goodness' with respect to stabilisation of collaboration, trust, in a community was demonstrated in the model study of Schmid *et al.* (2023). The authors were not surprised about the results from a psychological point of view. Dichotomous thinking may be, on the one hand, beneficial for quick decision-making in situations of emergency, but, on the other hand, it may be the basis for personality disorders, like a borderline one. It also is a hallmark of anti-social behaviour, which may become detrimental for a community, if present at a larger scale. Do we, therefore, need to characterise the Western, individualistic, societies as borderline already, as the Belgian psychiatrist Dirk de Wachter suggested in some newspapers? We do not try to answer this question here. It is up to psychologists and sociologists. The statements of de Wachter, however, point to the necessity of including societal structure in management of epidemics, in general, and appropriate, human, management, specifically. Management should maintain, or increase even, societal cohesion to be both, human and appropriate. And, management thus starts with a proper dissemination of a risk profile to people. A challenge as dissemination should be specific for the societal structure of a country and the target pathogen. In addition, country- and pathogen-specific dissemination may be over-ruled by circulating opinions, statements, on the world-wide web. But, should that hamper governmental bodies to disseminate an honest risk profile, including all the inherent uncertainty?

REFLECTIONS

We may recognise that appropriate, human, management of epidemics needs dissemination of proper risk profiles rather than simple communication. We may reflect about a proper dissemination of risk profiles of the five pathogens presented in Chapter 7 in countries that differ in societal structure, like China, USA, Russia and Nigeria. We may also think about coping with the potential 'disturbance' of the dissemination caused by the common use of the world-wide web.

8.4 Outlook

We feel surveillance of emerging epidemics is in its infancy still. The urgency of a proper surveillance is well-understood. Various initiatives and actions pop-up, but a clear concept is missing still. We envisage development of such a concept. If so, the focus will, likely, on the stage of epidemic source rather than that of epidemic seed, as we see it from a point of view of efficacy. Optimisation of the sensitivity and specificity of the surveillance will be inevitably, neither to miss incipient epidemics that may have a relatively high impact, nor to get societies on a permanent state of emergency. Latent Class Analysis, or Latent Profile Analysis, of initial data of epidemic sources will aid the set-up of a proper health risk assessment, including the assignment of risks to specific groups of people, where applicable.

Health risk assessments need to serve management of epidemics. Such assessments, however, are not fully exploited by a 'one size fits all' management. A type of management that we may see generally. We, therefore, feel that appropriate, human, management of epidemics will get to the forefront the more and more. We have provided a first step towards such a management in this chapter. We will work it out further in the next chapter.

References

Aarestrup, F. M., Bonten, M. and Koopmans, M., 2021. One Health preparedness for the next. The Lancet Regional Health – Europe 9: 100210. DOI: 10.1016/j.lanepe.2021.100210.

Arita, M., Karsch-Mizrachi, I. and Cochrane, G., 2021. The international nucleotide sequence database collaboration. Nucleic Acids Research 49: D121-D124. DOI: 10.1093/nar/gkaa967.

Asadgol, Z., Mohammadi, H., Kermani, M, Badirzadeh, A. and Gholami, M., 2019. The effect of climate change on cholera disease: the road ahead using artificial neural network. PLoS ONE 14: e0224813. DOI: 10.1371/journal.pone.0224813.

Baker, R.E., Mahmud, A.S., Miller, I.F., Rajeev, M., Rasambainarivo, F., Rice, B.L., Takahashi, S., Tatem, A.J., Wagner, C.E., Wang, L-F, Wesolowski, A. and Metcalf, C.J.E., 2022. Infectious disease in an era of global change. Nature Reviews Microbiology 20: 193-205. DOI: 10.1038/s41579-021-00639-z.

Dron, L., Kalatharan, V., Gupta, A., Haggstrom, J., Zariffa, N., Morris, A.D., Arora, P. and Park, J., 2022. Data capture and sharing in the COVID-19 pandemic: a cause for concern. The Lancet Digital Health 4: e748-756. DOI: 10.1016/S2589-7500(22)00147-9.

Huang, C., Wang, Y., Li, X., Ren, L., Zhao, J., Hu, Y., Zhang, L., Fan, G., Xu, J. Gu, X., Cheng, Z., Yu, T., Xia, J., Wei, Y., Wu, W., Xie, X., Yin, W., Li, H., Liu, M., Xiao, Y., Gao, H., Guo, L., Xie, J., Wang, G., Jiang, R., Gao, Z., Jin, Q., Wang, J. and Cao, B., 2020. Clinical features of patients infected with 2019 novel coronavirus in Wuhan, China. The Lancet 395: 497-506. DOI: 10.1016/S0140-6736(20)30183-5.

MacLean, E.L., Kohli, M., Köppel, L., Schiller, I., Sharma, S.K., Pai, M., Denkinger, C.M. and Dendukuri, N., 2022. Bayesian latent class analysis produced diagnostic accuracy estimates that were more interpretable than composite reference standards for extrapulmonary tuberculosis tests. Diagnostic and Prognostic Research 6: 11. DOI: 10.1186/s41512-022-00125-x.

Schmid, L., Ekbatani, F., Hilbe, C. and Chatterjee, K., 2023. Quantitative assessment can stabilize indirect reciprocity under imperfect information. Nature Communication 14: 2086. DOI: 10.1038/s41467-023-37817-x.

Van Wees, D.A., Heijne, J.C.M., Heijman, T., Kampman, K.C.J.G., Westra, K., de Vries, A., de Wit, J., Kretschmar, M.E.E. and den Daas, C., 2019. A multidimensional approach to assessing infectious disease risk: identifying risk classes based on psychological characteristics. American Journal of Epidemiology 188: 1705-1712. DOI: 10.1093/aje/kwz140.

Wang, Z., Liu, Y., Li, Y., Wang, G., Lourenço, J., Kraemer, M., He, Q., Cazelles, B., Li, Y., Wang, R., Gao, D., Li, Y., Song, W., Sun, D., Dong, L., Pybus, O.G., Stenseth, N.C. and Tian, H., 2022. The relationship between rising temperatures and malaria incidence in Hainan, China, from 1984 to 2010: a longitudinal cohort study. Lancet Planet Health 6: e350-358. DOI: 10.1016/S2542-5196(22)00039-0.

Weller, B.E., Bowen, N.K. and Faubert, S.J., 2020. Latent Class Analysis: a guide to best practice. Journal of Black Psychology 46: 287-311. DOI: 10.1177/0095798420930932.

Wille, M., Geoghegan, J.L. and Holmes, E.C., 2021. How accurately can we assess zoonotic risk? PLoS Biology 19: e3001135. DOI: 10.1371/journal.pbio.3001135.

SOCIETAL RISK PERCEPTION

We introduced the first step towards appropriate, human, management of epidemics in the previous chapter, a proper health risk assessment and its dissemination. This chapter provides us insight into the second step, the receipt of the risk assessment across a community, the societal risk perception, as we will call it. We will highlight three, correlated, determinants of societal risk perception, (1) social cohesion within a community, (2) level of basic health across a community, and (3) societal resilience of a community.

9.1 Social cohesion

The term 'trinity of life' expresses the profound interaction between, body, mind and environment, as we have seen in Chapter 2. We use our body to sense our non-social and social environment. The sensory impressions are processed by our mind to fit in our own narrative of life, or not. If not, mental disorders may result. If so, we are able to assimilate impressions, or to accommodate our narrative of life accordingly. Our narrative of life, in turn, determines our actions, in a physical and mental sense, with respect to the environment. We have also recognised that we may look at our-selves from some 'virtual' distance, say, we become part of the social environment and see us as an 'object'. In addition, we see the body separating as much as possible from the environment to protect it against foreign, adverse, agents, like pathogens. It is a primary, bodily, reaction as supported by the mind. In contrast, fulfilling desires of others, besides the own ones, may be seen as a basic mental attitude of humans. We like to become close to others. It may be seen as an expression of, being part of something 'transcendental', it may be a recognition of the own fragility that needs the protection of others, or just the mastering of the own solitude. Whatever it may be, we look actually for social cohesion from the start of our life on. Social cohesion is sought at various spatial levels and within various groups of people. Space scales up from, neighbourhood to town, region, county, state, federal state, and finally the world. Groups include, close relatives, scholars, teammates, colleagues, chatters, co-citizens, supporters of the same sport club, members of the same political party, and so on. We focus here on social cohesion among residents of a (federal) state, as it is commonly the primary entity to which management of epidemics is directed.

The multi-dimensional phenomenon of social cohesion got attention from the start of social sciences on. It was the French Émile Durkheim (1858-1917), who addressed the topic of social cohesion back in the 19th century already. He may be regarded as the formal founding father of the academic discipline of sociology. He studied at the prestigious French 'École Normale Supérieure', which did not include social sciences at that time. Durkheim graduated in philosophy, but he went to Germany for an additional study in sociology. Germany, where two other founding fathers of social sciences were at home, Karl Marx (1818-1883) and Max Weber (1864-1920). Durkheim returned to France and he established sociology as a scientific discipline. He provided the first definition of social cohesion (Schiefer and van der Noll, 2017). It reads as: "Interdependence between members of society, shared loyalties and solidarity". This definition of 1863 was followed by a series of definitions, especially, as social cohesion became a hot topic by the end of the 20th century. The process of globalisation in, economics, migration, and communication, triggered concerns about the maintenance of social cohesion, especially in high-income, market-driven, countries. Concerns arose not among scientists only, but also among governments and politicians. Concerns thus resulted in various definitions of social cohesion, even as all kinds of approaches to determine and govern social cohesion popped up. The aim of a literature review of Schiefer and van der Noll (2017), therefore, was to determine the essentials of social cohesion rather than coming up with a catch all concept, a definition. Their review was a bit deviant from common ones. One, they included publications that were directed to both, the academic and policy discourse. Two, publications in German were included, besides the ones in English. We may notice here that German was the leading language in sociology until the mid of the 20th century, which is understandable from a point of view of history.

Three dimensions of social cohesion may be considered as essential (Fig. 9.1). The first, major, one is the degree of orientation of people towards a common good. It constitutes of feelings of responsibility, solidarity, with the common good and, voluntary, compliance with the social order and rules. Feelings and behaviour that are measurable and these, therefore, provide an estimation of the degree of orientation towards the common good. The second dimension of social cohesion is called 'social relations'. It encompasses four components, which are actually measurable as well. One, the quantity and quality of social networks, which are also called social relations. Two, participation, engagement, in politics and socio-cultural activities. Three, trust among people, which we call horizontal trust, and trust in institutions, which we call vertical trust. Four, mutual tolerance among sub-communities. Sub-communities emerge inevitably at the level of states, and certainly in larger countries. Mutual tolerance is needed to counteract potential enforcement of social cohesion, uniformity, by way of repression of sub-communities. The third, minor, dimension of social cohesion is the one of attachment, belonging, to the community. It is the self-perception of the individual of being part of the community and mutatis mutandis considering the community as part of the own identity. The three dimensions show some overlap. If you are, for example, a self-confident person with trust in others, it is easier to have commitment to a community and its common good. We should also notice that social cohesion per se cannot be expressed in one score. It is expressed in a profile of the components of the dimensions. The profile, subsequently, provides the clues to strengthening social cohesion, if desired.

We identified trust earlier as a major component in the adoption, and adaption, of risks. Trust among people, trust in institutions. We see trust back now as a component of social cohesion. We, therefore, elaborate on trust here. We turn to data of the World Value Survey (WVS, https://www.worldvaluessurvey.org/WVSContents.jsp). Data is collected as part of an academic social survey programme, which is executed worldwide. National teams collect the data using a long list of questions. The programme runs since 1981. The use of data is open for anybody and free of charge. Data are available in several, easily accessible, formats, like the one of Excel. Collection of data is executed in waves of five years. We use here data of the seventh wave, which was executed between 2017 and 2022 (Haerpfer *et al.* 2022). The majority of questions was similar to those included in previous waves, which enables longitudinal analyses. The list of questions was extended with other questions to cope with novel societal topics. In addition, national teams were free to pose some country specific questions. About 300 questions were compiled in total. A representative sample of residents aged 18 years and older was composed by a national team. A sample needed to include 1000 – 1500 residents at least, depending on the size of a country. In addition, data collection by way of interrogation in a local language was required, at least for 95% of the residents included in a sample. We present here some results of the survey, as collected in 64 countries world-wide. The largest samples of the survey were used in Canada (4,018), Indonesia (3,200) and China (3,036). The sample of Northern Ireland was an outlier with 447 residents only.

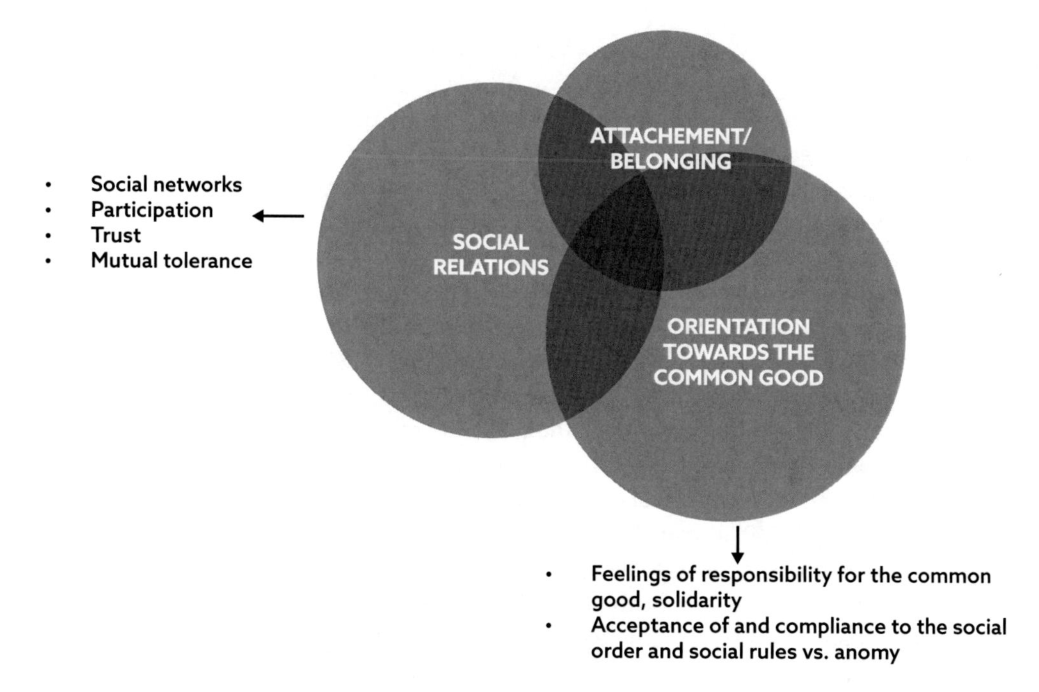

Figure 9.1. Three, partially overlapping, dimensions of social cohesion (circles), of which the constituting, measurable, components are indicated by arrows. See text for further explanation.

Questions 57-63 of the survey dealt with horizontal trust. The first one was a quite general one: "Generally speaking, would you say that most people can be trusted or that you need to be very careful in dealing with people? A majority of people of 75%, overall, opted for the option 'need to be very careful' instead of 'most people can be trusted'. In contrast, a majority of people in, China (64%), Netherlands (55%), and New Zealand (57%) indicated the second option. They expressed a general trust in people. The question of general trust was, subsequently, specified with respect to particular groups. The degree of trust in each group was reflected in four options of answers, (1) trust completely, (2) trust somewhat, (3) do not trust very much, and (4) do not trust at all. A majority of 77%, overall, trusted the own family completely and less than 1% not at all. Neighbourhood was trusted 'somewhat' by a majority of 54%. Seventeen percent of the people had, overall, a complete trust in the neighbourhood, whereas 7% had no trust at all. The scores on complete trust in neighbourhood were relatively high in Iraq (36%), Kyrgyzstan (30%) and Pakistan (34%). No trust at all scored less than 30% in all the countries. The answers turned, overall, from 'some trust' to predominantly distrust with respect to people that are met for the first time. A majority, overall, of 42% indicated 'do not trust very much' and 30% 'do not trust at all'. About a quarter indicated trust by choosing the options 'trust completely' (3%) and 'trust somewhat' (24%) , respectively. In contrast, a small majority indicated 'somewhat trust' in people met for the first time in, The Netherlands (56%) and Great Britain (51%). Half of the people of Northern Ireland also indicated somewhat trust. We do not deal with the answers to trust, or distrust, with respect to other specific groups here and we turn to vertical trust. The trust, confidence, in institutions.

Questions 64-89 of the survey covered confidence in a wide range of institutions, some very specific for a region, like the one directed to the Arab Ligue. We present here some questions that are widely applicable and relevant, as we feel, with respect to epidemiology. The questions, and options of answers, are all the same, except the addition of the specific institution. The core of the question is "I am going to name a number of organizations. For each one, could you tell me how much confidence you have in them: is it a great deal of confidence, quite a lot of confidence, not very much confidence or none at all?", and then a specific institution. The first one concerns religious institutions. Overall, 60% of the people indicated to have a great deal (29%), or quite a lot (31%), of confidence in a religious institution. A great deal dominated, especially, in Bangladesh (93%), Pakistan (87%) and Philippines (73%). In contrast, 12%, overall, indicated none confidence at all. The lack of confidence was, especially, abundant in Japan (40%), Andorra (34%) and Uruguay (34%). The following results were obtained with respect to confidence in the government. Overall, 45% of the people indicated to have confidence in their government, either a great deal (14%), or quite a lot (31%). A great deal of confidence dominated, especially, in, Tajikistan (53%), China (48%), and Myanmar (36%). In contrast, 23%, overall, indicated none confidence at all. The lack of confidence was dominant in, especially, Peru (58%), Tunisia (56%) and Mexico (54%). Confidence in universities, say science, was, overall, higher than the one in governments and religious institutions. Sixty-seven percent of the people indicated to have confidence in universities, either a great deal (20%), or quite a lot (47%). A 'great deal' dominated, especially, in Uruguay (52%), Pakistan (51%), and Puerto Rico (48%). In contrast, 7% indicated none confidence at all. The lack of confidence was, relatively, frequent in Iraq (30%), Jordan (26%) and Nicaragua (18%). Confidence in the press was, overall, rather low, as 41% of people had not very much and 19% none at all confidence. This lack of confidence was indicated by a relatively high proportion of residents, 86%, in Armenia, Libya and United Kingdom. Confidence in television showed a similar pattern. Confidence in social media, the world-wide web, was not determined. Confidence in the World Health Organization was expressed by 55% of the people, overall, as indicated by answering 'a great deal' (17%) or 'quite a lot' (38%). A great deal was indicated, especially, in Myanmar (59%), Kenya (43%) and Nigeria (39%). None confidence at all was, overall, indicated by 11% of the people, and it was indicated by a relatively high proportion of people in, Lebanon (35%), Serbia (32%), and Iraq (28%).

We highlighted here the data of some questions directed to horizontal and vertical trust in 64 countries. We do not have a single parameter expressing trust and, therefore, we need some reference to interpret the data of a specific country. The WVS-data provide such reference, which is a major strength of this repository. The reference is not provided by a comparison among countries only. We may also analyse the answers to the trust-directed question taking into account the answers to the other 200-300 questions. It enables a profound societal analysis of the residents of a country. It is another strength of the WVS-data, especially, as any scientist, researcher, or layman, may execute such an analysis. We did it not here, but we may understand from the data presented that trust, confidence, may appear in various expressions. Some trust their family, only, others do also trust unknown people. Trust may be especially in, science, religious institutions, or just a government. Does it matter for the social cohesion of a community, how trust appears? It does, whenever, people that have, for example, confidence in science do not tolerate those co-residents who have it in a religious institution, and *vice versa*. Confidence should be accompanied with mutual tolerance, as we indicated it in figure 9.1. In addition, we may notice that appropriate, human, management cannot be based on a one-size-fits-all approach. Likely, it takes into account the heterogeneity in horizontal and vertical trust of a society, as much as possible. If so, a managing body may use the channels of trust of specific groups to communicate about epidemics and it avoids interventions that may impair the trust of any group too strongly, or unnecessarily even. If not, the managing body may trigger intolerance among groups within a society. We feel maintenance of trust within a society is a basic condition in appropriate, human, management of epidemics.

We have described societal trust in a horizontal and vertical direction, respectively. Online platforms fit hardly this concept of trust in two directions. Online platforms may be defined as "programmable architectures designed to organize interactions between users online" (van Dijck, 2021). A platform is not a simple, neutral, technology. The hardware is first of all fuelled by data, which is generated by the users of the platform themselves often. Secondly, the platforms are automated and organised by closed algorithms that are, in general, not disclosed to third parties. Thirdly, the platforms are governed by global tech companies that actually escape from any public control. Nevertheless, digital platforms did, and do, direct the more and more our daily activities like, communication, purchase of goods, and decision-making with respect to, for example, health. We may turn to so-called 'platform societies' and, if so, this turn has consequences for, first of all, our vertical trust. It seems to shift, at least partially, from an institutional-professional one to a computational-corporate one. Likely, such a turn to 'platform societies' also affects horizontal trust.

We have zoomed in on trust. We will zoom out, back to social cohesion now. It is considered as a determinant of health in both, a positive and negative sense (Oberndorfer *et al*., 2022). Social cohesion may reduce, and buffer, chronic stress. Stress that may induce immune dysregulation by way of persistent inflammation. Societal cohesion may also protect from loneliness and its adverse effects on health. It has an ambiguous character with respect to diffusion of health information. It may enable dissemination of a healthy behaviour, but also the reverse is possible, depending on which institutions and social contacts are trusted. Social cohesion may also facilitate spread of pathogens and it turns then really into a health risk. We face, subsequently, the question whether the benefits of social cohesion outweigh such a risk. The answer will be a rather qualitative one, as we do not have a simple measure for social cohesion. In addition, the answer depends certainly on the target pathogen. Social cohesion will hardly contribute to the risk of spread of a pathogen that is common in an environment, like *Plasmodium* species for malaria, or a pathogen that disperses as a dispersive wave, like SARS-CoV-2 for COVID-19. It changes certainly for a pathogen like the Ebola-virus, of which the spread is especially among relatives, trusted people thus. So, we need understanding of the epidemic spread of a pathogen to estimate the epidemic risks imposed by social cohesion. If the risk is negligible, we may focus on the health benefits of social cohesion only. Health benefits that may be seen in the perspective of 'basic health', as we will do in the next section.

REFLECTIONS

The desire of social cohesion seems inherent to man. Social cohesion is a multi-dimensional phenomenon and it is, therefore, not easily to determine, or to predict. Trust is an essential component of social cohesion in both, a horizontal and vertical sense. We see heterogeneity in trust among, and within, countries. We may reflect about the role of trust in appropriate, human management of epidemics. How may we use it? How can we maintain, or even strengthen, it?

9.2 Basic health

Health is a multi-dimensional, subjective, phenomenon, as we have seen in Chapter 2. The WHO-definition of health is flawed and we do not have a one-size-fits-all alternative definition, or concept. We all, however, have some sense, some feeling, about health, and so, we may deal with the lack of a proper definition at the individual level. It is the basis of the common healthcare, as we know. We may also cope with this lack focusing on the control of a single disease at the population level, and thus, equalising absence of that disease in a community with (public) health. It changes as soon as we take a broader view. A broader view may be needed to determine side effects of interventions employed to control a specific disease. It also is needed to determine the effects of societal activities, like traffic, on health. Public health thus. Public health that we may see in legislation as a reason of exemption from specific rules quite often. Access of a product to the market may, for example, be denied due to its potential, and detrimental, impact on public health. Potential damage to public health may also trigger exceptional interventions by a government. So, we see some common sense with respect to (public) health at the community level. A common sense that is not expressed in clear definitions and concepts. It is more an overall feeling, a culture, within a community. And yes, we see differences in health policy resulting from such a common sense among, and within, countries. Subjectivity is inherent to our notion of public health. Subjectivity that we incorporated already in our concept of appropriate, human, management of epidemics, which we introduced in Chapter 8. Human, as it needs to address the feelings, concerns, of the whole community. We also included the term 'appropriate' in our management concept. We read it as being protective with respect to people at relatively high risk of suffering from an epidemic, while re-assuring those at relatively low risk that no specific interventions are necessary for them. Risks that can be estimated, relatively, objectively. Risks that vary among categories of people, as well. A variation that may, amongst others, depend on, (i) availability

and quality of nutrition, (ii) quality of the environment, (iii) grade of physical exercise, and (iv) social cohesion. We may see these as the determinants of 'basic health', as we call it here. Basic health that determines, in terms of infectious diseases, the relative level of resistance against, and tolerance of, pathogens. It is health as related to the style and environment of one's life. So, a relatively low basic health implies, in general, a relatively high risk of suffering from infectious diseases and a relatively high basic health is related to a relatively low risk.

The level of basic health does not affect the risk of suffering from infectious diseases only. It is also related to non-communicable diseases of both, mental and physical order. We may see again that the higher the basic health, the lower the probability of mental and physical disorders. We may, therefore, add an additional component of appropriateness to the management of epidemics. It needs to include interventions that impair the basic health of a community as less as possible to prevent increase of the incidence, or worsening, of non-communicable diseases and other infectious diseases. In addition, we may prevent epidemics, or reduce the potential impact of these, by way of health policies that increase the level of basic health in a community.

We will zoom in on the determinants of basic health in the following, except social cohesion. We have dealt with social cohesion in the previous section already. We will go along, nutrition, environment, and physical exercise, to estimate the effects of these on pre-disposition to non-communicable and infectious diseases. We will do this for communities under conditions of relatively low income and relatively high income, respectively. We feel income expresses fairly well the conditions of human life, especially focusing on the extremes of it, poor and rich thus.

Nutrition

Nutrition, in which we include water, is essential for human life. An essential that is not evident for millions of people worldwide, due to, poverty, climate disasters, and (inter)national conflicts. This is one extreme of the availability of nutrition. The other extreme is abundance of nutrition. And, we have all grades of nutrition in between. Availability of food, and water, is one aspect of nutrition as an essential of human life. Quality of it is another aspect. Water should be clean and food should have sufficient nutritional value to maintain a minimum of basic health at least.

Let us look now at the Dutch painter Vincent van Gogh (1853-1890). He painted 'The Potato Eaters' in 1885. It reflects the common daily pattern of eating among the poor people in The Netherlands. Potatoes at breakfast, potatoes at noon, if any at all, and potatoes in the evening. Potatoes were grown abundantly and these were relatively cheap. This nutritional habit of poor people could be observed in other countries, like Ireland and Germany, as well. And yes, potato itself is a relatively healthy food, but it is not healthy being the one and only food each day. We may state this for other staple foods, like rice and tapioca, as well. The implicit message of van Gogh's 'The Potato Eaters', *i.e.*, poor people who need to rely primarily on unilateral staple foods and to cope with a shortage of clean water, is reality for millions of people nowadays, still.

We stay at potatoes. We may imagine that Vincent Gogh could also have painted 'The Chips Eaters'. Chips as a symbol of the well-fed people, who eat just for fun, a pleasant feeling, or just to have something to do. The imaginary painting may illustrate, on the one hand, the surplus of food in some parts of the world and, on the other hand, the luxury of reducing the health value of potatoes by way of frying these. In addition, abundant consumption of chips may substitute relatively healthy components of the daily nutrition, like fruits and fresh vegetables. If so, nutrition becomes unilateral, like it was in the times of the potato eaters. Similarly, water may turn into beverages that may impair our health, like the alcoholised ones. So, we may characterise the daily diet of people nowadays, according to four major types:

(i) a shortage of nutrition;
(ii) malnutrition;
(iii) healthy nutrition;
(iv) a surplus of nutrition.

The first and last type express both the nutrition in a rather quantitative sense, say in calories. Malnutrition and healthy nutrition express additionally the composition of the diet with respect to the lack and presence, respectively, of all the essential nutrients we need. Quality thus. We see a shortage of nutrition especially among people with a low income and a surplus among those with a high income. In contrast, malnutrition may be observed among both, poor and rich people. We observed the paradox of an increase of food security that was accompanied by an increase of malnutrition in several countries during the last century.

Malnutrition has been recognised as a major risk factor of various diseases (Downer *et al.*, 2020). It has triggered the idea of positioning nutrition more prominently in healthcare systems. The slogan 'food is medicine' has been coined. Three major types of interventions and corresponding target groups are foreseen. One, preparation of medically tailored meals by a professional for people who are not able to shop and cook due to their medical state. We may think about cases of AIDS, cancer and heart failure. It is a target group that is closely related to the current core business of medicine. It is a relatively small target group. Two, distribution of medically tailored groceries, which are selected by a professional, to cases who are able to prepare food at home. People suffering from diabetes are a typical example of this target group. A group that is larger than the preceding one and, actually, we are at the border of the common healthcare systems, serving patients in this way. We are passing the border of healthcare systems certainly with respect to the third intervention. Delivery of produce prescriptions, of which the ingredients may be retrieved at discount, to people at risk, or those in an initial stage of pathogenesis. Obese people are, for example, at higher risk of, severe, COVID-19. They are either less resistant against the SARS-CoV-2 virus or, more likely, less tolerant for disease associated with the operation of their immune system. This third type of intervention targets the largest group of people, like elderly people and people with a low income. All of the three interventions executed by way of the common healthcare system may be more convincing to people, patients, than all kinds of public campaigns and projects. A substantial return on investment is, therefore, expected in terms of both, health and finances for such food interventions coupled to a healthcare system.

We may agree the concept of 'food is medicine', or not. It stresses, anyway, nutrition as a determinant of basic health, as we call it here. We turn now to another determinant of it, the environment and its quality, or lack of it.

Environment

Human life is a trinity of body, mind and environment. We distinguish these three entities, whereas we can actually not. In addition, the interactions of body and mind, respectively, with the environment are bilateral. The environment affects us and we are affecting the environment. An environment that encompasses social and non-social components. It may be characterised, especially, by stressors posing a health risk, or by opportunities to build up health resilience. The perception of an environment also varies from individual to individual, as each has its own physical constitution and narrative of life. We may talk about a personalised, subjective, environment of each of us at microscale. A micro-environment that we cannot separate from the environment at a larger scale. We all life in a neighbourhood, a community, a state, a climate zone, and we all live on one earth. We also see an intense, dynamic, interaction between the social and the non-social environment. A non-social environment that encompasses biotic and abiotic components. Our environment is a multi-dimensional, subjective, phenomenon. We do, however, reflect about it in a rather one-dimensional way in general.

The environment, social and non-social, is a determinant of (basic) health without doubt. We talk about 'environmental health' and we may associate environment with health equity (Zota and Shamasunder, 2021). Health equity is a societal state, in which everybody has opportunities to attain full health potential, irrespectively her, or his, social position, or any other socially determined condition. The opportunities are, however, not equally distributed among people due to, amongst others, variation in the quality of the environment in combination with socio-demographic inequalities (Ganzleben and Kazmierczak, 2020). Relatively deprived people, in any society, suffer more from environmental threats, and they benefit less from the opportunities of the environment, than the privileged people. The number of deprived ones may be substantial. It is estimated at about 6 million people in a high-income country like The Netherlands even, which is about 35% of the Dutch population. These people may be over-exposed to air pollution in their private, or professional, life, they may life in a noisy environment, they may not have easily access to green spaces, the housing may be poor, they may feel socially not accepted, and they may not have a perspective to improve their living conditions. We may extend this list endless and we may apply it to any country world-wide. Deprived people have to deal with one, or more, of these adverse health conditions. Irrespectively the number they have to deal with, the message remains the same. A low-quality environment impairs the basic health of relatively large proportions of people due to socio-demographic inequalities. These people do not have the capabilities, in general, to cope with epidemics adequately by way of, neither resistance against pathogens, nor tolerance of the resulting diseases. This also holds for another determinant of basic health, the degree of physical exercise, to which we turn now.

Physical exercise

Too few and too much food impairs health, respectively, as we elaborated above. It is similar for physical exercise. Poverty may force people to walk kilometres and kilometres to get water, or food. They may also walk to escape from, war, famine, or any other life-threatening condition. Long exercises exhaust the body and mind, especially under adverse conditions, and people may get in a vicious circle of, a need to move, exhausting their body, additional enforcement of the need to move, and so on, and so on. Once you start to move, you have to keep on moving, like refugees do. Too much physical exercise may also be caused by heavy, physical, jobs in combination with a lack of opportunities to relax. We also observe this commonly just among deprived people. In contrast, a lack of physical exercise is typical for people with a sedentary occupation and those who use predominantly motorised means of transport. Diabetes and cardiovascular diseases are examples of non-communicable diseases associated with a lack of physical exercise. Mental stress, and the resulting mental and physical disorders, also is related to absence of physical exercise. Such disorders may affect our resistance against, and tolerance of, pathogens and the diseases related to these. A lack of physical exercise may affect (basic) health of both, deprived and privileged people. We may see it as a kind of equity. Privileged people, however, have easier access to facilities that enhance physical exercise in a pleasant setting, *e. g.*, fitness clubs, green space for jogging, water sport, national parks for hiking, and so on. In addition, they may recognise the benefits of physical exercise due to their social environment in combination with access to health information. We may notice here that people are, in general, stimulated by a natural environment to exercise (*cf.* Remme *et al.*, 2021). So, privileged people may relatively easily escape from a sedentary, unhealthy, lifestyle in comparison with deprived people.

Degree of physical exercise is the last determinant of basic health. We noticed the heterogeneity in this determinant among people, as well as we observed it in the other three determinants of basic health, *i.e.*, social cohesion, nutrition and environment. All in all, we encounter a fairly large heterogeneity in basic health in each society. This heterogeneity is likely to affect the societal resilience with respect to epidemics. We will turn to that topic in the next section.

REFLECTIONS

Basic health cannot be expressed in a single value. It is a multi-dimensional phenomenon, of which we need to assess the major determinants, *i.e.*, social cohesion, nutrition, personalised environment, and physical exercise, to get grip on it. If so, we may determine quite a large heterogeneity in basic health among people. We may reflect about this heterogeneity with respect to appropriate, human, management of epidemics. How may we deal with it, if we take into account both, the group of deprived and privileged people?

9.3 Societal resilience

We introduced the term 'resilience' at the individual level already. It indicates the physical and metal capability to bend back after exposure to environmental 'stressors', like pathogens. The term 'resilience' refers to, the state before exposure to a stressor, the way of dealing with it, or the state after exposure. Resilience is based on the mechanisms of resistance and tolerance. We elaborated the difference between resistance against, and tolerance of, pathogens in Chapter 5. Resistance is directed to inhibition of pathogens, minimising the impact of these. Massive resistance against pathogens might, however, serve as a selecting factor in the evolutionary dynamics of pathogens, stimulating the build-up of pathogenicity. If so, pathogens may break the resistance. We see such a kind of selection with respect to anti-biotics resistance of bacteria, which became abundant due to the massive use of anti-biotics world-wide. In contrast, tolerance of pathogens does not result in selection, as it does not inhibit pathogens. Tolerance is directed to the 'damage' caused by pathogens, directly or indirectly. So, we need to keep in mind these different mechanisms, dealing with resilience at the population level, societal resilience, thus.

Resilience means, preparing, using, and restoring reserves. It holds at the individual level, but also at the population level. Reserves that are needed to keep on running daily life of a community, as much as possible, while dealing with a disaster (Wulff *et al.*, 2015). And we see an increase in, frequency, damage, and complexity of disasters, because people tend to aggregate in communities of a relatively high complexity. Build-up of societal resilience, community resilience, seems mandatory to deal with all the disasters, crises, nowadays. Health may be seen as the pivot in the build-up of resilience of a society to any disaster, whether it is a hurricane, an attack of terrorists, periods of extreme heat, epidemics, or whatever. In addition, everyday resilience may support disaster resilience and we also may connect people and systems. If so, we see the differences between disaster preparedness and build-up of resilience (Table 9.1). Disaster preparedness is governed by authorities, in which citizens have a minor role. A role that is clearly described in a plan. In contrast, resilience may be based on the inherent desire of people to aid each other, especially, in times of a disaster. It is based on social relations within a family, a neighbourhood, or just people unknown to each other. We may notice that the pool of volunteer citizens is, in general, far larger than a pool of professionals only. In addition, their aid is delivered faster than it is for professionals that need to come to the site of emergency, or disaster. A plan of emergency is also directed to restoring a pre-disaster state rather than strengthening a community. A strengthening that reduces the impact of consecutive disasters. Build-up of resilience is sustainable development of a community.

The COVID-19 pandemic may be seen as an outstanding stressor of testing resilience of societies. National responses of 28 countries to COVID-19 were investigated focusing on health systems in a broad sense (Haldane *et al.*, 2021). Countries were selected purposively with respect to, (i) relatively high, moderate, and low incidence of mortality attributed to COVID-19, respectively, (ii) representation of various regions worldwide, and (iii) reflecting a variety of health systems and economic status. A literature review was executed among peer-reviewed publications and public reports, extracting data of responses to the COVID-19 epidemic. Sixty-one items were included in the review. The review was supplemented with transcripts of semi-structured interviews and written submissions of experts in each country. It resulted in 45 interviews and submissions in total, which is less than two per country on average. Country-specific experts validated the data of each of the 28 countries. In addition, in-depth case studies of six countries were executed to support the data of all countries. Roundtable discussions with 35 national and international COVID-19 experts finalised the study. A health system resilience framework of the World Health Organization was adopted, and adapted, for the analysis of data. Within that framework, all sub-systems pivot around community engagement and health equity is a guiding principle. The study did not rank countries with respect to the degree of resilience of the health system. Trends and examples of the responses were provided only. We may summarise, and interpret, these as follows.

One, provision of healthcare services was limited in professionals. The lack of professionals could be a steady state due to, a lack of healthcare funding, under-payment of professionals in healthcare in comparison with other sectors, or efficiency in market-driven healthcare systems. It could also result from a relatively high rate of absenteeism of professionals due to COVID-19, as they were commonly exposed to the SARS-CoV-2 virus in combination with stressful working conditions. The lack was tackled by way of re-allocation of staff within healthcare systems. The re-allocation resulted, of course, in a lack of professionals to care for people suffering from other morbidities and disabilities, or being at risk for these. Re-allocation of staff does certainly not fit the concept of resilience. In contrast, engagement of community health workers and volunteers did fit resilience. They were engaged in, provision of essential services, risk communication, and provision of feedback on national strategies. The essential services encompassed home visits for, a check of the mental status of people, supporting vulnerable people in their daily care,

Table 9.1. Comparison of preparedness only and societal resilience of disasters. See text for explanation.

COMPONENT	PREPAREDNESS OF DISASTERS	RESILIENCE TO DISASTERS
Basis	Plan	Social relations
Executers	Governmental bodies	Whole community
Outlook	Short term	Long term
Focus on	Disaster	Daily life
Orientation	Risk	Social strenghts
Development	Not, bend back to initial state	Sustainable development

especially those with chronic medical conditions, and organising and promoting adherence to public health interventions. The degree of community engagement varied among countries. Thailand was quite exceptional in employing over one million community health workers.

Two, facilities and equipment were insufficient, overall, at the onset of the epidemic, to protect people, to screen on the virus, and to treat patients. The shortage in Intensive Care Units to treat a, severe, acute, respiratory, syndrome was striking. It resulted from a (too) strong efficiency policy in health care in various countries, like The Netherlands. Additional facilities were constructed massively in some countries, like China. In other countries, facilities were turned in ones dedicated to COVID-19. Countries like Singapore were prepared by way of stockpiles of personal protective equipment for months, but most countries were less well prepared. Countries without stockpiles were limited in purchase of equipment due to Corona-interventions that obstructed trading world-wide. The obstructions were especially detrimental with respect to purchasing in China, a major producer of medicinal equipment. All in all, we may conclude that countries, in general, were rather poorly resilient with respect to facilities and equipment to deal with the COVID-19 epidemic.

Three, governments provided substantial funding for, treatment of COVID-19, diagnostic testing, and relief of people and business. The degree and goals of funding varied among countries. The sources of funding were not indicated in the study of Haldane *et al.* (2021). It may have taken out of, national reserves, loans, or funding of other public tasks than health and well-being. The latter source does certainly not fit the concept of resilience, as it affects the execution of those public tasks. We may, overall, conclude that substantial, public, investments were required to compensate for a lack of societal resilience. High-income countries do, in general, have such a financial resilience, whereas low-income countries do not have. A country like Niger, for example, requested support from donors to deal with the COVID-19 epidemic.

Finally, the study of Haldane *et al.* (2021) did not investigate the mental resilience among countries. Did societies vary with respect to tolerance of a certain damage caused by a COVID-19 epidemic? We may imagine that societies struck frequently by epidemics may tolerate these at a higher degree, or these are forced to do so, than societies that are not exposed frequently to, severe, epidemics. We may have noticed a certain 'panic' in high-income, complex, societies upon onset of COVID-19, as these are not familiar anymore with such a kind of epidemics. We may refer here to the ongoing pandemic of cholera in low-income countries without massive interventions, like those we have seen for COVID-19.

REFLECTIONS

Societal resilience, in a narrow sense, is the ability of a community to return to its social state that preceded an epidemic. The return is facilitated by both, minimising the impact of an epidemic by way of resistance, and tolerating it to a certain degree. The latter may be beneficial from a point of view of evolutionary dynamics of pathogens. We may reflect about strategies to optimise societal resilience using both mechanisms.

9.4 Outlook

Social cohesion, basic health, and societal resilience have each been investigated extensively. Each of it is a complex on its own, which is rather impossible to abstract in a simple score, a parameter. The lack of such an abstracting score may explain that a link between each of these three and management of epidemics is uncommon. We did here and we went a step further even. We integrated the three into 'societal risk perception', as we called it. Such an integration makes sense, as a stronger social cohesion may result in sufficient basic health across a whole community and this may, subsequently, enforce societal resilience. In addition, sufficient societal resilience may keep us away from massive, invasive, social responses, and the drawbacks of these, as we will outline in Chapter 10.

We cannot improve simply societal risk perception, if necessary, as it would need a transformation of a whole society. We, however, envisage that epidemiologists take the status of societal risk perception of a country into account by drafting their strategies to cope with an epidemic. They will need a strong input of humanities, social sciences, to have a proper view on the actual status of societal risk perception in a country. If so, strategies may fit appropriate, human,

management of epidemics, as we call it. And, perhaps, experiences with epidemics, like the COVID-19 one, may provide a trigger to societies to invest in, social cohesion, basic health, and societal resilience. This would be worthwhile, not from a point of view of appropriate, human management of epidemics only.

References

Dijck van, J., 2021. Governing trust in European platform societies: Introduction to the special issue. European Journal of Communication 36: 323-333. DOI: 10.1177/02673231211028378.

Haerpfer, C., Inglehart, R., Moreno, A., Welzel, C., Kizilova, K., Diez-Medrano J., Lagos, M., Norris, P., Ponarin, E. and Puranen, B. (eds.), 2022. World Values Survey: Round Seven - Country-Pooled Datafile Version 5.0. JD Systems Institute & WVSA Secretariat, Madrid, Spain and Vienna, Austria. DOI:10.14281/18241.20.

Haldane, V., De Foo, C., Abdalla, S.M., Jung, A-S., Tan, M., Wu, S., Chua, A., Verma, M., Shrestha, P., Singh, S., Perez, T., Mieng Tan, S., Bartos, M., Mabuchi, S., Bonk, M., McNab, C., Werner, G. K., Panjabi, R., Nordström, A. and Legido-Quigley, H., 2021. Health systems resilience in managing the COVID-19 pandemic: lessons from 28 countries. Nature Medicine 27: 964-980. DOI: 10.1038/s41591-021-01381-y.

Downer, S., Berkowitz, S.A., Harlan, T.S., Olstad, D.L. and Mozaffarian, D., 2020. Food is medicine: actions to integrate food and nutrition into healthcare. The British Medical Journal 369: m2482. DOI: 10.1136/bmj.m2482.

Ganzleben, C. and Kazmierczak, A., 2020. Leaving no one behind – understanding environmental inequality in Europe. Environmental Health 19:57. DOI: 10.1186/s12940-020-00600-2.

Oberndorfer, M., Dorner, T.E., Leyland, A.H., Grabovac, I., Schober, T., Šramek, L. and Bilger, M., 2022. The challenges of measuring social cohesion in public health research: a systematic review and ecometric meta-analysis. SSM - Population Health 17: 101028. DOI: 10.1016/j.ssmph.2022.101028.

Remme, R.P., Frumkin, H., Guerry, A.D., King, A.C., Mandle, L., Sarabu, C., Bratman, G.N., Giles-Corti, B., Hamel, P., Han, B., Hicks, J.L., James, P., Lawler, J.J., Lindahl, T., Liu, H., Lu, Y., Oosterbroek, B., Paudel, B., Sallis, J.F., Schipperijn, J., Sosic, R., de Vries, S., Wheeler, B.W., Wood, S.A., Wu, T. and Daily, G.C., 2021. An ecosystem service perspective on urban nature, physical activity, and health. Proceedings of the National Academy of Sciences 118: e2018472118. DOI: 10.1073/pnas.2018472118.

Schiefer, D. and van der Noll, J., 2017. The essentials of social cohesion: a literature review. Social Indicators Research 132: 579-603. DOI: 10.1007/s11205-016-1314-5.

Wulff, K., Donato, D. and Lurie, N., 2015. What is health resilience and how can we build it? Annual Review of public Health 36: 361-374. DOI: 10.1146/annurev-publhealth-031914-122829.

Zota, A.R. and Shamasunder, B., 2021. Environmental health equity: moving toward a solution-oriented research agenda. Journal of Exposure Science & Environmental Epidemiology 31: 399-400. DOI: 10.1038/s41370-021-00333-5.

10

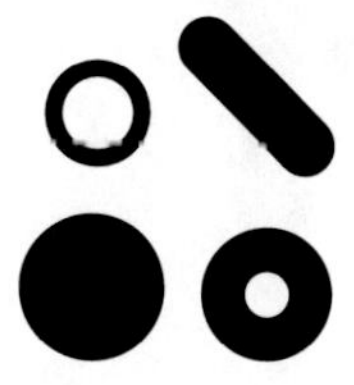

SOCIETAL RESPONSES

We have seen in the previous chapter that the societal risk perception of an epidemic determines actually the type, and extent, of interventions to manage an epidemic. We will present the major categories of interventions in this chapter, which are, social distancing, vaccination, and medication. We will provide the main characteristics of each briefly and we will estimate the efficacy, in general, with respect to the various types of epidemics that we have distinguished. In addition, we will point to potential drawbacks of the interventions from both, a bio-medical and social point of view.

10.1 Social distancing

Social distancing is a primary, relatively primitive, reflex to an external threat, which is based on our dual, bodily, closure, as we have seen in Chapter 2. Such a reflect of separating from the environment is essential in bodily survival as an organism. In contrast, we need to open our body to take up resources and to get rid of waste. This also opens the door for entry of pathogens and other detrimental substances. We unlock our body for reproduction as well. Love, in general, is a stimulus that does not open our body only, but also our mind. We look each other in the eyes, shake hands, embrace, kiss, and so on. In short, we need love to survive as human organism. So, we have two conflicting essentials of human life, separation versus attraction of each other. We focus on bodily separation, while ignoring the aspect of attraction, if we use the term 'physical distancing' instead of 'social distancing' (*e. g.*, Islam *et al.*, 2020). In contrast, use of the term 'social distancing' (*e. g.*, Lewnard and Lo, 2020) acknowledges that we need to suppress our feelings of attraction as well, if we keep distance from each other. The two essentials of human life cannot be treated independently, as the term 'physical distancing' suggests. We, therefore, refer further to social distancing rather than physical distancing.

People employ social distancing since ancient times to cope with infectious diseases. The bible is full of examples of isolation of lepers, or just the reverse, holy people who embrace lepers as an act of mercy, an act of love. The sexually transmitted disease of gonorrhoea is another historical example. Our history is full of celebrities suffering from gonorrhoea, or other sexually transmitted diseases. Abstention of intercourse was the only option to prevent such diseases in the past. A striking example of social distancing. Nowadays, we may use condoms, we may test on the causal pathogens before intercourse, and we have improved medication to treat the diseases. Promiscuity and abstention remain, of course, as proven means of prevention of sexually transmitted diseases.

Avoidance of bodily contact is one expression of social distancing. Avoidance of contact with blood and bodily excretions is another. Pathogen-contaminated blood may be transferred by way of wounds from a donor to a recipient. It may result from accidently and intentionally actions by people, but also by blood sucking vectors. We know, for example, the protective clothes that are worn by people, who take care of Ebola patients, and the sterilisation of equipment used in surgery, respectively, to avoid transfer of pathogen-contaminated blood. The use of a mosquito net is an example of prevention of pathogen transfer by way of a vector, in this case mosquitos transmitting *Plasmodium* species that cause malaria. Contact with faeces and urine may be reduced by way of proper sanitation and hygiene, like hand washing. This is the strategy to prevent infectious diseases of the digestive tract, like cholera. We keep out contact with excretes of the respiratory tract, *i.e.*, sputum, droplets, and aerosols, by way of, coughing and sneezing behind our hand, face-masks, and keeping distance from each other. Coverage of the mouth and nose by a hand is a way of social distancing that intends to protect others from you as a, potential, source of pathogens. It became actually part of our social norms, our civilisation, as we may call it. In contrast, face masks are, in general, not done, except at specific celebrations and health emergencies. Face coverage is, especially, attributed to criminals and terrorists. So, we have a rather negative association with face masks. We may notice that face masks have a dual function. These may protect the wearers as well as others, if the wearers serve as sources of pathogen dispersal. Similarly, keeping distance may be protective with respect to both, people serving as sources and those being receivers. We see heterogeneity in the social norms with respect to keeping distance. Crowding is, in general, detested by the high society, while it is highly appreciated by the 'common' people who, for example, visit a dance festival. The use of distancing to cope with epidemics is an old intervention, especially, with respect to the isolation of cases.

Social distancing became a topic of primary interest during the COVID-19 pandemic. We follow here the path of its implementation world-wide, including all the uncertainties about the fate of the pandemic and its management. A comment in The Lancet of March 2020 may illustrate both, the uncertainties in an early stage of an epidemic and the call for action by way of social distancing (Anderson *et al.*, 2020). They pointed to four uncertainties at that time. Uncertainties about SARS-CoV-2 and COVID-19 that were discussed from a point of view of the common, relatively well-known, seasonal influenza A. One, the case fatality rate was quite uncertain. A best guess of it seemed to be 3 to 10 times higher than the one of influenza. We know now that the case fatality rate of both viral diseases is, overall, quite similar (*cf.* Ioannidis, 2021). Two, onset of the infectious period after the incubation period was doubtful. The latent period of SARS-CoV-2 seemed to be shorter than the one of SARS-CoV-1 and rather similar to the one of influenza. If so, dispersal of the virus from pre-symptomatic, and symptomless, cases would be possible. It is indeed and it is quite essential in the epidemic spread of the virus, as we know meanwhile. Three, doubt about the proportion of a-symptomatic and relatively mildly-ill cases existed as well. Initial estimates indicated a proportion of about 80% of the cases. If so, symptom-based management of an epidemic would not be efficacious. Unnoticed dispersal of the virus turned out as Achilles heel of the management of COVID-19, indeed. Four, the length of the infectious period of the virus was unknown. It seemed relatively long in comparison with the one of influenza A, which has been demonstrated later on, indeed. All in all, the authors concluded that the size of a COVID-19 epidemic would exceed the common ones of seasonal flu. In addition, the lack of efficacious drugs and vaccines forced them to state: "So what is left at present for mitigation is voluntary plus mandated quarantine, stopping mass gatherings, closure of educational institutes, or places of work, where infection has been identified, and isolation of households, towns, or cities". We see here various types of social distancing suggested, from the relatively small scale of households up to the relatively large scale of lockdowns of cities. A lockdown means literally that people are not allowed to freely, enter, leave, or move around in an area.

The idea of social distancing was picked up further by a comment in the Lancet in June (Lewnard and Lo, 2020). They pointed to the necessity of evidence-based interventions of social distancing. Evidence that was provided predominantly by simulations of epidemics, which were based on the basic reproductive number R_0, at that time. We described the pitfalls of such modelling in the Chapters 5 and 6 already. The authors also requested attention for justice and appropriateness in implementing interventions of social distancing to avoid a disproportional burden of deprived people. In contrast, a systematic, rapid, review of 25 modelling studies and four observational studies had been indicated the available evidence of the efficacy of social distancing as low to very low back in April 2020 (Nussbaumer-Streit *et al.*, 2020). A 'natural' experiment in 149 countries' added practice-based evidence of social distancing by July (Islam *et al.*, 2020). We turn to it now.

Evidence, or not, governments had started to intervene as soon as the first case of COVID-19 was detected within a country, the natural experiment, as referred to by the authors. A team of policy and governance experts at the University of Oxford (UK) recorded public policy interventions directed to COVID-19 around the world. Data is stored at the Oxford COVID-19 Government Response Tracker (www.bsg.ox.ac.uk/research/covid-19-government-response-tracker). Data of governmental interventions directed to social distancing, which was collected until the 30th of May 2020, was used in the study presented here. Five types of interventions were distinguished, (i) closure of schools, (ii) closure of workplaces, (iii) restriction on mass gathering, (iv) closure of public transport, and (v) lockdown. A lockdown was defined as a combination of regulations for people to stay at home and restrictions on moving around in a country. Data of COVID-19 cases in the various countries in the period until the 30th of May was retrieved from the European Centre for Disease Prevention and Control. An interrupted time series analysis was employed to estimate the effects of interventions of social distancing on the incidence of COVID-19. The set-up of such an analysis is depicted in figure 10.1, in which COVID-19 incidence was the variable of interest reported here. It is also called a quasi-experimental time series analysis. We may also see it as an observational cohort study, in which the cohort becomes exposed to an intervention at a certain moment. Anyway, the slopes before and after exposure, respectively, were determined by way of a logistic regression model. The regression coefficients were, subsequently, presented as incidence rate ratios to express the effects of the interventions. A lag period of seven days was included. It seems a bit long taking into account the knowledge we have now.

One-hundred and eighteen countries employed all the five interventions of social distancing in the period of study. Twenty-five countries established four interventions, three interventions were applied in four countries, and two and one intervention in each of one country, respectively. The pooled incidence rate ratio determined for the 118 countries was 0.87 with a 95%-confidence interval between 0.85 and 0.90. So, the incidence of COVID-19 was, overall, reduced by 13% after implementation of the interventions. This estimate remained by including the 31 countries with fewer types of interventions. The lowest value of the incidence rate ratio of a country was determined for Slovenia and China. It was 0.65. The one of Slovenia was significant with a 95%-confidence interval of 0.54 – 0.79. It was not for China, showing a rather broad interval of 0.65 – 1.06. We may notice that insufficient compliance may have reduced the efficacy of the interventions, whereas other, concomitant, measures may have amplified the effects. Confounding cannot be ruled out in such an uncontrolled study. Wearing facing masks is a likely confounding factor increasing, or just decreasing, the efficacy of the interventions. We turn to face masks as another type of social distancing now.

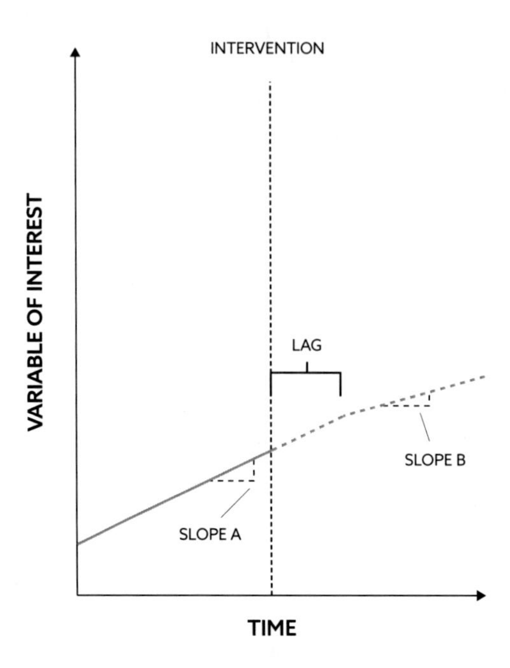

Figure 10.1. Schematic representation of an interrupted time series analysis of the variable of interest versus time. The variable of interest versus time is represented before and after an intervention directed to the variable. A delay of the effect of the intervention, as indicated by 'lag', may be included, or not. The effect of an intervention is here represented in a difference between the slopes of the line before and after the start of the intervention.

The use of face masks is common in the clinical setting, especially in the operating theatre. The wearing of masks became also popular among residents of air-polluted cities, like Beijing. The use of these was also recommended, or mandated even, by governments to cope with COVID-19. Research on the efficacy of such public interventions is limited in the heterogeneity of wearing face masks. Various types of face masks exist and the periods, and sites of, wearing vary relatively strongly among, and within, countries. Face masks may be worn, for example, in public transport only, or just for any out-of-doors movement. People may frequently refresh a mask, or not. They may wear it properly, or not. And, are people exposed to the SARS-CoV-2 virus the more, or the less? So, it is hard to obtain evidence of the efficacy of face masks in daily life. We may, however, investigate the efficacy of various types of masks under strictly-controlled conditions. We present an example of such research here.

The study was directed to both, the performance of the fabrics with respect to capture of ultra-small particles and the sealing performance of the masks (Duncan *et al.*, 2021). Masks representing five categories were included, (i) fabric 2-layer, which were constructed from, quilt batting, cotton, nylon, polyester, or silk, (ii) multi-layer, in which a cotton inner and outer layer enclose a layer of, furnace filter, quilt batting, electret filter membrane, disposable procedure mask, non-woven shopping bag material, or non-woven polypropene craft material, (iii)

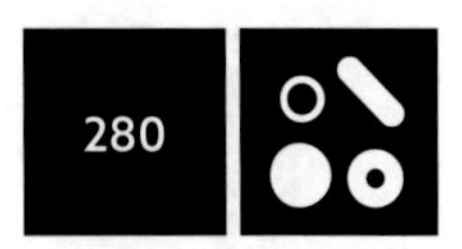

procedure/surgical of six different disposable face coverings commonly used in the clinical setting, (iv) two models of KN95 and (v) seven models of N95 Filtering Facepiece Respirators (FFR). KN95 and N95 refers to certification in China and the United States of America, respectively. Performance of fabrics was determined using an aerosol swatch penetration-set-up. An aerosol mixing chamber was used to create aerosols in a range of about 0.023 to 5 micrometres. Aerosols contained sodium chloride particles of sizes representative of bacteria and viruses. Standardised air flows passed first a charge neutraliser and then a fabric sample holder. Aerosols were quantified before and after passing the holder, the fabric. Temperature was 25 °C ± 5 °C and Relative Humidity 30% ± 10%. The set-up of the experiment enabled measurements of a filtration efficiency larger than 99.99%. The penetration of a fabric was determined for each aerosol size. A geometric mean of penetration was, subsequently, calculated including all the aerosol sizes, fabrics in a category, and replicates. The number of replicates varied.

The sealing performance of masks was determined by way of 11 subjects who wear the masks of the various categories. The subjects were employed at the research institute and they were familiar with wearing face masks. Five subjects were of smaller face size, four of medium size, and two of a larger size. A subject was placed in a closed, plastic, cabin, in which the same aerosol conditions were created as in testing the fabrics solely. A sampling probe was inserted in a mask between the mouth and nose of the subject to determine the aerosol concentration behind the mask. The concentration inside the cabinet was measured as well. The difference in concentration of aerosols between the chamber and behind the mask expresses the total inward leakage penetration by way of fabric penetration plus insufficient fit of a mask to the face. A geometric mean of total inward leakage penetration was calculated including the various mask types within a category and replicates. The number of replicates varied among types and categories.

The penetration of the fabric of a mask by particles decreased from 56% for those of the category 'fabric 2-layer' to about 1% for the categories of KN95 and N95 FFR (Table 10.1). The total inward leakage was higher for all categories of masks, except those of N95 FFR, which are well-known for a tight fit to the face. The lower value of N95 masks, in comparison with the one of fabric penetration only, may be attributed to experimental error. Anyway, the N95 FFR masks were the only ones achieving a protection that might be sufficient to cope with a virus like SARS-CoV-2. We may, therefore, question the efficacy of the face masks worn commonly in the public domain with respect to both, individual protection and inhibition of the COVID-19 epidemics. The wearing may have facilitated epidemics even, as people felt safe to use, for example, public transport and to go out for shopping. It is like the 1-3 metre distance rules that provided an unjustified feeling of safety.

An unjustified feeling of safety is one potential, adverse, side effect of social distancing. Social distancing may also have detrimental effects on health of people, as it results inevitably in a reduction of social contacts. Contacts with beloved ones, neighbours, colleagues, or just people you meet for the first time. No hand-shakes, hand on the shoulder, embraces and all those kinds of physical touches expressing compassion, signs of living together. Contacts,

Table 10.1. Overview of performance of various types of face masks under strictly-controlled conditions according to Duncan *et al.* (2021).

CATEGORY OF MASKS	FABRIC PENETETRATION*	TOTAL INWARD LEAKAGE PENETRATION**
Fabric 2-layer	56%	71%
Multi-layer	28%	56%
Procedure/surgical	10%	43%
KN95	0.7%	16%
N95 FFR	1.4%	0.6%

* Based on the geometric mean of, all sizes of particles, fabrics in a category, and replicates.

** Based on the geometric mean of, all types of fabrics of a category and replicates.

or the underlying emotions, that are known to have a protective role with respect to the detrimental effects of, pain, stress, and inflammation (Morese and Longobardi, 2022). Physical contacts provide people a feeling of well-being and social acceptance. If not, people experience loneliness. It is defined as the subjective perception of feeling socially isolated. Lonely people sense the discrepancy between desired and actual social relations. Social distancing increases loneliness and it, therefore, predisposes people to mental and physical disorders. Basic health is impaired. Deprived people are especially at risk of loneliness and they may, therefore, suffer relatively severely from side effects of interventions of social distancing. In addition, they may experience the economical drawbacks of such interventions (*cf.* Barnett-Howel *et al.*, 2021). Equity is a topic of concern in interventions based on social distancing. A lack of equity may also pop up with respect to vaccines, as we will see in the next section.

REFLECTIONS

The beneficial effects of, some degree of, social distancing are clear for some pathogens, whereas these are not for others. The detrimental effects may even exceed the beneficial ones. We may reflect about appropriate, human, social distancing with respect to various types of pathogens, various types of epidemics. We may need some 'gut-feeling' as the positive, and negative, effects of social distancing cannot be captured in simple calculations, simple models.

10.2 Vaccination

The human body has three major lines of defence against pathogens, as we have elaborated in Chapter 2. The first line is a physical barrier, of which the epithelium is the major component. Pathogens, or their excretes have to penetrate this barrier and its related, physical, components, like the muco-ciliary layer of the respiratory tract. The first line of defence is our constitutive resistance to pathogens, as we called it in Chapter 5. It prohibits thus the entry of pathogens. The second line of defence gets in operation as soon as pathogens are able to pass the first line. It is the innate immune system, of which the 'guards' are permanently on surveillance. The second line belongs to our induced resistance. The defence is operationalised as soon as a pathogen, or its exudates, enters the body. The innate immune system prevents establishment of pathogens in our body, or not. If not, the third line of defence is alerted, the adaptive immune system, which is especially characterised by the operation of anti-bodies. The second type of induced resistance we have. It needs to prevent outgrowth of pathogens. An outgrowth that may, directly or indirectly, result in disease. Vaccination is primarily directed to strengthening this third line of defence, the adaptive immune system. It is, therefore, an intervention to prevent (severe) disease rather than prevention of entry and establishment of a pathogen (Pollard and Bijker, 2021).

Development of vaccines was a matter of trial and error, so far. Research was actually following the practice of vaccines rather than providing the basis for these. We utilise in vaccination the ability of the adaptive immune system to 'remember' previous immune responses to specific pathogens. Immune responses that are induced by antigens of the pathogens, as we have outlined in Chapter 2. The 'memory' of the immune system enables a relatively quick immune response to a subsequent outgrowth of a pathogen, as compared with the preceding one. The response may be less strong, as the pathogen has less time to grow out, and the resulting disease may, therefore, be 'milder' due to vaccination.

Vaccines mimic actually an outgrowth of a pathogen in our body by introducing the whole pathogen, parts of it, or toxic compounds produced by it (Pollard and Bijker, 2021). The activity of a whole pathogen, which is used as vaccine, may be halted, or attenuated. Parts of a pathogen may be transferred into our body by way of various carriers, like liposomes. The parts may be, polysaccharides, proteins, a single gene, or nucleic acid. Genes and nucleic acid need to be expressed to serve as an antigen. Molecular mechanisms of a carrier may be used for the expression, or our own cells, like the mRNA-vaccine does with respect to COVID-19 prevention. Toxic compounds may serve as vaccines as long as these have an antigenic nature, like the antigenic one of whole pathogens, and parts of these, that are used as vaccines. Antigens that trigger the production of anti-bodies by B-cells. Anti-bodies that may directly interfere with a pathogen. If so, it may result in a so-called humoral immunity. Anti-bodies may also activate T-cells to interfere with a pathogen. If so, cellular immunity may be the result, as we call it.

Dosing is essential in vaccination. The dose should be high enough to trigger an immune response that results in the generation of sufficient anti-bodies, enabling a swift response to a pathogen later on. In contrast, it may be not too high to avoid a too strong a response that results in (severe) disease. We may notice here that disease, illness, may be strongly associated with immune responses, like it is for COVID-19. If so, we may experience the severer the illness, the stronger the immune response.

The generation of anti-bodies by way of vaccination is a challenge among children aged zero to five years. The immune system needs to develop still and vaccination may, therefore, not too early. It may not too late as the maternal protection wanes, which may leave the newborn, the child, unprotected. The maternal protection may, however, interfere with the generation of anti-bodies by the child itself upon vaccination. Timing is crucial in vaccination of children. Repetitions of vaccination, the so-called booster vaccinations, may aid to achieve the required levels of anti-bodies among young children. Boosters that may also be needed among old people, as the immune system wanes by ageing. The efficacy of vaccination-induced resistance may also be limited in a relatively fast outgrowth of a pathogen upon establishment, say within hours, or a couple of days, like it may be observed for bacteria. An adaptive immune response needs some time to develop, irrespectively the abundance of anti-bodies resulting from vaccination. Application of exogeneous anti-bodies may circumvent a too slow a response of the adaptive immune system. A way of vaccination that is needed for people with inherent deficiencies in the generation of anti-bodies, anyway. An immuno-competent donor offers then the required anti-bodies. Exogeneous anti-bodies are also used for maternal vaccination to protect new-borns against the typical paediatric diseases immediately after birth. A relatively novel approach to vaccination, of which the efficacy and safety has to be demonstrated unequivocally yet.

The term 'herd immunity' pops up frequently with respect to vaccination. It refers to protection of non-vaccinated people in a community, or those who do not respond to a vaccination, by those who are vaccinated. The idea of herd immunity is derived from a relatively simple SIR, compartmental, model, which hinges completely on the basic reproductive number R_0. We presented it in Chapter 6 already. The fraction of people that needs to be vaccinated to protect a whole community is given by the following equation,

$$S_{vac}=N(1-\frac{1}{R_0}),$$

in which S_{vac} is the number of susceptible people that needs to be vaccinated and N is the total number of people of a community. We may see that about 70% of the people needs to be vaccinated to achieve herd immunity, if the R_0 approaches a value of 3. A value that has been suggested for COVID-19. We have indicated the drawbacks of the model, and specifically those of R_0, already in the Chapters 5 and 6. An underlying assumption, for example, is a rather uniform distribution of the target pathogen and susceptibility of man, respectively. An assumption that may, in general, be met by general epidemics only. In addition, herd immunity assumes that the reproduction of a pathogen is completely inhibited by a vaccine,

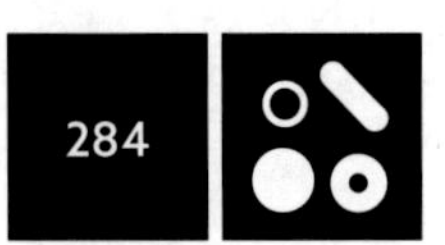

or at least for those responding to the vaccine. An assumption that was certainly not met by the COVID-19 vaccines. In contrast, the efficacy of vaccination against measles may go up to 98%, depending on age of first vaccination (Fu *et al.*, 2021). The coverage of vaccination needs to be rather high with 94% to achieve herd immunity, according to the model and a, presumed, relatively high value of R_0 of 16. If so, travelling of people who are not vaccinated, or who did not respond sufficiently to a vaccination, may break down the achieved herd immunity. Travelling is Achilles heel of achieving herd immunity by way of vaccination, besides the efficacy of, and the coverage by, the vaccination itself. Herd immunity, or not, vaccination remains efficacious in reducing mortality and morbidity due to measles (Fu *et al.*, 2021).

Epidemics may show a quite strong clustering of cases, especially at the onset and the termination of these. We see it especially in epidemics progressing as a travelling and dispersive wave, respectively. Clustering may also result from specific groups of people who are especially susceptible, like children, or just elderly. It may also result from a specific pattern of dispersal of a pathogen. We dealt with such a clustering in Chapter 6 already. In addition, emergence of clusters may also show substantial heterogeneity, say emergence of these is quite unpredictable. An unpredictable clustering of infectious diseases may affect both, execution of clinical trials to test interventions and the, subsequent, implementation of these. It affects these whenever limitations are present with respect to availability of, a vaccine, (diseased) subjects, time to act, or funding. Vaccination against Ebola, for example, was limited in a timely availability of sufficient vaccine. We may cope with such an unpredictable clustering of an infectious disease by way of so-called ring interventions. Interventions that may target relatively swiftly hotspots, clusters, of an epidemic. Vaccination may be such an intervention, if a proper vaccine is available.

Ring trial designs have been developed for the evaluation of ring interventions that are directed to infectious diseases (Butzin-Dozler *et al.*, 2022). Cases of the target disease are identified by way of passive, or active, surveillance of a catchment population. The cases identified, the so-called index cases, provide the basis of a ring trial. Subjects free of disease are enrolled and these constitute a ring around each of the index cases (Fig. 10.2). The ring may be constituted by, the household of an index case, a social network, people identified by contact tracing, and so on. The number subjects in a ring may vary within a single trial. The intervention of interest is randomly assigned to one part of the rings in a study. The other part serves as a control. We may notice that non-enrolled subjects are present between the rings. They also are, from a physical point of view, present within rings, if we assume that subjects of a ring move freely around in daily life. The presentation of a 'ring' as a circle (Fig. 10.2) is thus quite conceptual. Anyway, we turn to a practical example of a ring vaccination against the Ebola-virus during a flare-up of an epidemic.

A vaccine against Ebola passed clinical ring trials and it was, subsequently, implemented as a candidate vaccine under expanded access (Gsell *et al.*, 2017). It was implemented as a ring intervention during a flare-up of an Ebola epidemic in Guinea in 2016. Four index cases were selected and rings around these were defined by way of contacts of contacts with the index

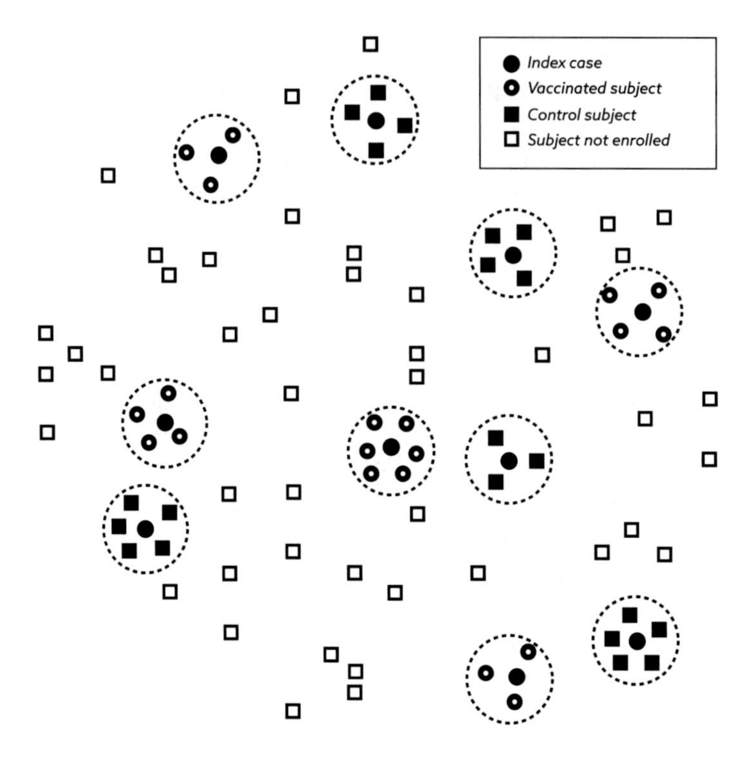

Figure 10.2. Schematic drawing of the design of a ring vaccination trial. Rings are depicted as circles for ease of presentation. See text fur further explanation.

case. Two index cases had been passed away at the start of the intervention. The four rings enclosed, 715, 75, 484 and 385 contacts of contacts, respectively. Three rings were situated in a rural area and one in an urban area. Ninety-one percent of the subjects was overall vaccinated. The major reason of non-vaccination was age. It was not allowed to vaccinate children younger than 6 years. In addition, five people did not consent finally, five women were pregnant at the time of vaccination, and one enrolled subject did not show up for vaccination. No one of the vaccinated subjects suffered from Ebola, nor the minority of non-vaccinated subjects, during a follow-up of 10 days, and later. The term 'later' was not specified, but we may assume that it means until the end of the flare-up of the epidemic.

We have an ongoing debate about the safety of vaccines (Pollard and Bijker, 2021). Pre-registration trials allow the detection of the quite common side effects, which are related to immune responses. These are, in general, transient and relatively mild. Detection of relatively rare side effects is not possible in relatively small pre-registration trials. Post-registration surveillance at larger scale is needed to detect, and attribute, these to a specific vaccination. The extent and quality of the surveillance system thus determines the likelihood of detection. In addition to the authors, we may notice that acceptance of side effects may be higher among people at relatively high risk of (severe) disease than those at relatively low risk. In addition, vaccination is directed to healthy people, in the sense that they do not suffer from the target disease yet. So, these people have a certain risk to become ill due to side effects of the vaccination. Vaccination may cause illness.

Vaccination is, in general, intended to prevent infectious diseases. It is efficacious in recurrent epidemics caused by indigenous pathogens, like the ones of measles and malaria. The recurrence may guarantee a return on investment of the time-consuming development of vaccines. A return on investment that may also result from a world-wide vaccination, like we have seen with respect to COVID-19. Return on investment also is a limitation of development of vaccines, besides challenges in the development itself (Pollard and Bijker, 2021). Anyway, we do not have vaccines for the majority of infectious diseases. Drugs provide an alternative to vaccination for recurrent, and other, types of epidemics. We turn to these in the next section.

REFLECTIONS

Vaccination is an old-fashion way to cope with infectious diseases. It has demonstrated efficacy, especially in paediatric infectious diseases. We see a shift of focus of vaccination from children to adults, and especially, elderly, and vaccination of complete communities even nowadays. We may reflect about novel challenges in vaccination due to this shift. How to cope with the waning of immunity with age? How to deal with the huge heterogeneity of pathogens and the resulting epidemics? And, is vaccination a timely intervention to manage an emerging epidemic?

10.3 Drugs

We presented five categories of pathogens, and the resulting diseases of these among man, in Chapter 3. These categories may also be used to distinguish drugs, which are used to cope with infectious diseases, *i.e.*, anti-prion agents, anti-viral drugs, anti-biotics, anti-protozoal drugs and anthelmintics, to inhibit outgrowth of, prions, viruses, bacteria, protists/chromists, and helminths, respectively. All these drugs support our resistance to pathogens. Another category of drugs, the anti-inflammatory ones, support us to tolerate pathogens, infectious diseases. We will briefly describe each of the six categories in short below.

Anti-prion agents

Prions are body-own proteins that are mis-folded. The mis-folding may be 'transmitted' to the correctly folded ones. Prions are, therefore, called to be 'infectious', as we know. Prions initiate a kind of chain reaction that results in aggregates, plaques. Potential targets for drugs result from this typical process of pathogenesis, as illustrated for the PrP prion (Shim *et al.*, 2022). The first target is at the transformation from the correctly-folded protein to the misfolded one. Various substances have been tested to stabilise the protein, which are, therefore, called chaperones. Another target is the oligomerisation of the mis-folded protein. Substances may also target the aggregation of the prion. Aggregates that cause finally (neuro-degenerative) disease. Elimination of the aggregates by way of autophagy may also be a target. Substances need then to promote this bodily process of autophagy. Substances may be directed to one specific target, or to several targets. Novel anti-prion agents are investigated, besides the re-purposing of existing drugs to anti-prion ones. Some of the drugs considered for re-purposing are directed to infectious diseases already. We see, the anti-viral drug efavirenz, various anti-biotics like doxycycline, the anti-protozoal one of quinacrine, and the Non-Steroidal Anti-Inflammatory Drug (NSAID) celecoxib. Celecoxib, which is an inhibitor of inflammation in general. The development of doxycycline and quinacrine got stuck in the stage of a phase II trial. These two agents were the only ones of all the substances under investigation that arrived in this relatively advanced stage of drug development, so far. Others are in a stage of pre-clinics still, or these were abandoned in an earlier stage, except one. The compound anle138b belongs to the class of diphenyl-pyrazoles. It inhibits the oligomerisation and aggregation of various types of prions. The compound has been followed up in a phase Ia study directed to the α-synuclein prion (Levin *et al.*, 2022). It had been turned out as efficacious with respect to Parkinson's disease in a murine model. The results of the phase Ia study demonstrated the safety and tolerability of anle138b. In such Phase Ia trials, a single dose of the agent of interest is delivered to healthy volunteers. Phase Ia trials may be followed up by a phase Ib trial, in which multiple doses of the agent of interest are tested among healthy volunteers. Likely, the results of the Ia trial of anle138b will be confirmed in a phase Ib trial, which is actually running. If so, efficacy needs to be demonstrated in subjects with Parkinson disease by way of a phase II trial. A stage of drug development that was detrimental to potential anti-prion drugs, so far.

We may conclude that R&D of anti-prion drugs is a real challenge still. We have to wait for a first agent that passes the stage of a phase II trial. It seems inherent to the very specific nature of prions, *i.e.*, body-own proteins that lose track. In contrast, we have several anti-viral drugs on the market, although multiplication of viruses is also bound intimately with our physiological functioning. The intimately binding, however, remains a major hurdle in the development of efficacious anti-viral drugs. We will turn to the anti-viral drugs now.

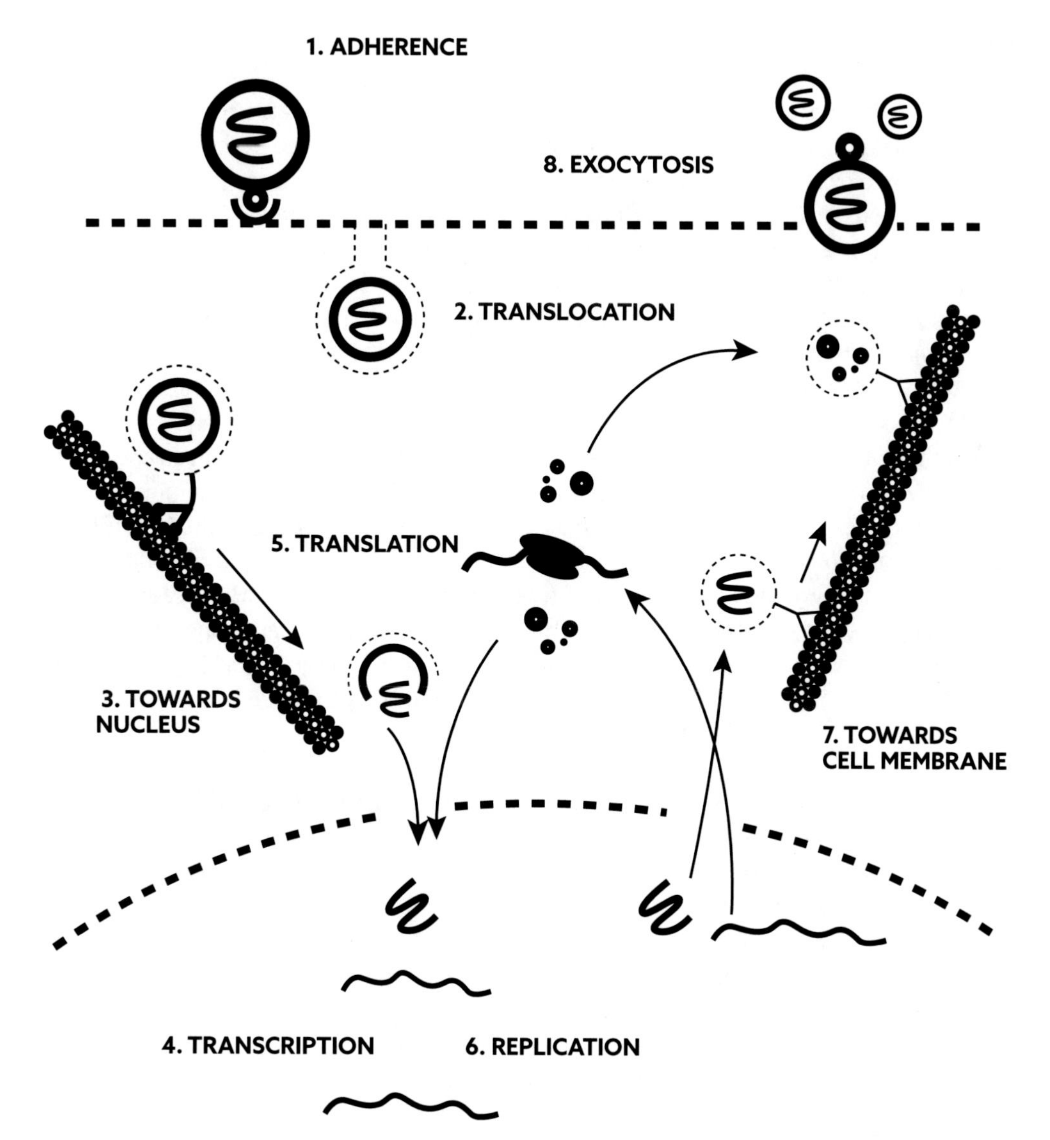

Figure 10.3. Schematic drawing of the process of uptake of a virus by a host cell, the subsequent multiplication of it by way of the molecular pathways of the host, and the release of the virus particles by way of exocytosis. Viral DNA is replicated in the nucleus, as indicated here, and viral RNA is replicated in the cytosol (not shown here). Virus particles may be released by way of exocytosis, as indicated here, or by lysis of the whole host cell (not shown here).

Anti-viral drugs

All viruses need a host cell to exist and multiply. Viruses are neither able to maintain for a long time outside a cell, nor to multiply without it. A virus is actually in a permanent mode of multiplication to exist, besides a rather short time of dispersal from one host cell to another, or from one host individual to another. Physical barriers protect our epithelial cells, in general, from a direct contact with viruses, as we have elaborated in Chapter 2. Once virus particles have passed these barriers, the infection process runs roughly in three stages, (i) adherence to, and translocation into, the cell, (ii) multiplication of the virus within the cell, and (iii) and exit of the novel virus particles (Fig. 10.3). Each of these are composed of several sub-processes. Once the epithelial cells have been passed, other types of cells may be infected by a virus, as well.

Our body is limited in recognition of viruses as foreign. Viruses are rather small-sized and these mimic body-own substances quite well. Viruses may, therefore, utilise the common cellular entrance. It may explain that a relatively small proportion of the approved anti-viral drugs is directed to inhibition of viral entrance of a cell (*cf.* De Clercq and Li, 2016; Tompa *et al.*, 2021). Three of these are targeted at the Human Immunodeficiency Virus (HIV), which causes AIDS. The peptide enfuvirtide blocks the fusion of a HIV-particle with the cell membrane. Maraviroc is a relatively small molecule that operates as an antagonist of HIV by way of blocking chemokine receptors on T-lymphocytes. Ibalizumab is a monoclonal anti-body that serves as an antagonist of HIV by way of binding to CD4-receptors of T-cells.

The majority of anti-viral drugs is thus directed to the multiplication process of a virus within a host cell. It ranges from viral un-coating via, transcription, translation, replication to assemblage and release of the novel particles from the cell. Anti-viral drugs may be directed to the virus itself, like idoxuridine. This inhibitor of the viral DNA-polymerase of the Herpes Simplex Virus was the first approved anti-viral drug. It was approved back in 1963. The pegylated interferon alfa 2a is a protein that targets indirectly virus by way of stimulation of the immune system, especially with respect to the Hepatitis B virus and Hepacivirus C. It may be used in combination with the anti-viral drug ribavirin for Hepacivirus C. Ribavirin targets at viral RNA-polymerase. Anti-viral drugs may, in general, be used alone, or in combination.

Anti-viral drugs are efficacious in containing viral infections, but these are not able to eliminate viruses completely from our body, either alone, or as a cocktail. Anti-viral drug treatments of HIV are a well-known example of containment without cure of the resulting disease AIDS. Application of an efficacious anti-viral agent is, in general, limited in the severe side effects that may result from the quite complete integration of viruses in our molecular pathways. DNA of viruses may be integrated in our genome even. So, the search of an efficacious anti-viral agent with a relatively low toxicity and well-tolerated by patients is a real challenge. In addition, viral genomes are rather unstable, which may result is a relatively rapid loss of the efficacy of an anti-viral agent. All in all, development of anti-viral drugs is a major task requiring long-time research. Research on anti-viral agents that may wane after a relatively short period of high activity upon emergence of a novel viral pathogen (Bobrowski *et al.*, 2020).

Prions and viruses are non-organisms intimately bound to our bodily processes. We turn now to drugs directed to pathogenic organisms, which are easier to recognise as foreign by our body in comparison with the non-organisms. We start with anti-biotics. These are directed to control of bacteria.

Anti-biotics

The bio-chemist and micro-biologist Selman Abraham Waksman (1888-1973) defined an anti-biotic as "a compound made by a microbe to destroy other microbes" (Hutchings *et al.* 2019). We have elaborated the vital function of such compounds in bacteria in Chapter 3. Waksman was borne in Little Russia, which we call Ukraine nowadays. He emigrated to the United States of America as he could, as Jewish, not attend a university in Little Russia. He investigated microorganisms systematically on the production of anti-microbial compounds. He discovered, amongst others, streptomycin as an agent active against *Mycobacterium tuberculosis*. The anti-biotic is produced by *Streptomyces griseus*, a soil-inhabiting bacterium. The discovery of streptomycin was contested by a PhD-student, Albert Schatz, who did actually the isolation in the lab of Waksman. Waksman received the Nobel Prize in Physiology, or Medicine, in 1952. Various bacterial diseases are treated with streptomycin still.

The genus *Streptomyces*, and more broadly the order of Actinomycetales to which it belongs, offers most of the clinically relevant anti-biotics, so far. Other classes of natural anti-biotics are provided by other bacteria and by fungi. Penicillin is the most well-known one among the anti-biotics derived from fungi. It was discovered by the Scottish physician and microbiologist Alexander Fleming (1881-1955). The discovery was one of serendipity. Fleming forgot a culture in a Petri dish and it got contaminated with the fungus *Penicillium rubens*. He, subsequently, detected the anti-biotic activity of the fungus. He called the crude extract of the fungus thus penicillin. Fleming received the Nobel Prize in Physiology, or medicine, together with the Australian pharmacologist and pathologist Howard Florey (1898-1968) and the German/British bio-chemist Ernst Boris Chain (1906-1979), in 1945. They isolated the specific compound and called it penicillin F, which was, subsequently, used by Fleming to treat streptococcal meningitis. That was back in 1942. Penicillin, and its semi-synthetic derivatives, is in use against a variety of bacteria still.

We may trace the origin of synthetic anti-biotics back to a winner of a Nobel Price as well. It is the German physician and scientist Paul Ehrlich (1854-1915). We presented his research om (adaptive)immunity in Chapter 2 already. He developed the anti-bacterial pro-drug salvarsan. The name indicates the major compound of the pro-drug already. It is a contraction of the Latin word 'salvatio', which means saved from a harm, and the word of the active-compound, arsenic. Ehrlich synthesised it in 1907 and Bayer introduced it onto the market in 1911 to treat syphilis. Syphilis is caused by the spirochaetal bacterium *Treponema pallidum*. Ehrlich was aware of the requirement of targeting specifically the pathogen in order to reduce detrimental side effects of anti-biotics. He was looking for the 'Zauberkugel', as he called it in German. Ehrlich was a real forerunner in the search of 'magic bullets' to treat diseases.

His pioneering work on anti-biotics was preceded by that of the German bacteriologist Rudolph Emmerich (1856-1914) and the German chemist and plant physiologist Oscar Löw (1844-1941). They observed that bandages of wounded patients coloured green and they identified *Bacillus pyocyaneus*, which we call now *Pseudomonas aeruginosa*, as the source. It was back in 1899. The extract of the bacterial culture was worked up to an anti-biotic drug, which was called pyocyanase. It was used in hospitals to treat various diseases, like cholera and diphtheria, until 1910. The efficacy and safety were questionable and the use of pyocyanase was abandoned.

Anti-biotics may inhibit synthesis, or break-down of, bacterial proteins, nucleic acids, or cell walls. Destruction of bacterial cell walls also is typical for the use of ordinary soap. A very efficacious method to cope with bacteria, so far these are located on our skin. It is, of course, no way to treat bacterial infections of, for example, our respiratory tract, or intestine. We then rely on other types of anti-biotics. Anti-biotics that are essential in medicine. Bacteria may, however, cope with such a kind of anti-biotics. First of all, bacteria aggregate in colonies that are protected by a bio-film, as we have seen in Chapter 3. In addition, mutations and genetic recombination may result in novel bacterial strains that are insensitive to anti-biotics. Such strains may gain a competitive advantage under conditions of massive use of anti-biotics and, if so, these may start to dominate bacterial populations, as a result of the process of natural selection, which we outlined in Chapter 5. And yes, we had a massive use of anti-biotics world-wide the last decades, and yes, we see a global, bacterial, resistance against the common anti-biotics, in short antimicrobial resistance. Antimicrobial resistance was indicated as an, indirect, leading cause of mortality world-wide in 2019 (Antimicrobial Resistance Collaborators, 2022). We need to be very cautious, in general, to attribute mortality to a specific cause, but we cannot ignore that bacterial resistance to anti-biotics affects the more and more a proper treatment of, bacterial, infectious diseases. The current sandwich approach to reducing the frequency of anti-biotics resistance pairs a responsible, minimal, use of current anti-biotics in treatment of patients with a quest for novel, efficacious, anti-biotics.

The topic of resistance against drugs also pops up in anti-protozoal drugs. We may notice that the term 'anti-protozoal' also includes treatment of diseases caused by chromists, which were seen as protists as well in the past. We elaborated on this in Chapter 3. Anyway, we continue with anti-protozoal drugs in a broad sense now.

Anti-protozoal drugs

Development of drugs to treat chromists/protists received relatively little attention, so far. Twenty-five anti-protozoal drugs are currently available to target the top ten of diseases caused by chromists/protists (Supuran, 2023). Malaria is at pole position. Relatively many drugs are directed to this disease and its causal chromists belonging to various *Plasmodium* species. All of the drugs were, however, developed in the period 1930-1990 and the efficacy is reduced meanwhile due to expansion of drug-resistant types of the pathogens. An expansion that seems to be supported by the chemical similarity of the available drugs. In addition, the drugs

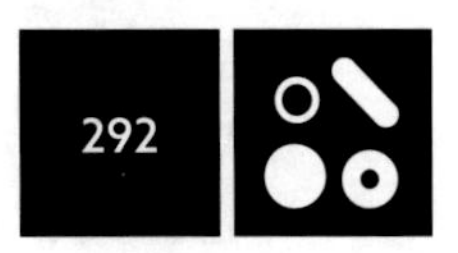

show a rather high toxicity. The pipeline of novel, efficacious and safe, anti-protozoal drugs looks quite empty. Technical challenges may be too serious, or the market is not profitable for big pharma, to fill the pipeline. We may make a similar statement with respect to anthelmintics. We turn to these now.

Anthelmintics

The pipeline of anthelmintics shows a similar emptiness, as the one of anti-protozoal drugs (Nixon *et al.*, 2020). Treatment of diseases caused by worms hinges exclusively on drugs approved before 2000, like praziquantel for schistosomiasis. We described it in Chapter 3. The efficacy of the anthelmintics, in general, decreases, whether, or not, caused by drug resistance among the target worms. Development of novel anthelmintics is limited in, (i) a lack of basic research, (ii) proper strategies to screen compounds on efficacy, and (iii) strong requirements of efficacy and safety in the stage of pre-clinical development, as helminths resemble relatively strongly humans in comparison with other categories of human pathogens. We may state it in another way, the probability is rather low that a lead compound results in an approved drug. The return on, a relatively high, investment is far from guaranteed for anthelmintics. Return on investment is less of a concern with respect to anti-inflammatory drugs.

Anti-inflammatory drugs

We have positioned inflammation and cytotoxicity quite centrally in our presentation of the immune system in Chapter 2. Inflammation, and the accompanying cytotoxicity, is inherent to an appropriate response of our immune system to the outgrowth of a pathogen in our body. Appropriate also means that an inflammatory response turns off timely to minimise collateral side damage. Damage that is inevitable. The process of inflammation incorporates a mechanism of feedback to switch off properly. We may interpret the term 'proper', 'on time', as the moment that a pathogen is controlled sufficiently, while the accompanying bodily damage is tolerable still. We may see here the subjectivity of the process of inflammation. Subjectivity that is related to, the type of pathogen, the amount of pathogen that passed the physical barriers, the internal and external conditions that affect the strength of the immune response, and the internal and external conditions that determine tolerance of the immune response, the disease. The timely switch-off is rather personally, rather subjective. Anti-inflammatory drugs are for those people who cannot switch-off inflammation on time, taking into account their tolerance of inflammation, of disease, for whatever reason.

We focus here on anti-inflammatory drugs that are in use to inhibit inflammation that results from outgrowth of a pathogen. We may distinguish two major, and one minor, category of drugs (Kuchar *et al.*, 2022). The first category is the one of Non-Steroidal Anti-Inflammatory Drugs (NSAIDs). The use of NSAIDs dates back to ancient times, at which people discovered the anti-inflammatory properties of the bark of willow. It is in use in the form of acetyl-salicylic acid still. We know it better as aspirin, the branding it got by the pharmaceutical company Bayer by the end of the 19th century. NSAIDs inhibit Cyclo-OXygenase enzymes (COX), which are involved in immune responses. The use of NSAIDs to turn off the process of inflammation may result in an uncontrolled outgrowth of a pathogen, which may cause complications of a minor infection even. Pre-clinical use of ibuprofen, for example, increases the risk of complicated pneumonia among children. The use of NSAIDs should, therefore, be accompanied by sufficient control of the causal pathogen by way of, for example, anti-biotics. It is similarly for the use of the second category of anti-inflammatory drugs, the glucocorticoids. Glucocorticoids may increase the expression of anti-inflammatory genes, as well as, decrease the expression of pro-inflammatory genes. The expression may be directly, or indirectly, affected. Glucocorticoids may be indispensable as an adjuvant therapy of drugs directed to a pathogen itself to reduce morbidity and mortality. The addition of, for example, dexamethasone resulted in a significant reduction of mortality due to tuberculous meningitis in comparison with no addition of it. In contrast, glucocorticoids as adjuvants did not reduce mortality in children with acute bacterial meningitis caused by *Haemophilus influenzae* type b, but fewer neurological sequelae were observed in those who were treated with glucocorticoids. These beneficial effects were, however, observed among children in high-income countries only and not among those in low-income ones. A difference that may be explained, amongst others, by malnutrition and abundant co-morbidities of children in the low-income countries. The third, minor, category of anti-inflammatory drugs encompasses actually one drug only, colchicine. It is indicated for some non-infectious diseases, but we have some evidence of efficacy in treatment of infectious diseases as well, like malaria and COVID-19.

Anti-inflammatory drugs are, in general, recommended for people at risk of severe disease due to inflammation and cytotoxicity. The benefits of the treatment should thus outweigh the potential, severe, side effects. The drugs are, therefore, directed especially to the late stage of inflammation (Tu *et al.*, 2022). Bio-materials may modulate the process of inflammation in the early and middle stage, avoiding severe side effects. In the early stage, biomaterials may scavenge various compounds that actually initiate the process of inflammation and cytotoxicity. Biomaterials may also block the migration of leucocytes to the site of infection in the middle stage. Finally, biomaterials may serve as carriers of anti-inflammatory drugs in the late stage of inflammation, delivering these efficiently at the site of action. Various types of biomaterials may be used to modulate the process of inflammation rather specifically. Biomaterials may thus offer ample opportunities of regulation of inflammation, if the promising results of pre-clinical research can be confirmed by results of clinical research.

REFLECTIONS

Medication is directed to the outgrowth of pathogens, once these have passed the physical barriers of the epithelium. Anti-prion agents, anti-viral drugs, anti-biotics, anti-protozoal drugs, and anthelmintics target directly at the pathogens. These fit the mechanism of resistance against pathogens. In contrast, anti-inflammatory drugs enable us to tolerate pathogens. We may reflect about the pros and cons of supporting resistance and tolerance, respectively, from a point view of efficacy and safety in the short term, and evolutionary dynamics in the long term.

10.4 Outlook

The ratio behind appropriate, human, management of epidemics, as we outlined it here, is one of a minimisation of the need of massive societal interventions by way of a proper risk assessment (Chapter 8) and a proper societal risk perception, which is based on social cohesion and a relatively high, and uniform, level of basic health within a community (Chapter 9). We may, however, need societal responses to deal with epidemics still.

Social distancing may be mandatory with respect to epidemic spread of relatively aggressive pathogens as travelling waves. The Ebola virus is a striking example of such a pathogen. The social distancing can be applied at a quite local scale, *i.e.*, the social contacts of cases. Vaccination may support the management of such epidemics, but it may also provide an unjustified feeling of safety among people at risk, who may, subsequently, ignore the necessary social distancing with respect to cases.

Relatively aggressive pathogens may also disperse from a multitude of sources, the general epidemics, like those of the *Falciparum* species that cause malaria. Social distancing is not an option as societal response. Elimination of all sources is an option, but it may be limited in deprived socio-economic conditions, or biological constraints posed by the target pathogen. If so, vaccination would be a likely societal response. Development of a vaccine may, however, face technical challenges. In addition, the return on investment may be too low for pharmaceutical companies, especially for diseases that occur in low-income countries only. A hurdle that exists with respect to the development of medication, as well.

Relatively non-aggressive pathogens may disperse exponentially as a dispersive wave, like the SARS-CoV-2 virus that causes COVID-19. Social distancing and vaccination are, actually, no options to manage such epidemics on time. The focus of management should then be on persons at high risk and severe cases, *i.e.*, those in need of hospitalisation. Existing anti-pathogen drugs may be employed and, if not efficacious, anti-inflammatory drugs and other supportive care to tolerate the pathogen, the disease.

Interventions directed to, social distancing, vaccination, and medication need to be efficacious, first of all. We provided here a rough indication with respect to the, expected, efficacy of the three major categories of interventions. Estimations of the efficacy need, of course, to be specified with respect to the target epidemic. Specification, or not, we are dealing with estimations, and the accompanying uncertainties, still. We feel over-estimation of the efficacy of an intervention is less desirable than an under-estimation from a point of view of maintenance of confidence of the general public in health authorities.

Safety is a second, major, topic of interest with respect to the implementation of interventions. We have a general framework of safety assessment available for vaccines and drugs. An early assessment is limited in the relatively small size of clinical trials. The assessment needs to be complemented by large-scale surveillance post registration. We have, nevertheless, a common framework at hand to estimate the safety of vaccines and drugs. In contrast, we do not have it for social distancing. Interventions have been done massively without a proper, commonly accepted, framework to estimate the safety of these. We can, in general, not assume the safety of social distancing. We observed the negative impact of social distancing on public health during the COVID-19 pandemic in, an increase of the incidence and severity of mental disorders, solitude among relatively large groups of people, insufficient care for other diseases than COVID-19, and impairment of basic health due to a lack of physical exercise and social contacts. We may foresee that the development of a safety assessment framework for social distancing will get high priority among scientists and policymakers. The development will be a real challenge, as it requires a strong collaboration of bio-medical sciences and humanities, social sciences. This textbook may provide a basis for such a collaboration.

Human rights constitute a third, major, topic of interest with respect to interventions directed to social distancing. The Universal Declaration of Human Rights declares in clause 13 that "everyone has the right to freedom of movement and residence within the borders of each State" and "everyone has the right to leave any country, including his own, and to return to his country" (United Nations, 1948). In addition, "everyone has the right freely to participate in the cultural life of the community", as it is stated in clause 27. The statements of clause 29 restrict these rights as "everyone has duties to the community in which alone the free and full development of his personality is possible" and "everyone shall be subject only to such limitations as are determined by law solely for the purpose of securing due recognition and respect for the rights and freedoms of others and of meeting the just requirements of morality, public order and the general welfare in a democratic society". We feel this clause is a call for prudence with respect to social distancing rather than a license for it. Interventions need to pass a proper process of decision-making. Proper means that the process is democratic and non-prejudiced, enabling a deliberate consideration of efficacy and safety of an intervention of social distancing versus the impairment of human rights. A well-developed society is able to secure such a process in times of emergency, crises, even.

References

Anderson, R.M., Heesterbeek, H., Klinkenberg, D. and Hollingsworth, T.D., 2020. How will country-based mitigation measures influence the course of the COVID-19 epidemic? The Lancet 395: 931-934. DOI: 10.1016/S0140-6736(20)30567-5.

Antimicrobial Resistance Collaborators, 2022. Global burden of bacterial antimicrobial resistance in 2019: a systematic analysis. Lancet 399: 629-655. DOI: 10.1016/S0140-6736(21)02724-0.

Barnett-Howell, Z., Watson, O.J. and Mobarak, A., 2021. The benefits and costs of social distancing in high- and low-income countries. Transactions of the Royal Society of Tropical Medicine and Hygiene 115: 807-819. DOI: 10.1093/trstmh/traa140.

Bobrowski, T., Melo-Filho, C.C., Korn, D., Alves, V.M., Popov, K.I., Auerbach S., Schmitt, C., Moorman, N.J., Muratov, E.N. and Tropsha, A., 2020. Learning from history: do not flatten the curve of antiviral research! Drug Discovery Today 25: 1604-1613. DOI: 10.1016/j.drudis.2020.07.008.

Butzin-Dozier, Z., Athni, T.S. and Benjamin-Chung, J., 2022. A review of the ring trial design for evaluating ring interventions for infectious diseases. Epidemiologic Reviews 44: 29-54. DOI: 10.1093/epirev/mxac003.

De Clercq, E. and Li, G., 2016. Approved antiviral drugs over the past 50 years. Clinical Microbiology Reviews 29: 695-747. DOI: 10.1128/CMR.00102-15.

Duncan, S., Bodurtha, P. and Naqvi, S., 2021. The protective performance of reusable cloth face masks, disposable procedure masks, KN95 masks and N95 respirators: filtration and total inward leakage. PloS ONE 16: e0258191. DOI: 10.1371/journal.pone.0258191.

Fu, H., Abbas, K., Klepac, P., van Zandvoort, K., Tanvir, H., Portnoy, A. and Jit, M., 2021. Effect of evidence updates on key determinants of measles vaccination impact: a DynaMICE modelling study in ten high-burden countries. BMC Medicine 19: 281. DOI: 10.1186/s12916-021-02157-4.

Gsell, P-S., Camacho, A., Kucharski, A.J., Watson, C.H., Bagayoko, A., Danmagjii Nadlaou, S., Dean, N.E., Diallo, A., Honora, D.A., Doumbia, M., Enwere, G., Higgs, E.S., Mauget, T., Mory, D., Riveros, X., Thierno Oumar, F., Fallah, M., Toure, A., Vicari, A.S., Longini, I.M., Edmunds, W.J., Henao-Restrepo, A.M., Paule Kieny, M. and Kéïta, S., 2017. Ring vaccination with rVSV-ZEBOV under expanded access in response to an outbreak of Ebola virus disease in Guinea, 2016: an operational and vaccine safety report. Lancet Infectious Disease 17: 1276-1284. DOI: 10.1016/S1473-3099(17)30541-8.

Hutchings, M.I., Truman, A.W., and Wilkinson, B., 2019. Antibiotics: past, present and future. Current Opinion in Microbiology 51: 72-80. DOI: 10.1016/j.mib.2019.10.008.

Ioannidis J.P.A., 2021. Infection fatality rate of COVID-19 inferred from seroprevalence data. Bulletin of the World Health Organization 99: 19 - 33F. DOI: 10.2471/BLT.20.265892.

Islam, N., Sharp, S.J., Chowell, G., Shabnam, S., Kawachi, I., Lacey, B., Massaro, J.M., D'Agostino Sr., R.B. and White, M., 2020. Physical distancing interventions and incidence of coronavirus disease 2019: natural experiment in 149 countries. British Medical Journal 370: m2743. DOI: 10.1136/bmj.m2743.

Kuchar, E., Karlikowska-Skwarnik, M. and Wawrzuta, D., 2022. Anti-inflammatory therapy of infections. Encyclopedia of Infections and Immunity 4: 791-797. DOI: 10.1016/B978-0-12-818731-9.00181-6.

Levin, J., Sing, N., Melbourne, S., Morgan, A., Mariner, C., Spillantini, M.G., Wegrzynowicz, M., Dalley, J.W., Langer, S., Ryazanov, S., Leonov. A., Griesinger, C., Schmidt, F., Weckbecker, D., Prager, K., Matthias, T. and Giese, A., 2022. Safety, tolerability and pharmacokinetics of the oligomer modulator anle138b with exposure levels sufficient for therapeutic efficacy in a murine Parkinson-model: a randomised, double-blind, placebo-controlled phase 1a trial. eBioMedicine 80: 104021. DOI: 10.1016/j.ebiom.2022.104021.

Lewnard, J.A. and Lo, N.C., 2020. Scientific and ethical basis for social-distancing interventions against COVID-19. Lancet 20: 631-633. DOI: 10.1016/S1473-3099(20)30190-0.

Morese, R. and Longobardi, C., 2022. The impact of physical distancing in the pandemic situation: considering the role of loneliness and social brain. Frontiers in Psychology 13: 861329. DOI: 10.3389/fpsyg.2022.861329.

Nixon, S.A., Welz, C., Woods, D., Costa-Junior, L., Zamanian, M. and Martin, R., 2020. Where are all the anthelmintics? Challenges and opportunities on the path to new anthelmintics. International Journal for Parasitology: Drugs and Drug Resistance 14: 8-16. DOI: 10.1016/j.ijpddr.2020.07.001.

Nussbaumer-Streit, B., Mayr, V., Dobrescu, A.I., Chapman, A., Persad, E., Klerings, I., Wagner, G. Siebert, U., Christof C., Zachariah, C. and Gertlehner, G., 2020. Quarantine alone or in combination with other public health measures to control COVID-19: a rapid review. Cochrane Database of Systematic Reviews 2020: CD013574. DOI: 10.1002/14651858.CD013574.

Pollard, A.J. and Bijker, E.M., 2021. A guide to vaccinology: from basic principles to new developments. Nature Reviews Immunology 21: 83-100. DOI: 10.1038/s41577-020-00479-7.

Shim, K.H., Sharma, N. and An, S.S.A., 2022. Prion therapeutics: lessons from the past. Prion 16: 265-294. DOI: 10.1080/19336896.2022.2153551.

Supuran, C.T., 2023. Antiprotozoal drugs: challenges and opportunities. Expert Opinion on Therapeutic Patents 33: 133-136. DOI: 10.1080/13543776.2023.2201432.

Tompa, D.R., Immanuel A., Srikanth, S. and Kadhirvel, S., 2021. Trends and strategies to combat viral infections: a review on FDA approved antiviral drugs. International Journal of Biological Macromolecules 172: 524-541. DOI: 10.1016/j.ijbiomac.2021.01.076.

Tu, Z., Zhong, Y., Hu, H., Shao, D., Haag, R., Schirner, M., Lee, J., Sullenger, B. and Leong, K.W., 2022. Design of therapeutic biomaterials to control inflammation. Nature Reviews Materials 7:557-574. DOI: 10.1038/s41578-022-00426-z.

United Nations, 1948. Universal Declaration of Human Rights, Paris.

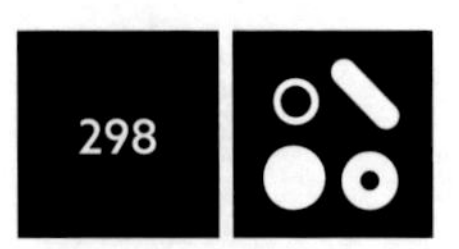

Epilogue

Our quest for appropriate, human, management of epidemics passed across 10 chapters, 10 major topics of epidemiology. A symbol characterised each. Did the symbols characterise the topics well? The symbols all together may reflect the multi-faceted, the multi-dimensional, phenomenon of epidemics. Additional facets, dimensions, may be added to enrich our view on epidemics and the management of these. This textbook may provide the framework to continue the quest for appropriate, human, management of epidemics.

My passion for epidemics dates back about forty years. I investigated and observed these, in general, without a strong personal involvement. This changed completely upon emergence of COVID-19 and the massive interventions executed at national and international level. I noticed the lack of epidemiological knowledge among health authorities and experts. I also experienced the impact of the epidemic, and the associated interventions, on the Dutch society in daily life. The idea of a trans-disciplinary textbook was born. I spent a whole year to arrive at a proper, viable, set-up of the book. The writing took another two years. It was, all in all, to the limits of my abilities. Fortunately, I had great support of various people.

Renate Smallegange of Wageningen Academic Publishers supported me at maximum in the tough start-up of the process of writing. Unfortunately, she had to step out. I am very grateful to her. Mike Jacobs took over her job. He was willing to listen to my disappointments and excitements during the process of writing. He continued his support after the merge of Wageningen Academic Publishing and Brill, although he decided to leave. Mike, many thanks to you! Suzanne Mekking of Brill managed the book, and me, through the final stage of writing and printing. She knew to cope with all my worries about the finalisation, and presentation, of the book. Thank you very much Suzanne!

Some people were willing to spend their spare time on, reading, commenting, and discussing drafts of various chapters. I am really grateful for the indispensable input of, Theo, Esther, Iris, Christopher, Hans, Marc, Els, Bernadette and Lex! Our son Lucas also provided an indispensable input by designing the book graphically. Lucas, thank you very much! My wife Kirsten accepted all the private time that I spent on writing. She comprehended that this book had to be written. Thanks, and love!

Finally, my parents taught me to serve the public interest, the common good. My mother expressed this explicitly at the time I went to university: "you do not go to university for a great salary, but you have to serve society". My parents passed away already. Hopefully, I have accomplished my duty to them in writing this book.

Index